Image Structure

Computational Imaging and Vision

Volume 10

Image Structure

by

Luc Florack

Department of Computer Science,
Utrecht University,
Utrecht, The Netherlands

KLUWER ACADEMIC PUBLISHERS
DORDRECHT / BOSTON / LONDON

A C.I.P. Catalogue record for this book is available from the Library of Congress.

ISBN 978-90-481-4937-7

Published by Kluwer Academic Publishers,
P.O. Box 17, 3300 AA Dordrecht, The Netherlands

Sold and distributed in the U.S.A. and Canada
by Kluwer Academic Publishers,
101 Philip Drive, Norwell, MA 02061, U.S.A.

In all other countries, sold and distributed
by Kluwer Academic Publishers,
P.O. Box 322, 3300 AH Dordrecht, The Netherlands

Printed on acid-free paper

Gedanken ohne Inhalt sind leer, Anschauungen ohne Begriffe sind Blind.

—IMMANUEL KANT

Contents

Foreword

Because of reasons best known to their authors I have already written forewords to various books on the general topic of "scale space": Why yet another one? Well, the present book is *different*. Most developments in scale space theory and practice have been due to scientists who are either best classified as "computer scientists" or as "(applied) mathematicians". Whereas the mathematician rightly pursues art for art's sake, the computer scientists tend to pursue craft for craft's sake, they are—in the first place—*engineers*. This places both disciplines at most at the margin of the empirical sciences in the sense that neither is censored much by exposure to the physical world. (The bulk of computer science is of this nature, in this sense it is different from civil engineering: At least bridges may collapse.) The present book has been written from the perspective of the *physicist* and this shows through at every page.

Indeed, it is possible to view scale space theory as essentially a mathematical theory and thus *arbitrary* from the standpoint of physics (I will write "physics" when "empirical sciences" would serve equally well). This is in fact the common attitude and it explains the many *ad hoc* "generalizations" and embellishments that appear indeed attractive from the vantage point of the mathematician. Despite the often admirable virtuosity displayed, such exercises often fail to excite me because I can't perceive them as being *about* something. Even the "hard results" in computer vision and image processing often leave me in a quandary because we only too often seem to lack the touchstone needed to size up the result.

Luc Florack bases the theory firmly as a (very general) description of the process of observation. This is much more intricate than it might sound at first blush since when one thinks of "observation" one tends to assume the existence of an entity that is *being observed*. Such is certainly the case in the context of signal detection and communication theory, *i.e.*, the communication between agents. However, nature is not an agent and whatever it may be that is "being observed" can only be known through the observation itself! *Observations are indeed all we have.* Florack takes this seriously in that he defines the elementary process of observation in terms of the structure of the observer, rather than the structure of the entity being observed. The latter is only (indeed: *can* only be) implicitly defined in terms of the structure of the observer. (Notice how this is the exact opposite of the standard procedure of signal detection and communication theory and also of most image analysis.) Different observers, when confronted with the same slice of nature, will come up with different observations: Reality is observer–relative and not "God given". Indeed, "man is the measure of all things": This goes so far that even the spatiotemporal framework is seen as a property of the structure of the observer, rather than of the world! Of course this is not really that novel (few concepts are) and Florack finds himself in (among more) the varied company of Kant (philosopher), Mach (physicist) and Whitehead (mathematician).

What makes the book fascinating to me is that such general and important concepts underlying human perception (and thus cognition) and empirical science are shown to lead to a very beautiful, firmly conceptually based and computationally powerful theoretical framework. The mathematics involved is the calculus in its widest sense, involving topology, measure theory and continuous groups. Much of the backbone is Schwartz's notion of "tempered distributions". This theory—sorely neglected by the natural sciences—provides the natural tool for *differentiation* in the real world. Notice that perceptual or image analysis necessarily involves the differentiation of observations, or rather, the observation of differences. Here the "infinitesimal domain" becomes tangible. In fact the theory provides rigorous definitions of such geometrical entities as "point" or "tangent vector" that are also fully *operational*: "Rigorous" will appeal to the mathematician, "operational" to the physicist and operational in the guise of "computational" to the computer scientist.

The conceptual and formal groundwork leads to an edifice that contains all of "linear scale space theory", the theory of local image operators and differential invariants, by now generally recognized as the natural tools in the front end of image processing and analysis. Florack shows generalizations to arbitrary spatial dimensions including space–time. Well chosen examples include the deblurring of pictures, multiscale representation in real time, and various image flow estimators.

It is my hope that the book will find a wide audience, including *physicists*— who still are largely unaware of the general importance and power of scale space theory, *mathematicians*—who will find in it a principled and formally tight exposition of a topic awaiting further development, and *computer scientists*—who will find here a unified and conceptually well founded framework for many apparently unrelated and largely historically motivated methods they already know and love. The book is suited for self–study and graduate courses, the carefully formulated exercises are designed to get to grips with the subject matter and prepare the reader for original research.

Jan Koenderink

Utrecht, July 7th 1997

Preface

Being still in its toddlerhood, image analysis has not yet found a balance between theorisation, experimentation and engineering. But the syncretistic role of experimentation is clear from the fact that it straddles the other fields, and aims to reconcile two opposing tenets: that of the theoreticist employing abstractions for reasons of *genericity*, versus that of the engineer sharply focused on *specificity*.

Relative to existing image literature this monograph is clearly biased towards theorisation. As such it does not claim to be a review covering conventional models and techniques, but rather reflects many of the ideas on image structure and front-end vision pioneered by Koenderink, and pursued by many others (the "Koenderink school"). For this reason it will appeal to the mathematician, who will appreciate that image analysis triggers many challenging problems of a fundamental nature. However, practically minded readers should not put this book aside *prima facie*. On the contrary, as the book deals with basic matters of *image structure* it lies at the core of virtually all (potential) applications that require an analysis of image content.

Like any observable an image comprises a "physical picture" (pixel and header data, say, in a digital image) as well as a "mental picture" (a model). The physical picture is a matter of public evidence, but there is no guarantee that mental pictures coincide. Indeed, the very term "image analysis" is akin to the fact that "image synthesis", i.e. an *operational definition* of an "image", more sophisticated than its mere physical format, remains unstressed. This is quite unfortunate, as conceptual mistakes at the axiomatic stage may well conspire to produce wrong results in the final analysis with one hundred per cent confidence. For instance, a naive application of results from standard analysis or differential geometry—probably *the* tools for handling image structure—will almost certainly lead to failure.

Apart from mathematical *rigour*, the need for *robustness* despite noise poses an additional, equally fundamental demand. However, the latter is usually of little or no concern in pure mathematics, simply because the objects of investigation are not imposed as facts of life, but are rather arbitrarily defined. The *principle of duality* allows one to manifestly combine these requirements by stating/solving problems in terms of robust machine concepts defined within a rigorous mathematical framework. It is the guiding principle adopted in this book, and is more constrained than, for instance, the traditional "Marr paradigm" [207], which allows more leeway at the algorithmic stage. For example, Marr's philosophy does not prohibit us to formulate an ill-posed problem as a viable theoretical model, as long as one accounts for some kind of *regularisation* in the algorithms. The Marr paradigm is thus somewhat indirect, because clearly one *has to* regularise ill-posed problems [12, 270]. However, a regularised model is a *new* model, different from the original as well as from any alternatively regularised one. This leaves no room for regularisation as a degree of freedom in algorithmic design. Regularisation must be an integral part of the theory; "incorrectly formulated problems" must be rejected from the outset, or considered incomplete.

As for the technical content, most of the mathematics used in this book is fairly elementary. However, a basic skill in analysis (differentiation, integration, etc.) and algebra (linear spaces, linear operators, etc.) is an absolute prerequisite. Geometrical expertise is helpful but not necessary. Many mathematical concepts are explained in this book to make it self-contained. Two levels of reading are supported. Sections marked with * contain technical details, and should cause no difficulties if skipped. Numbered boxes are provided as standalone figures; these contain side material that may be of interest, but likewise does not interfere with the main text. Problem sections are included at the end of each chapter, except the first, as an aid to acquire active skill; again these can be safely ignored on first reading. Solutions to a selection of problems have been provided. Finally, there is a glossary one may consult to get the gist of a few central concepts as they are used in this book.

Acknowledgement

This monograph is a product with a truly international flavour. The Computer Vision Research Group in Utrecht, in which I participated as a Ph.D. student, continued to provide practical support after my departure in 1994. Max Viergever, Jan Koenderink, Bart ter Haar Romeny, and Peter Johansen have scrutinised preliminary versions of the manuscript, and have provided useful comments and suggestions for improvement. An ERCIM/HCM fellowship, financed by the Commission of the European Communities, enabled me to continu research at INRIA Sophia-Antipolis (France) and at INESC Aveiro (Portugal). The hosting professors, Olivier Faugeras and Antonio Sousa Pereira, are gratefully acknowledged for providing excellent scientific infrastructures. The same credit goes to Peter Johansen, who invited me on behalf of the Danish Research Council for a one-year position as an assistant research professor at DIKU, the Computer Science Department of the University of Copenhagen (Denmark). The opportunity to use a draft of the manuscript as lecture notes for a Ph.D. course has provided me with useful feedback from graduate and undergraduate students, which has contributed to the clarity of presentation. The manuscript has been finalized at the Computer Science Department of Utrecht University (together with the Computer Vision Research Group participating in the Image Sciences Institute), where I am currently working as a scientific researcher with professor Mark Overmars and professor Max Viergever.

I would like to acknowledge Mads Nielsen, Jon Sporring, Ole Fogh Olsen,

Niels Holm Olsen, Koen Vincken, Alfons Salden, Ruud Geraerts, Stiliyan Kalitzin, Armando Pinho, Andrea Koenderink van Doorn, Tony Lindeberg, and colleagues, secretaries and students—their sheer number prevents me from mentioning all by name—in the various institutes for discussions, illustrations or other material, or assistance otherwise. Special thanks goes to Wiro Niessen, who carried out the experiments described in Chapter 6, and to Robert Maas for his help with some nontrivial mathematics. Paul Roos from Kluwer Academic Publishers is gratefully acknowledged for his help with the editorial procedure. Last but not least I am indebted to Annemieke Wolthuis for bearing the brunt of my "European tour", and for joining me whenever she had the opportunity.

Needless to say I am solely to blame for the errors the careful reader will almost certainly discover.

Luc Florack

Utrecht, July 7th 1997

Der Verstand vermag nichts anzuschauen, und die
Sinne nichts zu denken.

—IMMANUEL KANT

CHAPTER 1

Introduction

A *scalar image* is the result of a physical observation of some scalar field configuration within some confined region of spacetime, i.e. a set of numbers ("scalars") that capture the field's *coherent* structure[1]. It may also represent a scalar aspect of another type of field, such as a "scalar density" (*density images*) or a vector field (*vector images*[2]). Typical examples are proton density in a magnetic resonance scan and grey-tone in a colour image.

Before going into further detail, let us look at some practical cases to get the intuitive picture.

1.1 Scalar Images in Practice

Medical imaging [263, 288] may serve as an example in which various image modalities are used in daily practice for the purpose of diagnosis and therapy planning. A non-exhaustive list includes Magnetic Resonance Imaging or Nuclear Magnetic Resonance [206] (MRI/NMR), Computed Tomography (CT), Ultra Sound (US), Single Photon Emission Computed Tomography (SPECT), Positron Emission Tomography (PET), Digital Angiography, and X-ray imaging.

Besides medicine there are many other fields in which physical measurements are converted into scalar images (a procedure called *image reconstruction*) to support particular tasks. The applications are abundant. Facilities handling nuclear waste may be interested in the migration and activity profile of this material when encapsulated in concrete barrels, the oil industry may want to study geological profiles from seismic data, satellite images can be used to monitor the earth's atmosphere, or to chart environmental pollution, various image modalities can be

---

[1]In the context of perception Koenderink proposes the term "picture"—"an *ordered* record"—instead of "image", reserving the latter term for an abstraction that is beyond the scope of image analysis [149]. The point of departure in this book is to consider it essentially undefined.

[2]Vector images are to be distinguished from multispectral images.

used for quality inspection of raw materials or end products, astronomical observatories produce spectral images of cosmological objects, and so on. One may even think of "retinal images" in terms of the firing rates of neurons in our visual system when probed by a light stimulus. As such, scalar images are relevant for both primary visual perception (disregarding alternative visual cues such as colour and stereo), as well as for image analysis, although the focus in this book will clearly be on the latter (see Box 1.1). Figure 1.1 shows several image modalities that fall within the scope of the book. From these examples it is clear that the meaning of an image is to be taken in a broader sense than in plain photography (which, by the way, leads us to yet another example: the police may want to take a picture of your car for speeding, and automatically extract the licence plate number). We shall account for static images as well as video sequences or movies.

Static images are typically two- or three-dimensional, and time adds one more dimension in the case of movies. 2D images are usually obtained from projections, such as perspective or orthographic projection in photography (irradiance image) and range imaging (depth map), or line-integral projection (e.g. X-ray). In many cases images are intended to represent *densities* considered to be of direct relevance, rather than scalars. Such images are often referred to as density images. A density is typically the kind of quantity one tries to reconstruct in medical imaging, with—ideally—a direct functional or anatomical relevance.

Quantitative reconstruction is not always achieved though. For instance, in nuclear medicine it has proven notoriously difficult to relate SPECT images quantitatively to the radio-activity distribution generated by radio-pharmaceuticals in a patient's body. Quantitative SPECT [15, 279] is sometimes referred to as the Holy Grail of Nuclear Medicine; state-of-the-art SPECT is still only a qualitative source of information.

Another problematic case is conventional photography (and likewise vision), which essentially deals with *fluxes* rather than densities. But this fact is not itself the core of the problem, because a projected flux is just a density in the projection plane. The real difficulty is that photon fluxes, captured by means of an ordinary, projective camera, are of a rather *indirect* relevance, since we are typically not interested in shading details.

However, it must be appreciated that relevance in all these cases only makes sense relative to a model or a task. Without any extrinsically imposed model, or *a priori* knowledge (image formation details, projective camera geometry and calibration [64], shading model [122], etc., the image as such is the only objective source of evidence we have. Even if its relevance is only indirect or merely qualitative, there is no essential difference between all image modalities mentioned above at the basic level of representation. For this reason we will assume throughout this book that by default we are dealing with density images (we will discuss the "scalar versus density" issue as soon as we have adequate mathematical tools for it). For the same reason one must expect that the theory in this book may well apply to front-end vision, despite the fact that autonomous vision can hardly be compared to supervised image processing.

There is one rather unfortunate observation to be made here: *densities cannot be*

☞ **Box 1.1 (Front-End Vision)**

> Page 2: "As such, scalar images are relevant for both primary
> visual perception (disregarding alternative visual cues such as
> colour and stereo), as well as for image analysis..."

Thinking of visual perception one could call this, taking into account the above-mentioned limitations, "front-end vision". One then has to think of the visual stages preceding cognitive processing in the visual cortex, from the retinal stage, say, via the lateral geniculate nucleus (LGN), somewhere up to the first layers of the visual striate cortex. It is, however, not the aim of this book to push the analogy with vision to the extreme. Let us simply take this analogy for what it turns out to be worth, given the assumption that an operational representation of image structure must also be essential for visual perception.

An excellent introduction to the biology and physiology of vision is the book by David Hubel [125]. Koenderink has pioneered the visual system as a "geometric engine", and discusses matters of implementation of basic topological and geometrical expertise in the visual front-end [149, 153, 154, 157, 161, 163, 165, 166]. Much of this can be assumed to be pre-established or is otherwise easily implemented in image analysis, since unlike visual systems, computers are *supervised* data processors. Although more or less trivial relative to an external user, who has the ability to *interpret*, the matter is highly nontrivial, and above all, essential in the context of unsupervised agents, as Koenderink points out.

---

measured. It is sad but it is true. The same is true for fluxes, which manifest themselves as densities upon projection, as well as all other so-called "k-forms" (things that need to be *integrated* over a k-dimensional volume [158, 211, 264, 265]), *except* scalars (0-forms). That is why in the end every image is a scalar image (or collection of such images). A scalar can be obtained from a density by volumetric integration, which is automatically enforced in a detector instrument (e.g. a CCD element). In physics variables like densities are also known as "intensive quantities"; although glibly defined as point mappings, they only make sense on neighbourhoods of finite extent. In modern geometry such quantities are handled properly; although attached to a base point, they are not conceived of as point mappings, but as multilinear mappings of vectors that could be said to span a "sensory element" at that point [158, 211, 264, 265]. Mapping *as well as sensor* determine a numerical signal at each point, which is indeed the physically most sensible picture. Any weighted ensemble of neighbouring sensory elements defines a *sampling aperture*. If we could conceive of such an aperture (also called *point spread function, impulse response, receptive field*, etc., depending on context) as a "mandatory argument" rather than a pre-established "configuration constant", then we would have an operational means to disclose some aspect of the "true" physical field of interest beyond sensor specific details. We will account for such a functional analysis approach explicitly in the next chapters. But let us first look

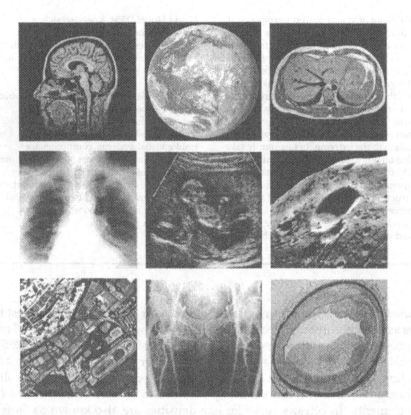

Figure 1.1: Nine images obtained from a variety of image modalities. Note that some of these are 2D slices taken from 3D images merely for the sake of display. In most applications, these should be dealt with in a genuine 3D way. Even 4D images, i.e. (3+1)D movies, are common practice in medical imaging nowadays. This is why we consider images of arbitrary dimensionality n.

at some general conceptual issues.

1.2 Syntax versus Semantics

Scalar image values are often referred to as "grey-values", "intensities", or "luminances". Sometimes the terminology refers to a specific physical field these numbers are intended to represent, e.g. "proton density". This implies a certain *interpretation* of the image data; the premise is a correct (quantitative) *reconstruction*, the conversion from measurements to grey-values proportional to the physical quantity of interest.

Interpretation belongs to the domain of *semantics*. In this book we study only the evidential structure of an image. That is to say, we do not want to anticipate its semantical purpose, but only account for its *syntactical* structure. Syntax is *pre-categorical* and purely *conventional* (it is imposed rather than inferred), whereas semantics is inherently *dialectical* (it demands consensus of interpretation relative to an explicitly formulated goal). Much as it is the case in linguistics, the convention of a syntax is a *conditio sine qua non* for image analysis, since it allows us to structure "raw data". In a way image analysis is indeed a language, in which information exchange is mediated by images.

Syntax subserves semantics and must therefore be *conveniently* defined. Subdividing a written sentence into characters has no syntactical relevance (this is rather explicit in the Chinese language). A character set is of mere practical relevance; for one thing, it keeps our keyboard size within manageable proportions without limiting our word vocabulary. Characters constitute the "grain" of a written language, but do not form the core of its syntax. If we indeed want to perceive of images as *structured* data, we must likewise distinguish between grain and syntax, and determine the kind of structure we may want to impose, as well as the operational means to implement that. To some extent this depends on what we want to do with the images.

Typically, grain structure is specified in terms of scan raster and pixel grid details ("raw data" versus "header data"). For the purpose of storage, display or transfer, this is indeed all the structure we need, much like a character set is all we need to produce a sentence on a type-writer. A pixel-based syntax is convenient relative to digital computer architecture. However, images are intended to communicate information subject to interpretation, not to fill up hard-disks (that is an unfortunate side effect). In *image processing* one aims to convert one image into another, such that properties that are most relevant for a certain *task* become more pronounced or more explicit. For example, smoothing, spatial enhancement, histogram equalisation, thresholding, etc. In *image analysis*, the input is an image, but the output is not (necessarily). The idea here is to arrive at a *symbolic* description of relevant image content. In any case, image handling is semantically inspired.

The focus in image analysis is primarily on the semantical sector, for obvious reasons one might argue; as an engineering discipline befits, it aims to solve practical tasks. Given the outstanding complexity of typical image analysis problems, it is nevertheless prudent not to underestimate structural aspects. Indeed, a misconception at the syntactical level is likely to propagate, causing overall failure or at least unsatisfactory performance of image algorithms, even if the *real* difficulty, viz. the semantical part of the problem, has been handled intelligently.

Typically, and *ideally*, the details of discretisation and quantisation are *irrelevant* in most image problems that are typically raised in the intermediate stage between reconstruction and visualisation. Of prime interest is the semantical content of an image, not how it fits into the slots of a computer register. An important consequence is that a syntactical representation, in order to provide a convenient basis for such applications, must be likewise independent of the grid. If input images are not represented accordingly, one will have to face the cumbersome job of unconfounding grid details from relevant image structure in each and every computational routine. It would also pose a heavy conceptual burden. A more sophisticated structural convention is thus an *a priori* necessity in image analysis.

The popularity of a pixel-oriented approach in the computer vision literature finds its cause in its inevitability. At some stage sooner or later, image data will have to be squeezed through the CPU registers of a computer. This is however no reason why we should maintain a pixel-based syntax in conceptual design strategies. On the contrary, discrete representations are more likely to be distracting than useful. The conception of images as discrete objects tends to overemphasize computational details relative to more serious conceptual problems. In electrodynamics one rarely encounters Maxwell's field equations sampled on a discrete grid, not because of tradition—that Maxwell himself was in the unfortunate situation of lacking a computer is no longer a handicap to us—but because it would only obscure their physical content. The apparent ability of a human observer to solve many tasks on the basis of an image rendered on *whatever* display device likewise argues for a grid-independent philosophy behind any image analysis program designed to assist in or take over such a task.

1.3 Synthesis versus Analysis

In this book we study the possibility of a structural image description explicitly decoupled from grid artifacts. It should not come as a surprise that mathematics plays an important role in image analysis precisely for its potential ability to provide structural representations. Much inspiration is also derived from physics, which has served as a purveyor of mathematical techniques to many related areas of endeavour, and from the physiology, psychophysics and electrophysiology of biological visual systems. One cannot play down the importance of these empirical sciences for image analysis, because the physics of image formation and the human visual inclination lie at the very basis of it. Last but not least image analysis is an arena for philosophy. Quite a lot of practical problems can be avoided if image analysis is not confused with "axiomatic vision". Axioms are objects of *analytical* reasoning, which by definition never exceeds beyond the scope already logically contained in these objects. Axioms *as such* cannot be justified by analytical considerations, they must necessarily be assembled by *synthetical* reasoning [248]; in this process, axioms—if any—typically come in *at the end*, whereas an analytical approach takes them as a point of departure.

Synthetical reasoning must be based on public facts and common sense. All the abovementioned empirical sciences contribute to the synthesis of axioms by a method known as *induction* [248]: facts of experience induce axiomatic principles (models) that are believed to explain our observations, but do not logically follow from them. Indeed, insight tends to change over time as newly acquired experience forces us to modify the axioms. Their self-consistency is never put to doubt though. Within a given axiomatic framework, the only self-consistent method for analysis is *deduction*. The reason why it is important in image analysis to make the distinction "synthesis versus analysis" explicit is not only conceptual—clearly, basic axioms need to be motivated—but is also of practical importance: only analytical reasoning, because of its logical structure, can potentially be converted into autonomous computer programs. If based on incorrect axioms, analysis will produce wrong results with one hundred per cent confidence. For this reason we will motivate *in extenso* the synthetical arguments that lead to our syntax axioms, so that these become a plausible, transparent, as well as convenient point of departure for semantical analysis.

One could call the line of approach adopted in this book "theoretical image analysis". This is not a *contradictio in terminis* as long as theory seeks to establish *operationally meaningful* concepts. The essence of an operational concept is that it is *self-contained*. An operational description or definition basically aims to capture a concept by means of a recipe or program that can readily be realized in hardware and/or software (or may have been realized in nature in biological "wetware"). An operational definition is essentially a *computation*, a protocol free of ambiguous key words. The ultimate goal of this book is to represent image structure in terms of operational concepts. In other words, the basic strategy is not the usual one of "theory *before* computation", but rather "theory *of* computation". The latter is more restrictive and more direct (it avoids non-implementable concepts from the outset). Indeed, adopting a strictly operational attitude, implementation issues will often be a matter of "spelling" hardly worth addressing.

An operational description of a problem essentially entails the solution of that problem (just "run the program"); that explains why it is generally difficult to come up with such a description. For example, we could define "image segmentation" as "the partitioning of an image into meaningful segments", and although this is rather vague, it does make some sense to the trained human eye in a lot of cases. It is also the kind of description one must expect to receive from people interested in industrial or medical applications. Yet this definition is far from operational; it does not allow us to sit down and start writing a segmentation program, at least not one that readily yields segments physicians would consider meaningful in, say, an MRI scan. It is clearly the word "meaningful" that is operationally meaningless here, unless one explains how to incorporate the "trained human eye" into the program.

Synthesis and analysis must concentrate on operational concepts. Indeed, whimsically defined concepts may be indispensable for our intuition, but should subserve attempts to make such concepts operational.

1.4   Image Analysis a Science?

Image analysis is an engineering discipline. It clearly finds its place somewhere "in-between" the sciences, but is it a science *proper*? In the introduction of his book, Faugeras has repeatedly raised this issue in the context of computer vision[3] [64]:

> "We strongly believe that the mathematics are worth the effort, for it is only through their use that computer vision can establish itself as a science. But there is a great danger of falling into the trap of becoming yet another area of applied mathematics. We are convinced that this trap can be avoided if we keep in mind the original goal of designing and building robotics systems that perceive and act. We believe that the challenge is big enough that computer vision can become, perhaps like physics, a rich source of inspiration and problems for mathematicians."

Solving the science issue to everyone's satisfaction is clearly beyond the scope of a monograph. The purpose of this section is to stress its relevance. Moreover, the suggestions made here reflect the general strategy along which concepts will be developed in the rest of this book.

The essence of a science *stricto sensu* is that it has a characteristic *method*, like induction in physics, and deduction in mathematics. Mathematics has a certain privileged status among the sciences; its methodology is the only one shared by all of them. But even prior to a method comes the *object* of methodological investigation, such as the assessment of natural laws in physics. A scientific approach is an oriented and systematic approach that enjoys public consensus. The orientation is what matters; unlike in engineering, in which the goal is *pragmatic* and *specific*, it makes little sense to pose a dead-line on actually fulfilling the ultimate scientific project. But apart from the egocentric desire to satisfy individual curiosity, science *does* have an indisputable practical justification: it provides *coherence* to a discipline, and serves engineering via its *spin-off*. A mature science is a well-dosed blend of pure and applied scientific inclinations that can be pushed to either extremum, from *l'art pour l'art* (an enjoyable hobby) to a *why bother it works* attitude (a disposable application).

The consensus on object and methodology in physics is enviable, and its scientific coherence is beyond doubt. Compared to this, image analysis is anarchy. Nevertheless image analysis poses rather explicit demands of synthesis and analysis subordinate to a functional goal. These demands are in some sense *opposite* to those in physics. And this very fact might well contribute to a discussion of the science issue.

Figure 1.2 illustrates this point. Physics concentrates on modelling assertoric state **A** (the hypothetical, unrevealed "world out there") by regressive induction based on observations **O**. Metamerism **M** is the obstacle (which, as we all know, prevents us from seeing Higgs bosons, gravitons, and other exotic particles), and the challenge is to remove it by building new detectors that might *disambiguate*

---

[3]Faugeras' concern regarding computer vision and its goal—relating to conventional imaging using projective camera(s) subserving robotics systems—clearly applies to image analysis in general.

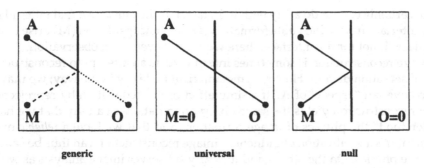

Figure 1.2: Operational versus assertoric point of view. Upon observation, the *assertoric state* A (the universe in the sense of physics) manifests itself in a detector specific disguise (the observer finds himself on the right of the semi-transparent window symbolising the detector interface). The degrees of freedom actually probing the detector determine the *operational state* O (i.e. the observation). The "ghost" degrees of freedom that do not pass the interface or do not segregate upon detection determine a *metameric*[5] state M $\sim$ A mod O [160, 161]. Assertoric state space is *assumed* to be governed by *natural laws*, whereas operational state spaceIndexstate space!operational is *designed* according to *conventional rules* (the details of the interface). Two non-generic situations arise as limiting cases, viz. the "universal" detector (middle): O $\sim$ A and M $\sim$ $\emptyset$, and the "null" detector (right): O $\sim$ $\emptyset$ and A $\sim$ M.

metameric states. From this point of view, the middle image shows the *ideal* situation (the Holy Grail of Physics); albeit a hypothetical goal, it is precisely this "point at infinity" that gives the field its characteristic orientation and keeps the physicist off the street.

Image analysis on the other hand concentrates on functionality; it too must be run on the basis of O (we are all inside Plato's cave). Yet the strategy employed here is to design the interface so as to *support its functional purpose* ("perceive and act"). Purpose is not at all served by attempts to nullify M. Quite the opposite, the state of the universe A is *irrelevant per se*, and metamerism M is a *blessing*: it is the trash can for all the garbage we *don't* need for the accomplishment of a specific goal[4]. Perhaps the greatest challenge is the synthesis of *relevant* information, and thus the *formation of metamerism*, which optimally supports a given task. The smarter the synthesis, the easier the analysis, and the better the task performance. The situation sketched in the middle is the computer scientist's worst nightmare: he would not know where to *begin*. As Faugeras puts it [64]:

> "The space of tasks is usually so different from the space of visual measurements that using the first one to guide the gathering of information in the second is an area that is still in its infancy."

Data acquisition is of the generic type sketched in the left image. The detector interface could be physical hardware, such as a gamma camera, but whatever it

---

[4]If this metaphor triggers the image of a physicist as someone eating from a garbage can, I apologise; one may think of organising one's desk by creatively using its drawers instead.

is, one certainly cannot do away with it. In fact, it is the *interaction* that takes place at the interface from which data formation (O) and encapsulation (M) arise in the first place. If not for the interface, there would not even *be* an observation.

Image reconstruction is sometimes interpreted as an attempt to reconstruct A from observation data O. However, metamerism is clearly in the way; we may at best recover an "aspect" of A. If a simulation of the acquisition by reprojection of the reconstruction yields the same observation data, we can say that we have "understood" the physics of image formation, and that we have a *reliable* measurement. As a verification of induction, image reconstruction can thus be said to subserve physics. On the other hand it clearly subserves image analysis as well.

Once a non-interactive program has been fed an input image, we might be inclined to think that we are in a situation as sketched on the right ("end of data"). This may be true from the point of view of the *user*, but let us not forget it is the *programmer*, faced with a specific task, who actually has to digest the image. In fact, we are back in the generic situation shown on the left. We may redefine A to be our input image (the rest of the universe does not exist anyway, since we went off-line), and we may want to design a detector device (as a software device it is usually called a *filter*) so as to recast the image data into a format more suitably adapted to our task. This is after all what image processing is all about.

Vision (artificial or biological) can be defined as "visually guided behaviour". The main difference with off-line algorithms is that functional behaviour in active vision *affects the state* A, and thus in turn M and O: "observer", or processing device, becomes "participator". Thus one must close the loop A $\longleftrightarrow$ O across the sensory interface (or modify the figure by embedding O and M into A, akin to the well-known Yin and Yang symbol). Note that the interaction A $\longrightarrow$ O is determined by natural laws, while *efficacious* action A $\longleftarrow$ O requires at least the relevant phenomenological laws to be embodied in the visual system (knowledge). Clearly one needs to understand relevant details of image formation in order to "invert the arrow", see e.g. [64, 121] in the context of machine vision.

The theory described in the rest of this book will be seen to mirror the generic picture on the left of Figure 1.2.

1.5 An Overview

Since this book is about image structure, it seems appropriate to start by posing the question of what we mean when we loosely speak of "an image" as an operational concept. The basic observation that an image exists by virtue of physical interaction between a source field on the one hand ("what one is looking at") and a detector device on the other is crucial. We shall refer to such an interaction as an *observation* or a *measurement*.

- In Chapter 2 we focus on a general paradigm that accounts for the dual role of sources and detectors in the realization of a measurement.

- In Chapter 3 we propose two equivalent image definitions based on this paradigm. The first definition will be related to general linear image processing, and could be called "dynamic" insofar that it requires one to allocate filters *after* one is given a particular task to solve. The second one is known by the name "scale-space image". It is a "static" definition, in the sense that it is suited for an image representation that can be computed *beforehand*. Apart from measurement duality, both definitions explicitly incorporate a *spacetime model*, as well as a *consistency requirement*, needed to reconcile basic image processing demands with the duality paradigm. The focus in Chapter 3 is on the synthesis of the (very few) basic axioms that form the point of departure in the rest of the book.

- Chapter 4 focuses on the analysis of the results established in the previous chapter, notably the *scale-space paradigm*. Further motivation is given to support and clarify the theory of Chapter 3. More specifically we discuss matters of *temporal causality*.

- In Chapter 5 we study *local image structure*. We investigate the *general* differential structure, as well as the *geometric* structure, that is, the structure of the image's iso-grey-level topography. This chapter also contains an introduction to tensor calculus, on which the subsequent theory heavily relies.

- The goal of Chapter 6 is to study spatiotemporal images from a different perspective. As opposed to the conventional scale-space representation, in which one views the image as a coherent set of local samples in spacetime, we consider its *kinematic structure*, captured by the notion of *optic flow*. We define a "syntactical optic flow field" as a fully equivalent reformatting of the conventional scale-space representation, based on a conservation principle. This is to be distinguished from any "semantical optic flow field", which requires an additional model independent of the data.

- Several appendices are included, and referenced where appropriate. Sections marked with a * contain details that are not necessary to appreciate the main text, and could be skipped.

Wenn wir aber auch von Dingen an sich selbst etwas durch den reinen Verstand synthetisch sagen könnten (welches gleichwohl unmöglich ist), so würde dieses doch gar nicht auf Erscheinungen, welche nicht Dinge an sich selbst vorstellen, gezogen werden können. Ich werde also in diesem letzteren Falle in der transzendentalen Überlegung meine Begriffe jederzeit nur unter den Bedingungen der Sinnlichkeit vergleichen müssen, und so werden Raum und Zeit nicht Bestimmungen der Dinge an sich, sondern der Erscheinungen sein: was die Dinge an sich sein mögen, weiß ich nicht, und brauche es auch nicht zu wissen, weil mir doch niemals ein Ding anders, als in der Erscheinung vorkommen kann.

—IMMANUEL KANT

CHAPTER **2**

Basic Concepts

In this chapter we discuss some basic physical and mathematical concepts relevant for the description of real-valued images. The focus is on images as instances of physical *measurements*. The considerations made here are necessary in order to appreciate the precise image definition that will be given in the next chapter.

Images are assumed to live on a connected subset Ω of a *spacetime manifold* M of dimension n. The explicit subdivision into space and time is irrelevant in this chapter, and we shall resort for convenience to the "spacetime" terminology to cover the general case.

2.1 A Conventional Representation of Images

It is conventional in the literature to model an image by means of a—possibly discretised—function $f(x)$. Quite often this function is taken for granted, and as such, it is not a satisfactory model in our operational approach. In particular it does not explicitly capture the fact that an image is a physical measurement. A serious weakness of writing $f(x)$ is that it is ambiguous and deceptive; it suggests that a measurement is a *point* mapping. This is clearly a physically non-realizable situation! One cannot confine a physical measurement to a single point due to its

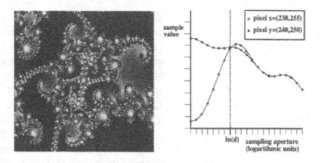

Figure 2.1: **Left:** Fractal image (originally sampled as 8-bit per pixel grey-values on a regular 512×512 pixel grid). **Right:** Grey-value samples at two nearby pixel locations, each displayed as a function of resolution. The samples are obtained by correlation with normalised Gaussian apertures of various widths σ, centred at the respective locations. Vertical axis: grey-values in arbitrary units; horizontal axis: aperture widths sampled on a logarithmic scale within the scale-range $\sigma \in [1, 128]$ in grid units. The open circles correspond to a "bright pixel" near the centre, with pixel coordinates $x = (238, 255)$. The filled circles correspond to a neighbouring "dark pixel" at $y = (240, 250)$ (scan order is from top left $(0, 0)$ to bottom right $(511, 511)$). Note that grey-values at a given point (i) depend significantly upon resolution, and (ii) become highly correlated between "neighbouring" points as soon as scale exceeds their mutual distance $d = \|x - y\|$ (the transient scale indicated by the dotted line): $f(x) \approx f(y)$ if $d \ll \sigma$, whereas $f(x)$ and $f(y)$ are highly uncorrelated if $d \gg \sigma$.

intrinsically finite resolving power. What is lacking in this representation is some notion of *scale* or *resolution*[1]: the physical realization of a grey-value f cannot depend only upon x, but must necessarily involve a *neighbourhood* of x. There is, however, not a single trace of this in the notation, and one is led to guess how "the value f at point x" came into existence. In fact, since the sampling aperture of the measurement has been left out, there is a lot of hidden ambiguity here: a mere change of sampling aperture at a given point x will generally result in a *totally different* value f. Figure 2.1 illustrates this ambiguity for a fractal image.

A related drawback is that by writing $f(x)$ one usually *assumes*—rather than *imposes*—coherence of neighbouring samples; this occurs for instance as soon as one has to rely on assumptions about regularity, even seemingly mild ones such as continuity. Virtually every image based model requires some form of regularity to hold. Indeed, it should be rather obvious that a sufficiently small spatiotemporal perturbation $x \mapsto y = x + \delta x$ (an insignificant change of localisation of the detector device, say) should have hardly any measurable effect, and yet we end up with two *a priori* uncorrelated values $f(x)$ and $f(y)$. It is desirable, and obviously necessary from the viewpoint of stability, to have an operationally well-defined notion of smoothness that enforces the desired coherence, i.e. that forces $f(x)$ and $f(y)$ to be close if x and y are close *relative to some fiducial level of*

---

[1]Not to be confused with limitations due to quantisation; quantisation details are ideally determined *in conjunction with* knowledge of resolving power to avoid "undersampling" and to limit redundant "oversampling".

resolution. If you come to think of it, the only sensible definition of x and y being "close" is that one refrains from resolving a lot of details in-between.

Note that smoothness in the conventional sense of $f \in C^{\infty}(\Omega)$ does not provide such a coherence and is *operationally ill-defined*: its definition relies on operationally non-realizable infinitesimal limiting procedures. Even worse, differentiation is *ill-posed in the sense of Hadamard*[2] (see Problem 2.1 for details). This basically means that even the most insignificant perturbations δf of f may randomise its derivatives (unless such perturbations are—unrealistically—subjected to very strong topological constraints, cf. Problem 2.1). In practice this leads to the paradoxical situation that discrete differences intended to "approximate" derivatives may yield *worse* results when refining the grid (e.g. by resorting to a higher resolution data acquisition), because seemingly insignificant noise tends to become prohibitive! People have been hacking around quite a bit to "fix" this, typically by preprocessing a digital representation of f so as to make it look "smoother" (by low-pass filtering, interpolation, or even by sophisticated nonlinear processing), followed by some kind of discrete differencing scheme. But it should be appreciated that ill-posedness of differentiation has nothing to do with lack of regularity of the operands, so this two-stage procedure is at best *ad hoc*. Besides, the results of discrete differencing will always and crucially rely on the details of the applied preprocessing method. Last but not least, any preprocessing such as smoothing applied prior to the formulation of a goal *corrupts the input data.* At the syntactical stage, in which we have not yet set out such a goal, one should never "tamper with the evidence". We will return to the ill-posedness phenomenon in more detail later, suffice it to say here that it makes "classical" differentiation utterly useless for practical purposes in image analysis.

All the abovementioned problems arise from a single, fundamental misconception: the representation of an image by means of a *function* f mapping a spacetime region onto a subset of the real numbers. If we want to dispense with the abovementioned deficiencies, then we should *explicitly* account for

(i) the notion of spacetime as a *coherent* structure, and

(ii) the notion of *measurement*.

Mathematicians typically introduce coherence on a space by assuming it is endowed with a *topology*: every point p has a neighbourhood of points that are "close" to p. For practical reasons this space is assumed to be coordinatisable by postulating the existence of smooth local coordinate charts[3]. Via such coordinate charts, one naturally obtains a notion of *differentiability* for functions on this topological vector space. The resulting space is *locally* nice and coherent (but may *globally* be quite nontrivial), and is called a *differentiable manifold*. It is one of the most basic concepts mathematicians start out from when they build their formidable theories in differential calculus.

---

[2]A problem is well-posed in the sense of Hadamard if it has a unique solution which depends continuously upon the input data. Otherwise it is ill-posed [106].

[3]Strictly speaking we have $p \in M$ with local coordinates $x \in \mathbb{R}^n$, but, encouraged by the existence of coordinate charts, we will be sloppy and simply write $x \in M$, identifying local neighbourhoods of M with corresponding ones of $\mathbb{R}^n$. The premise is of course that we agree on the charts.

Clearly it would be very helpful if we could just take the whole machinery of differential calculus and apply whatever is useful for the description of image structure. Unfortunately, our operational attitude forces us to admit that this remains wishful thinking: *none* of the local concepts of differentiable manifolds has any operational significance! We should not *assume* somebody has been so kind as to endow spacetime with a topology for us to start differentiating around, but rather *impose* such a structure. In other words, we should not make the claim that spacetime *is* coherent, but rather *enforce* coherence. It is clear that this can only be achieved *a posteriori* via the notion of observation[4], i.e. spacetime topology must be induced by a suitably defined and operationally meaningful topology for images, and not *vice versa*. Thus (i) and (ii) are intimately connected.

2.2 Towards an Improved Representation

We have seen previously that the representation of an image by means of some function $f(x)$ is, at best, ambiguous. We have argued that the reason for this lies in the fact that it hides the observation characteristics (the integration measures, to be more specific), which are nevertheless of paramount importance to the measurement results. Despite this ambiguity, we might hypothesise the existence of a unique physical system, the *state space* in which our scalar field configuration lives, Σ say, independent of the existence of measurement devices. We may then assume that we would be able to chart this system completely by considering the *joint* response of *all possible* detector devices—or filters—when exposed to Σ, collectively denoted by *device space* Δ for ease of reference (cf. Koenderink's *sensorium* [165]). In other words, that we would be able to resolve ambiguities inherent to any single measurement by collecting a huge bunch of independent measurements carried out on Σ. Of course, complete disambiguation can be no more than a *Gedanken* experiment, since we will never be able to carry it through *ad infinitum*. Anyway, collecting as much evidence as possible seems to be the closest we can get in order to disclose the "true nature" of Σ. Note that, apart from obvious practical limitations, we typically don't want an overkill of measurements, but rather restrict ourselves to some subset of relevant measurements. But, without anticipating final purpose, this subset could be anything, so we must not be too restrictive at this point.

To proceed, we would like to couple Σ and Δ so as to give rise to the notion of a measurement, and ultimately, to that of an image. Following our thought experiment, we would like the collection of all possible measurements to be, in some precise sense, a *faithful* representation of the underlying physical state. A convenient way to express this is by *topological duality*. In principle, we have two options here: we can either define Σ and take Δ to be its dual, or *vice versa*. These

---

[4] According to Kant [140], our very idea of "space" and "time", as well as our notion of "causality", are "forms of intuition", by which he means that they *precede* our experiences; they are so to speak part of our mind. Thereby he draws a clear distinction between an event that causes an experience and the spacetime context in which it is placed by the observer. Replacing "observer" by whatever "sensory device", it seems compelling to me to take this view seriously in image analysis when operationalising such basic concepts as space, time, and causality.

options may differ in more than a conceptual sense: taking the dual of the dual does not necessarily bring us back to our point of departure, while even if it does, there may be practical considerations to favour one over the other. The notion of topological duality as well as these two duality options are explained in the following two sections.

2.2.1 Device Space as the Dual of State Space

In this section state space will be considered as our paradigm. This approach enjoys great popularity in the literature; one usually aims for a *direct* model of a physical field of interest as having an existence of its own, de-emphasising observation details. It also seems an intuitively plausible point of departure; after all we are not interested in the details of our detector devices. Yet it will be argued that one cannot operationalise this paradigm. It is the purpose of this section to point out why. In passing, concepts will be introduced that will nevertheless be useful for future purposes.

Let us denote a particular scalar field realization in Σ, or *source*, by an integrable, almost everywhere defined function $f \in L^1(\Omega)$. This seems a plausible thing to do for the following reasons. $L^1(\Omega)$ is a *Banach space*, i.e. a *complete, normed vector space*. The norm is given by

$$\|f\|_1 \stackrel{\text{def}}{=} \int_\Omega dx \, |f(x)| \,. \tag{2.1}$$

A norm induces a translation invariant metric and hence a topology[5] (although a topology does not require a norm). So we actually have a complete, normed, topological vector space. This means first of all that, by virtue of the topology, we have captured the intuitive idea of "almost identical" states, which is of obvious practical importance. Moreover, by virtue of the norm, we can sensibly talk about suitably weighted integrals of f. Integration is exactly the thing one needs in order to obtain a measurement from f, the weight or "measure" corresponding to the sampling aperture. This amounts to replacing the *Lebesgue* measure in Equation 2.1 by a more general, so-called *Borel* measure. Finally, completeness guarantees that sufficiently small perturbations δf of f (small relative to the $\| \, . \, \|_1$-norm) will again yield an element of $L^1(\Omega)$. This is clearly desirable from a physical point of view: such perturbations should not matter, hence it would be rather artificial to exclude them from our theory. And we'd better make sure that they indeed do not mess up our models; there is something conceptually wrong with a model based on a descriptor f that falls apart after a physically insignificant, but otherwise unconstrained perturbation δf. If we cannot get it conceptually right, there is little hope for computational success. A typical thing that happens in practice is that a conceptually wrong model falls apart after discretisation or under circumstances of minor noise.

Other examples of Banach spaces include $L^p(\Omega)$, with $1 \leq p \leq \infty$. These are the spaces of measurable functions defined almost everywhere on Ω, such that

---

[5]Don't confuse topology on function spaces with spacetime topology.

$|f|^p$ is integrable. The norms are given by

$$\|f\|_p \stackrel{\text{def}}{=} \left\{ \int_\Omega dx \, |f(x)|^p \right\}^{1/p} . \tag{2.2}$$

In particular we have the following limiting case:

$$\|f\|_\infty \stackrel{\text{def}}{=} \text{ess} \sup_\Omega |f(x)| , \tag{2.3}$$

i.e. the "essential supremum" of f, or the smallest value M for which $|f(x)| \leq M$ almost everywhere.

A *Hilbert space* $\mathcal{H}$ arises as a special case of a Banach space $\mathcal{B}$, equipped with a strictly positive, bilinear[6] mapping $\mathcal{H} \otimes \mathcal{H} \to \mathbb{R}$, or *scalar product*. The norm can be deduced from this scalar product in a natural way. The canonical example of a (separable) Hilbert space of infinite dimension is $L^2(\Omega)$, the space of square-integrable functions on Ω. Hilbert spaces play a dominant role in *wavelet theory* (see Page 91 for references).

In the context of the above norms, a useful theorem is the following.

Theorem 2.1 (Hölder Inequality)
Let $1 \leq p, q \leq \infty$ with $1/p + 1/q = 1$, and let $f \in L^p(\Omega)$, $\phi \in L^q(\Omega)$, then

$$\|f\phi\|_1 \leq \|f\|_p \|\phi\|_q .$$

A special case of this theorem is the familiar Schwartz inequality in Hilbert space ($p = q = 2$). Using this theorem, one can proof the following *inclusion theorem*.

Theorem 2.2 (Inclusion Theorem)
Let Ω be a compact subset of $\mathbb{R}^n$, then

$$L^q(\Omega) \subset L^p(\Omega) \quad \text{whenever} \quad 1 \leq p < q .$$

So our choice for $L^1(\Omega)$ seems, at least for compact Ω—not a severe constraint for practically minded people—strong enough to cover all other potentially feasible Banach spaces of the form $L^q(\Omega)$. It is in fact the strongest of this type, since $L^p(\Omega)$ with $p < 1$ is *not* a Banach space: the expression for $\| \cdot \|_p$ in Equation 2.2 is not a valid norm if $p < 1$. A proof of the Hölder inequality and of a somewhat stronger version of the inclusion theorem can be found in [36].

What we effectively have now is a function $f \in L^1(\Omega)$ modelling the state of our physical system Σ; this state function has a neighbourhood comprising all practically indistinguishable variations $f + \delta f \in L^1(\Omega)$. But it is important to realize that, even in the hypothetical case that we would be able to probe the system in *all possible* ways, we would *still* be unable to reveal "the value of f at the point x". To see this, note that $\delta f(x)$ might be assigned *any* value at any isolated point x and yet be of zero norm $\|\delta f\|_1 = 0$. Such variations lead to *equivalence classes* of

---

[6]Strictly speaking the scalar product is bilinear on $\mathcal{H} \otimes \mathcal{H}'$, in which $\mathcal{H}'$ is the so-called dual of $\mathcal{H}$, and "sesquilinear" on $\mathcal{H} \otimes \mathcal{H}$. When dealing with real vector spaces this subtlety may be ignored.

functions that induce identical measurements: $f \sim g$ iff $f(x) = g(x)$ almost everywhere (one could call $0 \in L^1(\Omega)$ the "vacuum state" of Σ and any $\delta f \sim 0$ a "vacuum fluctuation"). Note also that the space of continuous functions with compact support (and suitable topology, cf. Definition 2.1 later on)—notation: $\mathcal{D}^0(\Omega)$—is dense in $L^1(\Omega)$, so that we can always model our Σ-equivalence classes up to any desired approximation by such functions.

Now we would like both the system Σ as well as the entire measurement apparatus Δ to be manifest in an operational definition of an image. Moreover, it would be desirable to account for the fact that the set of all possible measurements on a given physical state is a *faithful* description of that state. A convenient way to do this, as alluded to previously, is to consider the so-called *topological dual* Σ' of Σ, and to identify measurement devices on this state space, i.e. detectors, with elements of this dual: $\Delta \equiv \Sigma'$.

In general, the topological dual of a topological vector space is the space of all real or complex valued, linear functions on that vector space, also called *linear forms*, *1-forms*, *covectors*, or, in case of function spaces, *linear functionals*, which are *continuous*. It is itself a linear space, with a straightforward definition of addition and scalar multiplication. This may sound rather abstract at first sight, but it does capture the notion of observation in a quite practical manner; after all, the state vector f has to be probed in one way or another in order to evoke a numeric response. Evoking a *linear* response is merely a matter of convenience, since there is, at least at this point, no compelling reason for making things more complicated; everything is fine as long as our measurements provide us with a *faithful* description of the actual state (we will return to linearity later). Avoiding the abstract terminology we may think of Σ' as "probes of Σ" to capture the idea.

The *algebraic dual* Σ^* is similar to the topological dual, but without the requirement of continuity. In finite dimensional spaces linearity always implies continuity, so then we have $\Sigma^* = \Sigma'$, but in infinite dimensional function spaces, Σ^* and Σ' no longer coincide. Even though the algebraic dual Σ^* is more general, linear functionals that are not continuous are rather pathological, and since they undermine stability, we will exclude them from our considerations right from the start[7] (after all, why use a nice norm-topology in the first place?). Besides, the topological dual Σ' is generally rich enough, as we will see, and allows us to draw many conclusions that do not necessarily hold for the algebraic dual Σ^* (Problem 2.2).

Now there exists a famous theorem that allows us to identify dual spaces.

Theorem 2.3 (Riesz Representation Theorem)
The dual of $L^p(\Omega)$, $1 \le p < \infty$, is isomorphic to $L^q(\Omega)$, with $1 < q \le \infty$ given by $1/p + 1/q = 1$. Each element Φ of the dual of $L^p(\Omega)$ can be represented by some element $\phi \in L^q(\Omega)$ in the following way:

$$\Phi : L^p(\Omega) \to \mathbb{R} : f \mapsto \int_\Omega dx \, f(x)\phi(x) \, .$$

---

[7]Discontinuities of functionals are not to be confused with those of functions! Discontinuity of a functional pertains to the problem of "ill-posedness".

In particular we may identify the dual Σ' of our physical state space $\Sigma = L^1(\Omega)$ as the Banach space $\Delta = L^\infty(\Omega)$; this determines the nature of the linear detectors appropriate for probing Σ. Note that the case $(p,q) = (\infty, 1)$ falls outside the scope of Theorem 2.3. Indeed, the theorem fails to be true for this case; the dual of $L^\infty(\Omega)$ is *not* isomorphic to $L^1(\Omega)$. This is most easily appreciated by means of an example.

Example 2.1
Let $f \in L^\infty(\Omega)$ be a continuous and bounded function, and let $\Phi : L^\infty(\Omega) \to \mathbb{R}$ be the linear continuous functional given by $\Phi[f] = f(x)$ for some point x. (It can be shown that this functional can indeed be extended to all of $L^\infty(\Omega)$: for details cf. the Hahn-Banach theorem, e.g. [36].) This functional cannot be represented by an integral formula as in Theorem 2.3 for any $\phi \in L^1(\Omega)$, hence $L^1(\Omega)$ is not the dual of $L^\infty(\Omega)$.

The fact that $\Sigma \neq \Sigma' \equiv \Delta$ says that the roles of sources and detectors is not interchangeable in our model (as opposed to Hilbert spaces $\mathcal{H} = L^2(\Omega)$, for which we do have $\mathcal{H} = \mathcal{H}'$, leading to a source/detector symmetry phenomenon well-known in quantum physics as *crossing*). The thing that should really worry us though is the fact that $\Sigma \neq \Sigma'' \equiv \Delta'$; there may exist sources that trigger our detectors and yet are not accounted for in our model of $\Sigma = L^1(\Omega)$. Indeed, these are just the *point sources* or, more precisely, linear derivatives of so-called *Dirac distributions* (to be discussed later), an instance of which has been given in Example 2.1. It is nevertheless desirable to allow for point sources in order to explain the cause of a detector response; after all, relying on the response of linear continuous detectors apparently does not exclude their possible existence. A more practical reason why it is desirable to extend state space by admitting point sources is that it lies at the very basis of empirical methods in "reverse engineering". In electrophysiology, for instance, "approximate" point stimuli are routinely employed to probe retinal and cortical receptive fields [47, 125]. For this reason one often uses a functional notation similar to that of Theorem 2.3, even if f is not a locally integrable function, but a point source.

There is another artifact that we may want to reconsider, viz. the fact that, in order to *conceive* of Σ as the space $L^1(\Omega)$, we apparently have to *implement* Δ as the space $L^\infty(\Omega)$. This is the plain consequence of our paradigm $\Delta \equiv \Sigma'$. Such a device space is, however, far too large for actual (sufficiently dense) implementation. If only for pragmatic reasons, we may want to consider only a subset of $L^\infty(\Omega)$. Yet, *by construction, we cannot maintain a faithful description of a predefined state space when reducing device space!* Besides these, there are many more pathologies we would have to hack around if we would hold on to the assertoric state space $\Sigma = L^1(\mathbb{R}^n)$, such as the fact that there is a somewhat artificial discrepancy with its Fourier equivalent space, and that such an extremely powerful tool as differentiation is not even defined. All these apparent conceptual problems may pose huge obstacles at the ultimate stages of application and implementation; it is prudent to face them in advance.

In summary, setting up a seemingly plausible model for Σ, and defining Δ as its dual Σ' in order to circumvent the conceptual flaws of the conventional approach, we apparently find ourselves confronted once again with intricate problems.

2.2.2 State Space as the Dual of Device Space: Distributions

We could of course continue to pursue the paradigm $\Delta = \Sigma'$ based on assertoric state space laid out in the previous section, by replacing candidate function spaces for Σ until we have satisfactorily removed all potential problems. But we won't. In this section it will be argued that the intricacies we got entangled in as we followed the line of reasoning of the previous section are in fact the result of a wrong attitude. The line of approach thus far has been one of "first theory, *then* computation". But not all theories can be turned into computations. It is rather cumbersome and quite annoying to develop nice theories only to find out *afterwards* that they don't work in practice. A (more restrictive) "theory *of* computation", aiming for *theoretical concepts defined in terms of computational modules*, would be most welcome if only to avoid disappointments of that kind.

The thing is that, in the previous section, we started out to model our *sources*, rather than the only sensible thing we should have done, viz. to model our *detectors*. In the framework of topological duality these correspond to readily implementable linear filters. The sources are (at least in image analysis) beyond our control. One does not change input data before one knows which problem to solve; thus sources must be "read-only". In particular one cannot *a priori* exclude structural configurations that do not comply with any given model we might want to conjecture in order to describe the degrees of freedom of our "naked" source field. For example, holding on to our operational attitude consistently, we simply cannot deny the existence of point sources, such as the one in Example 2.1, if all we ever get to see is a bunch of numbers arising from a measurement. Another example is "band-limitedness"; as a viable model for a source field its meaning can be only semantical, i.e. relative to the performance of a task (e.g. can one "decode" a signal if band-limited such-and-such); purely syntactical it is either trivial (measurements *are* bandlimited) or does not mean anything at all (no way to verify it *without* a measurement). All we see are the responses of the detectors we have implemented. The inevitable conclusion of this is that we can identify sources only as far as our detector devices allow us to; "segregation of quality" is determined by our choice of device space!

For this reason we must give up our previous attempt and pursue the alternative option, in which we start out to define Δ instead of postulating a unique Σ, and then define Σ "via interaction" as Δ' rather than *vice versa*: Figure 2.2.

Paradigm 2.1
State space is the topological dual of device space, the latter is the paradigm:

$$\Sigma \stackrel{\text{def}}{=} \Delta'.$$

Thus we identify state space with its device specific manifestation. We decline to investigate physical degrees of freedom that do not interact with our devices, nor do we distinguish between physical degrees of freedom that do not segregate upon detection. In particular this shift of paradigm changes our conception of a "vacuum state"; if all detector devices remain silent, then for all operational pur-

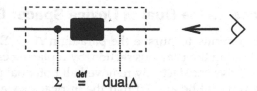

$$\Sigma \stackrel{\text{def}}{=} \text{dual}\,\Delta$$

Figure 2.2: Duality amounts to a "what-you-see-is-all-you've-got" philosophy.

poses, there is nothing out there. Box 2.1 discusses duality as a general principle in a broader context.

☞ Box 2.1 (Duality)

The reader should be aware of the distinct meanings of "duality" depending upon context (at an abstract level some may well coincide). Loosely speaking, two objects (operators, operands, spaces) are said to be dual if they are formally equivalent despite distinct interpretations. Duality principles are typically employed in practice if (at least) one of these interpretations has operational significance.

One employs duality principles in mathematical morphology in statements such as "the dual of an opening is a closing" [112], in projective geometry, "the pencil of planes through a point is the dual of the set of points on a projective plane" [64], etc.

In this book we consider the geometric notion of duality when applied to vector spaces—"the dual of a vector space is a covector space"—in particular to linear function spaces. When we say "state space is the dual of device space" we identify state space with the space of all possible measurement responses induced by that particular device space. Thus "formal equivalence" here means that a source (an element of state space) is identified with the read-out of all detector instruments to which it is exposed. Any source detail that *might* exist but does not manifest itself unambiguously in the measurement results (does not segregate from other source measurements or does not trigger any response at all) is purely *hypothetical*. In other words, a source *in operational sense* is *always* a metameric class of hypothetical configurations, because people may speculate *at liberty* about the latter (unless one has the devine knowledge of a universal device: Figure 1.2). Recall that assertoric state space is any set of hypothetical sources compatible with the evidence obtained through measurement (thus it is ambiguous by nature), whereas operational state space is just the evidence as such.

In principle this paradigm allows us to single out virtually any set of detector devices, although in practice we obviously have to single out a *finite* set. The price we pay for this is that we no longer have a unique description of our scalar field configuration: different Δ's lead to different Σ's: each such Σ allows only *aspects* of a physical scalar field. These aspects can be called "sufficiently rich" if they capture at least those degrees of freedom that are *relevant* for operational purposes. Thus we circumvent the problem of having to hypothesise the nature of the "true degrees of freedom" that constitute the "system beyond the measurements"; *by construction*, any fiducial "internal structure" of sources that cannot be disambiguated given our device space Δ ends up neatly encapsulated in equivalence classes, or *metamers* (cf. Koenderink's example of the "sextuplet image" and his exposition of "metameric black pictures" [160, 161], or, to take a more familiar example in the context of colour perception, the "tristimulus curves" that give rise to indistinguishable colours despite different underlying spectral sources [121]). Hiding the "internal structure" of a source need not be a problem, as long as we can make our computer programs run successfully on the basis of its "apparent structure" only. In fact, the greatest virtue of the model $\Sigma \equiv \Delta'$ is precisely its potential ability to hide whatever features that are *irrelevant* for solving a task! Thus specific tasks may vote for specific choices of Δ; as a consequence of Paradigm 2.1, the definition of physical state has become a "dynamic" rather than a "static" one.

There is a gamut of tasks in image analysis, which all have to be handled on the basis of such Δ's. In view of our desire for *flexibility* rather than *specificity*, let us try to set up a single, unifying model for Δ, plausible enough to encompass such a variety of tasks. The idea is to establish a unique representation suitable for *retrospective* interpretation (a semantical routine may "read" the output spawned by such a device space *selectively*, and *combine* it in whatever way it thinks fit; nonlinearities will be rule rather than exception at a task-oriented level). This implies that we should try to commit ourselves as little as possible. It is clearly beneficial for a general-purpose system to share a common format for the representation of its source data; it facilitates interactions between various routines that may need to share and compare evidence at will. A reasonable thing to do is to use our freedom of detector design to enforce a nice spacetime topology, for one can build quite a lot on the foundations of such a structure (see also Box 2.2).

The idea of the following sections is first to get some flavour of how one might proceed to construct device spaces. To this end we will consider the spaces $\mathcal{D}(\Omega)$ in Section 2.2.2.1 and $\mathcal{E}(\Omega)$ in Section 2.2.2.2, defined on some open subset Ω of $\mathbb{R}^n$. Then we will define the space $\mathcal{S}(\mathbb{R}^n)$ in Section 2.2.2.3, which will turn out to be the most convenient one for future purposes in this book. The reason for this deviation $\mathcal{D}(\Omega) \to \mathcal{E}(\Omega) \to \mathcal{S}(\mathbb{R}^n)$ is mainly didactical. The reader may safely ignore all mathematical details if he/she feels comfortable with Assumption 2.1 on page 27; its significance will be explained in sufficient detail in Section 2.3.

2.2.2.1 ✳ $\Delta \stackrel{\text{def}}{=} \mathcal{D}(\Omega)$

Let us first consider the case $\Delta \equiv \mathcal{D}^m(\Omega)$, i.e. the space of "test functions" having the following properties. First of all, functions in this space are "nice" in the sense that $\phi \in \mathcal{D}^m(\Omega)$ iff $\phi \in C_0^m(\Omega)$, i.e. the set of m-fold continuously differentiable functions of compact support in Ω, with m large enough for their intentional use in practice. Moreover—and this is the reason for writing $\mathcal{D}^m(\Omega)$ instead of $C_0^m(\Omega)$—the space $\mathcal{D}^m(\Omega)$ is endowed with a topology. One way to introduce topology is through a notion of *convergence*.

☞ **Box 2.2 (Local Sign)**

Page 23: "A reasonable thing to do is to use our freedom of detector design to enforce a nice spacetime topology, for one can build quite a lot on the foundations of such a structure."

Nature appears to have followed this design strategy in the evolution of biological visual (as well as other sensory) systems, although matters get seriously complicated here due to the *local sign* problem ("Lokalzeichen") [114, 201], roughly the absence of explicit spatiotemporal coordinate labels in the neural circuitry; local measurements appear to come in as pieces of a scrambled jigsaw puzzle. Topology—perhaps even a geometry—can nevertheless be restored by virtue of the unique correlation structure of the individual pieces (indeed, the nature of the game). Spatiotemporally uncorrelated detectors clearly do not support such topological expertise. A clear explanation of this has been given by Koenderink [149, 157]. In typical image analysis tasks the problem does not arise, since one always has the extrinsic pixel coordinates (though we still have to come up with a topology ourselves). It may well become a relevant issue when simulating visual perception [26, 276]; in one way or another, one will have to account for local sign.

Definition 2.1 (Convergence on $\mathcal{D}^m(\Omega)$)

Let $\phi_k, \phi \in C_0^m(\Omega)$, $k \in \mathbb{Z}_0^+$. Then

$$\lim_{k \to \infty} \phi_k = \phi \quad in \quad \mathcal{D}^m(\Omega), \quad if$$

- there exists a compact subset $X \subset \Omega$ such that supp $\phi_k \subset X$ for all $k \in \mathbb{Z}_0^+$, and

- all partial derivatives of ϕ_k of orders up to m (inclusive) converge uniformly to the corresponding partial derivatives of ϕ.

The spaces $\mathcal{D}^m(\Omega)$ and $\mathcal{D}(\Omega) \equiv \mathcal{D}^\infty(\Omega)$ are no longer Banach spaces, but something very close to that. In fact, the difference is so insignificant that it takes hardly any effort to turn them into Banach spaces again. If X is a compact subset of Ω, then we can construct similar test functions having their support confined to X; these do form a Banach space $\mathcal{D}_X^m(\Omega)$, of which the limiting case $\mathcal{D}_X(\Omega) \equiv \mathcal{D}_X^\infty(\Omega)$ is a particularly nice instance. The norm on $\mathcal{D}_X^m(\Omega)$ may be defined as

$$\sigma_{m,X}[\phi] \stackrel{\text{def}}{=} \sup_{x \in X} \sup_{|\alpha| \leq m} |\nabla_\alpha \phi(x)|, \tag{2.4}$$

in which $\alpha \equiv (\alpha_1, \ldots, \alpha_n)$ is a *multi-index*: see Box 2.3. Note that $\mathcal{D}^m(\Omega) = \cup_X \mathcal{D}_X^m(\Omega)$.

Although not normed, we can provide $\mathcal{D}^m(\Omega)$ with a *family of seminorms*, which is an alternative way to arrive at a topology compatible with that of Definition 2.1.

☞ **Box 2.3 (Multi-Index Notation)**

A *multi-index* α of dimension n is an n-tuple of integers $(\alpha_1, \ldots, \alpha_n)$, allowing us to use convenient shorthand notation. The *order* or *norm* of a multi-index α is defined as $|\alpha| \equiv \alpha_1 + \ldots + \alpha_n$. Most manipulations carried out with multi-indices are self-explanatory:

$$x^\alpha \stackrel{def}{=} x_1^{\alpha_1} \ldots x_n^{\alpha_n},$$

$$\alpha! \stackrel{def}{=} \alpha_1! \ldots \alpha_n!,$$

$$\binom{\alpha}{\beta} \stackrel{def}{=} \prod_{i=1}^{n} \binom{\alpha_i}{\beta_i},$$

$$\nabla_\alpha \stackrel{def}{=} \frac{\partial^{\alpha_1}}{\partial x_1^{\alpha_1}} \cdots \frac{\partial^{\alpha_n}}{\partial x_n^{\alpha_n}},$$

$$\vdots$$

etc.

Definition 2.2 (Family of Seminorms for $\mathcal{D}^m(\Omega)$)
Let $\sigma_{k,X}$ be given as in Equation 2.4 for compact X. Then

$$\mathcal{F}_\Omega^m \stackrel{def}{=} \{\sigma_{k,X} \mid k = 1, \ldots, m, \ X \subset \Omega\}$$

defines a family of seminorms for $\mathcal{D}^m(\Omega)$.

A seminorm is like a norm, except for the fact that nonvanishing elements may have zero seminorm. Indeed, for every compact $X \subset \Omega$, we can construct non-trivial elements $\phi \in \mathcal{D}^m(\Omega)$, the support of which is contained in $\Omega \backslash X$, so that $\nabla_\alpha \phi(x) = 0$ for all multi-indices α of norm $|\alpha| \leq m$ and all $x \in X$ (even if $m = \infty$). Such elements obviously have zero seminorm $\sigma_{m,X}[\phi] = 0$. It may be clear that in the context of seminorms one typically considers families, rather than single instances of seminorms, in order to get around potential ambiguities. This is accomplished by setting up the family in such a way that if *all* seminorms of a given element vanish, the element must necessarily be zero as well. Such a family of seminorms is then called *separated*. A linear space together with a separated family of seminorms is called a locally convex topological vector space [25].

If we are not too modest we can start with $m = \infty$. We then obtain the space $\mathcal{D}(\Omega)$ of $C^\infty(\Omega)$-functions whose topology follows from the limiting case of Definition 2.1 or 2.2. Of course we should convince ourselves that we are not overconstraining our test functions by requiring smoothness as well as compact support; people not previously exposed to these simultaneous constraints might wonder whether we haven't singled out the zero function as the only admissible test function in this way. Requiring *analyticity*[8] rather than smoothness would certainly have yielded only this trivial solution! It is left as an exercise to construct an example of a nontrivial, compactly supported, smooth test function (Problem 2.6).

---

[8] Recall that an analytic function is a smooth function with a convergent Taylor series.

In fact, $\mathcal{D}(\Omega)$ *is dense in* $L^p(\Omega)$ *for all* $p \geq 1$. This shows that $\mathcal{D}(\Omega)$ is indeed not an unreasonable choice to start with; it is certainly rich enough to convey such cases as the prototypical Banach spaces $L^p(\Omega)$.

2.2.2.2 $\ast \, \Delta \stackrel{\text{def}}{=} \mathcal{E}(\Omega)$

The spaces $\mathcal{E}^m(\Omega)$ are defined in a similar way as $\mathcal{D}^m(\Omega)$ (Definition 2.1), except that the constraint of compact support on the elements is now dropped.

Definition 2.3 (Convergence on $\mathcal{E}^m(\Omega)$)
Let $\phi_k, \phi \in C^m(\Omega)$, $k \in \mathbb{Z}_0^+$. *Then*

$$\lim_{k \to \infty} \phi_k = \phi \quad \text{in} \quad \mathcal{E}^m(\Omega),$$

if for all compact subsets $X \subset \Omega$ *all partial derivatives of* ϕ_k *of orders up to* m *(inclusive) converge uniformly to the corresponding partial derivatives of* ϕ *on* X.

We will henceforth only consider the limiting space $\mathcal{E}(\Omega) \equiv \mathcal{E}^\infty(\Omega)$, equipped with a straightforward topology. A family of seminorms compatible with Definition 2.3 is the same as that for $\mathcal{D}(\Omega)$, see Definition 2.2. The space $\mathcal{E}^m(\Omega)$ is a so-called Fréchet space, or a complete metrisable topological vector space. That is almost the same as a Banach space, except that the norm is now replaced by a translation invariant metric [36].

2.2.2.3 $\ast \, \Delta \stackrel{\text{def}}{=} \mathcal{S}(\mathbb{R}^n)$

We have seen two qualitatively different candidate spaces for Δ: $\mathcal{D}(\Omega)$, the most restrictive one, containing smooth functions of compact support, and $\mathcal{E}(\Omega)$, a relaxation with no constraint on function support. However, these spaces are not so different after all, for it can be shown that $\mathcal{D}(\Omega)$ *is dense in* $\mathcal{E}(\Omega)$. That means they are *practically* identical. So the question arises whether, and on the basis of which criteria we might want to prefer one over the other.

It is easy to get carried away by scrutinising all mathematical details involved in the construction of all these topological function spaces. But we should keep in mind that *in practical applications* such decisions as whether a set is open or closed, compact or non-compact, no matter how different these concepts are from a strict mathematical point of view, may hardly be relevant at all. If we are pragmatic, we should be content with *approximations*, and allow such mutually exclusive attributes to be interchanged whenever we feel this is an appropriate or convenient thing to do. Sooner or later we are going to make approximations anyway, so we had better turn them into a virtue and use them to facilitate our models right from the start. We have already put to virtue one such approximation by the smoothness constraint on Δ, motivated by the argument that it would not make much sense to do otherwise, since smooth functions are densely embedded in virtually any function space of practical interest in image analysis. Apart from being justified as an arbitrarily close approximation to whatever nonsmooth function spaces people might want to come up with, we may have gained a modest amount of foresight that, at least from a conceptual viewpoint, this is what we *need* to do in order to turn spacetime into a differentiable manifold. Anyway, since we are dealing with device space, not state space, we *are* in a position to pose structural demands. Note that we are in no way constraining the sources in state space to be smooth; that would make very little sense.

Since both $\mathcal{D}(\Omega)$ as well as $\mathcal{E}(\Omega)$ seem plausible choices for implementing Paradigm 2.1, let us try to find a compromise based on the above considerations. The idea is to come up with a space "somewhere in-between" that inherits the most desirable properties from both. In fact there exists precisely such a space, known as $\mathcal{S}(\mathbb{R}^n)$ (*Schwartz space*), after the mathematician Laurent Schwartz, who laid the foundations of modern *distribution theory* [253]. It is constructed such that

$$\mathcal{D}(\mathbb{R}^n) \subset \mathcal{S}(\mathbb{R}^n) \subset \mathcal{E}(\mathbb{R}^n) \quad \text{and} \quad \mathcal{E}'(\mathbb{R}^n) \subset \mathcal{S}'(\mathbb{R}^n) \subset \mathcal{D}'(\mathbb{R}^n), \tag{2.5}$$

with all dense topological embeddings, so that it is indeed a "compromise" in a precise mathematical sense. But the real point is that $\mathcal{S}(\mathbb{R}^n)$ has many desirable features that make life easy.

2.3 More on the Theory of Schwartz

For reasons explained in the previous section we make the following

Assumption 2.1 (Device Space)
Recall Paradigm 2.1. Unless stated otherwise, we shall assume that $\Delta \equiv \mathcal{S}(\mathbb{R}^n)$, whence $\Sigma \equiv \mathcal{S}'(\mathbb{R}^n)$.

In Section 3.8 we shall see that this assumption is in fact consistent with image processing demands. For future use, let us summarise the most important facts of the theory of Schwartz.

Although we have tacitly extended the support of the test functions to all of spacetime, $\Omega = \mathbb{R}^n$, it is clear that whatever happens at infinity is not of much practical interest; images cover only a finite field of view. In order to get some kind of confinement to a local region, the functions $\phi \in \mathcal{S}(\mathbb{R}^n)$ are chosen to "decrease sufficiently fast"—a good approximation to "compactly supported"—in the following sense.

Definition 2.4 (Class of Smooth Test Functions)
The class $\mathcal{S}(\mathbb{R}^n)$ of smooth test functions, or Schwartz functions, is defined by

$$\phi \in \mathcal{S}(\mathbb{R}^n) \quad \text{iff} \quad \phi \in C^\infty(\mathbb{R}^n) \quad \text{and} \quad \sup_{x \in \mathbb{R}^n} |x^\alpha \nabla_\beta \phi(x)| < \infty,$$

for all multi-indices α and β.

So "sufficiently fast" applies to all derivatives and means: faster than any polynomial, or any other *function of polynomial growth*, would be able to counteract:

Definition 2.5 (Function of Polynomial Growth)
An almost everywhere defined function $f : \mathbb{R}^n \to \mathbb{R}$ is said to be of polynomial growth, notation $f \in \mathcal{P}(\mathbb{R}^n)$, if there exist constants $c > 0$ and $m \geq 0$ such that for all $x \in \mathbb{R}^n$

$$|f(x)| \leq c \left(1 + \|x\|^2\right)^m .$$

Definition 2.4 implies a *very strong* topology on $S(\mathbb{R}^n)$. The idea is that the stronger this topology, the easier it becomes to create continuous mappings $F \in S'(\mathbb{R}^n)$, which is what we are looking for (Paradigm 2.1, Assumption 2.1). Alternative ways to look at the topology of $S(\mathbb{R}^n)$ is by considering convergence on $S(\mathbb{R}^n)$, or by introducing a family of seminorms (in fact even norms), as follows.

Definition 2.6 (Convergence on $S(\mathbb{R}^n)$)
Let $\phi_k, \phi \in S(\mathbb{R}^n)$ be as defined in Definition 2.4, $k \in \mathbb{Z}_0^+$. Then

$$\lim_{k \to \infty} \phi_k = \phi \quad \text{in} \quad S(\mathbb{R}^n),$$

if for all multi-indices α, β the linear derivatives with polynomial coefficients $x^\beta \nabla_\alpha \phi_k$ converge uniformly to $x^\beta \nabla_\alpha \phi$.

Definition 2.7 (Family of Norms for $S(\mathbb{R}^n)$)
Let $\phi \in S(\mathbb{R}^n)$ be defined as in Definition 2.4 and

$$\nu_{k,N}[\phi] \overset{\text{def}}{=} \sup_{|\alpha| \leq k, |\beta| \leq N, x \in \mathbb{R}^n} |x^\beta \nabla_\alpha \phi(x)| .$$

Then a family of norms for $S(\mathbb{R}^n)$ is given by $\mathcal{F} \equiv \left\{ \nu_{k,N} \mid k \in \mathbb{Z}_0^+, N \in \mathbb{Z}_0^+ \right\}$.

The class $S(\mathbb{R}^n)$ is very nice when it comes to differentiation, as opposed to $C^\infty(\mathbb{R}^n)$ endowed with a weaker topology.

Result 2.1 (Differentiation on $S(\mathbb{R}^n)$)
Linear partial differential operators with polynomial coefficients $\lambda_\alpha(x)$

$$P(\lambda, \nabla) \overset{\text{def}}{=} \sum_{|\alpha| \leq m} \lambda_\alpha(x) \nabla_\alpha$$

are continuous linear mappings from $S(\mathbb{R}^n)$ to itself.

The key-word is "continuous"; it is really topology, not smoothness, that makes this happen.

Another of those indispensable tools in image analysis is *Fourier transformation*.

Definition 2.8 (Fourier Transformation on $S(\mathbb{R}^n)$)
Fourier transformation $\widehat{\phi} = \mathcal{F}\phi$ is a continuous linear invertible mapping on $S(\mathbb{R}^n)$:

$$\widehat{\phi}(\omega) \overset{\text{def}}{=} \int dx \, e^{i\omega x} \, \phi(x),$$

with inverse

$$\phi(x) = \frac{1}{(2\pi)^n} \int d\omega \, e^{-i\omega x} \, \hat{\phi}(\omega) \, .$$

Note that Fourier transformation can be extended to the space of all Lebesgue integrable functions $L^1(\mathbb{R}^n)$. The idea behind Fourier transformation is that it diagonalises linear shift invariant operators (Problem 2.7), hence in particular all linear partial differential operators $P(c, \nabla)$ with *constant* coefficients c_α (cf. Result 2.1). It decomposes a spatiotemporal function into a continuous spectrum of harmonic planar waves $e_\omega(x) \equiv (2\pi)^{-n} e^{-i\omega x}$, which are precisely the eigenvectors to $P(c, \nabla)$, with eigenvalues $P(c, -i\omega)$ (Problem 2.8). For computational purposes there exists a *Discrete Fourier Transform* (DFT) and an optimised *Fast Fourier Transform* (FFT), with properties analogous to those of the integral version [40, 238].

Another linear shift invariant operator on $S(\mathbb{R}^n)$ is *convolution*. It is compatible with the nice structure of $S(\mathbb{R}^n)$ in the sense that it does not take us into yet another function space.

Definition 2.9 (Convolution on $S(\mathbb{R}^n)$)
The convolution product of two functions $\phi, \psi \in S(\mathbb{R}^n)$ is defined by

$$\phi * \psi(x) \stackrel{\text{def}}{=} \int dy \, \phi(y) \, \psi(x - y) \, .$$

Lemma 2.1 (Convolution on $S(\mathbb{R}^n)$)
For all $\phi, \psi \in S(\mathbb{R}^n)$ we have $\phi * \psi \in S(\mathbb{R}^n)$.

Correlation is similar:

Definition 2.10 (Correlation on $S(\mathbb{R}^n)$)
The correlation product of two functions $\phi, \psi \in S(\mathbb{R}^n)$ is defined by

$$\phi \star \psi(x) \stackrel{\text{def}}{=} \int dy \, \phi(y) \, \psi(y - x) \, .$$

Lemma 2.2 (Correlation on $S(\mathbb{R}^n)$)
For all $\phi, \psi \in S(\mathbb{R}^n)$ we have $\phi \star \psi \in S(\mathbb{R}^n)$. Unlike convolution, correlation is neither commutative nor associative; if P denotes the parity operator, $P(x) = -x$, then we have

$$\phi \star \psi = (\psi \star \phi) \circ P \, ,$$
$$(\phi \star \psi) \star \chi = \phi \star (\psi * \chi) \, .$$

So far for $\Delta = S(\mathbb{R}^n)$, now let us turn to its dual $\Sigma = S'(\mathbb{R}^n)$ to study the "apparent structure" of state space.

Definition 2.11 (Tempered Distribution)
A linear functional $T \in \mathcal{S}'(\mathbb{R}^n)$ is called a tempered distribution if there exist a constant $c > 0$ and multi-indices α, β such that

$$|T[\phi]| \leq c \sup_{x \in \mathbb{R}^n} |x^\alpha \nabla_\beta \phi(x)|.$$

A tempered distribution $T \in \mathcal{S}'(\mathbb{R}^n)$ is called *positive* if $T[\phi] > 0$ for all positive $\phi \in \mathcal{S}(\mathbb{R}^n)$. Another case of special interest is that of the so-called *regular tempered distributions*.

Definition 2.12 (Regular Tempered Distribution)
Let $f \in \mathcal{P}(\mathbb{R}^n)$ be a function of polynomial growth, then its associated regular tempered distribution $T_f \in \mathcal{S}'(\mathbb{R}^n)$ is defined by the tempered distribution

$$T_f : \mathcal{S}(\mathbb{R}^n) \to \mathbb{R} : \phi \mapsto \int dx \, f(x) \, \phi(x).$$

One may easily verify that a regular tempered distribution as defined by Definition 2.12 is indeed a tempered distribution according to Definition 2.11. Often, and for obvious reasons, a regular tempered distribution T_f is identified with the function f, or with an equivalent one that equals f everywhere except for a set of measure zero. In particular, for *continuous* functions f, the linear mapping $f \mapsto T_f$ becomes one-to-one. Thus the subspace of regular tempered distributions can be strictly identified with $C^0(\mathbb{R}^n) \cap \mathcal{P}(\mathbb{R}^n)$, and "loosely" with $\mathcal{P}(\mathbb{R}^n)$ itself, but the more general space of tempered distributions is larger, as the following definition shows (recall Example 2.1).

Definition 2.13 (Dirac Distribution)
Let $\phi \in \mathcal{S}(\mathbb{R}^n)$, and $x \in \mathbb{R}^n$ any given point. Then

$$\delta_x : \mathcal{S}(\mathbb{R}^n) \to \mathbb{R} : \phi \mapsto \phi(x)$$

defines a tempered distribution, called the Dirac point distribution at point x.

It is clearly *not* a regular tempered distribution. However, one likes to abuse the notation of Definition 2.12, writing *any* tempered distribution T in the form T_f, even if no such function $f \in \mathcal{P}(\mathbb{R}^n)$ exists. In particular, the Dirac point distribution δ_x is associated with the "function under the integral" $\delta_x(y) = \delta(y - x)$ (y being the dummy coordinate), and in turn this encourages the identification $f \equiv T_f$ for every such f. We will henceforth adhere to this widespread but convenient abuse (it does not severely offend mathematicians as long as one remains diplomatic about it).

Notation 2.1 (Function versus Functional)
Tempered distributions corresponding to "functions" $f, g, h, \ldots$ will henceforth be denoted by corresponding capitals $F, G, H, \ldots$

So small f maps points and capital F maps test functions. Note that the support of the Dirac distribution is $\operatorname{supp} \delta_x = \{x\}$, a set of measure zero. The Dirac point distribution at the origin $x = 0$ is often simply written as δ, or as $\delta(y)$ in function form.

There exists a natural definition of convergence on $\mathcal{S}'(\mathbb{R}^n)$, and hence a naturally induced topology.

Definition 2.14 (Convergence on $\mathcal{S}'(\mathbb{R}^n)$)
Let $F_k, F \in \mathcal{S}'(\mathbb{R}^n)$, $k \in \mathbb{Z}_0^+$. Then

$$\lim_{k \to \infty} F_k = F \quad in \quad \mathcal{S}'(\mathbb{R}^n) \quad if \quad \lim_{k \to \infty} F_k[\phi] = F[\phi] \quad for \ all \ \phi \in \mathcal{S}(\mathbb{R}^n) .$$

Of course, we can relate topology to a family of seminorms again.

Definition 2.15 (Family of Seminorms for $\mathcal{S}'(\mathbb{R}^n)$)
Let $F \in \mathcal{S}'(\mathbb{R}^n)$, $\phi \in \mathcal{S}(\mathbb{R}^n)$, and

$$\varsigma_\phi[F] \stackrel{\mathrm{def}}{=} | F[\phi] | .$$

Then a family of seminorms for $\mathcal{S}'(\mathbb{R}^n)$ is given by $\mathcal{F}^t \equiv \{\varsigma_\phi \mid \phi \in \mathcal{S}(\mathbb{R}^n)\}$.

Tempered distributions can be assigned a natural "order".

Definition 2.16 (Order of a Tempered Distribution)
See Definition 2.7. Suppose $F \in \mathcal{S}'(\mathbb{R}^n)$, $c > 0$ and $k, N \in \mathbb{Z}_0^+$ such that $| F[\phi] | \le c \, \nu_{k,N}[\phi]$ for all $\phi \in \mathcal{S}(\mathbb{R}^n)$. If k is the smallest such integer, then F is said to be of order k.

Measures constitute a particularly important class of distributions.

Definition 2.17 (Measure)
A (regular) measure is a (regular) tempered distribution of order 0.

The core of distribution theory is that we can define *arbitrary derivatives* of tempered distributions in a *well-posed* way.

Definition 2.18 (Derivative of a Tempered Distribution)
The derivative $\nabla_\alpha F \in \mathcal{S}'(\mathbb{R}^n)$ of a tempered distribution $F \in \mathcal{S}'(\mathbb{R}^n)$ is the tempered distribution defined by

$$\nabla_\alpha F : \mathcal{S}(\mathbb{R}^n) \to \mathbb{R} \ : \ \phi \mapsto (\nabla_\alpha F) [\phi] \stackrel{\mathrm{def}}{=} F \left[\nabla_\alpha^t \phi \right] ,$$

in which the transposed partial differential operator is given by $\nabla_\alpha^t = (-1)^{|\alpha|} \nabla_\alpha$.

The reason for the minus signs is that in the subspace of regular tempered distributions T_f for which f is smooth, the above definition essentially boils down to the classical definition of differentiation. This can readily be verified by repeated partial integrations using the integral expression on the r.h.s. of Definition 2.12

and the fact that f is smooth and of polynomial growth. One does not have to keep track of minus signs if one uses the "Hermitian" gradient $-i\nabla$ instead of ∇ itself (this operator should appeal to the physicist, see also Problem 2.8). Another way to get rid of the pesky minus signs, and most convenient for our purposes, is to insert *parity reversed* test functions $\tilde{\phi}$ defined as $\tilde{\phi}(x) \equiv \phi(-x)$. Clearly we have $\tilde{\phi} \in S(\mathbb{R}^n)$ if and only if $\phi \in S(\mathbb{R}^n)$, so it is really just a matter of convenience. This observation boils down to the self-explanatory triviality

$$\nabla_\alpha^\dagger = \tilde{\nabla}_\alpha. \tag{2.6}$$

Concerning its differential structure there is apparently no need to distinguish between a smooth function and its associated regular tempered distribution, and this justifies the terminology "derivative" in Definition 2.18. But note that, even if a tempered distribution has no such smooth counterpart, it is *still* infinitely differentiable. Note also that if F has order k, then the order of $\nabla_\alpha F$ equals $k + |\alpha|$.

Generalised differentiation follows the usual *Leibniz's product rule*:

$$\nabla_\alpha(\gamma F) = \sum_{\beta \leq \alpha} \binom{\alpha}{\beta} \nabla_{\alpha-\beta}\gamma \, \nabla_\beta F, \tag{2.7}$$

in which $\beta \leq \alpha$ means $\beta_i \leq \alpha_i$ for all $i = 1, \ldots, n$, and in which $\gamma \in C^\infty(\mathbb{R}^n) \cap P(\mathbb{R}^n)$ (the product γF has an obvious meaning: $\gamma F[\phi] \equiv F[\gamma \phi]$ for all $\phi \in S(\mathbb{R}^n)$). Moreover, the derivatives *commute*: $\nabla_\alpha \nabla_\beta F = \nabla_\beta \nabla_\alpha F$ for all multi-indices α, β.

In particular, we can now differentiate our Dirac point distribution.

Result 2.2 (Derivatives of Dirac Distribution)
Recall Definition 2.18. Let $\phi \in S(\mathbb{R}^n)$, $x \in \mathbb{R}^n$ any given point, and α a multi-index. Then
$$\nabla_\alpha \delta_x : S(\mathbb{R}^n) \to \mathbb{R} : \phi \mapsto \nabla_\alpha^\dagger \phi(x).$$

It is not difficult to show that a distribution $F \in S'(\mathbb{R}^n)$ whose support is a single point is a finite linear combination of derivatives of the Dirac distribution at that point.

Theorem 2.4 (General Point Distributions)
If $F \in S'(\mathbb{R}^n)$ is a general point distribution, i.e. if supp $F = \{x\}$ for some point $x \in \mathbb{R}^n$, then there exists an order $m \in \mathbb{Z}_0^+$ such that

$$F = \sum_{|\alpha| \leq m} c_\alpha \nabla_\alpha \delta_x.$$

with constants c_α given by $c_\alpha = F[\tilde{\psi}_x]$, in which $\tilde{\psi}_x(y) = (x-y)^\alpha/\alpha!$.

It follows from Theorem 2.4 and Definition 2.17 that point measures are of the Dirac type.

Result 2.3 (Point Measure)
If $F \in S'(\mathbb{R}^n)$ is a point measure with supp $F = \{x\}$ for some point $x \in \mathbb{R}^n$, then

$$F = c\,\delta_x\,,$$

with amplitude $c = F[1]$.

Point measures are useful for the representation of "rastered images" defined on a discrete grid[9]. In this book, however, we will continue to use a differentiable spacetime manifold instead of a discrete grid, since it is easier to work with, but also to deliberately de-emphasize the role of the grid. One typically does not care about the grid (although one obviously has to think about matters of *accuracy*); images are intended to support various dedicated tasks independent of how they are discretised. Moreover, we can always perceive of spacetime as a differentiable manifold even in the discrete case, since we are free to model grid data by a tempered distribution defined on all of spacetime $f_S(x) = c_S \delta_S(x) \equiv \sum_{s \in S} c_s \delta_s(x) \in S'(\mathbb{R}^n)$, with a discrete number of amplitudes $\{c_s \in \mathbb{R} | s \in S\}$ (the "raw data", or the pixel values typically stored in a file). The grid itself can thus be defined as the tempered distribution δ_S (the information of which one typically stores in the "header" accompanying the raw data). See also Problem 2.13.

Fourier transformation extends naturally to the dual space of $S(\mathbb{R}^n)$.

Definition 2.19 (Fourier Transformation on $S'(\mathbb{R}^n)$)
Recall Definitions 2.8 and 2.12. Fourier transformation $\widehat{F} = \mathcal{F}F$ is a continuous linear invertible mapping on $S'(\mathbb{R}^n)$. If $F \in S'(\mathbb{R}^n)$, then $\widehat{F}$ is the tempered distribution defined by

$$\widehat{F}[\phi] \overset{\text{def}}{=} F[\widehat{\phi}] \quad \forall \phi \in S(\mathbb{R}^n)\,.$$

By abuse of notation, writing $F = T_f$ for an appropriate "function under the integral" f, we may write

$$\widehat{f}(\omega) \overset{\text{def}}{=} \int dx\, e^{i\omega x}\, f(x)\,, \tag{2.8}$$

with inverse

$$f(x) = \frac{1}{(2\pi)^n} \int d\omega\, e^{-i\omega x}\, \widehat{f}(\omega)\,. \tag{2.9}$$

Caution: *never assign any significance to $f(x)$ or $\widehat{f}(\omega)$ for any fixed point x, repectively ω*; both forms derive their proper meaning only as distributions. See also Problem 2.15. We have the following inclusions: $S(\mathbb{R}^n) \subset L^1(\mathbb{R}^n) \subset S'(\mathbb{R}^n)$, so Definition 2.19 is a—subtle but essential—extension of Definition 2.8 as well as of the more general one for the space of Lebesgue integrable functions.

It is possible to define convolutions and correlations of distributions, most easily with the help of sloppy notation.

---

[9]However, by virtue of nice topology, it is not crucially important to do this; one may just as well use any *regular* tempered distribution, i.e. function of polynomial growth sufficiently close to such a point measure.

Definition 2.20 (Convolution on $S'(\mathbb{R}^n)$)
The convolution product of a tempered distribution $F \in S'(\mathbb{R}^n)$ with $\phi \in S(\mathbb{R}^n)$ is defined by

$$F * \phi(x) \stackrel{\text{def}}{=} \int dy\, f(y)\, \phi(x - y)\,.$$

And we have indeed a kind of "closure" property under convolution on $S'(\mathbb{R}^n)$.

Lemma 2.3 (Convolution on $S'(\mathbb{R}^n)$)
For all $F \in S'(\mathbb{R}^n)$, $\phi \in S(\mathbb{R}^n)$, we have $F * \phi \in S'(\mathbb{R}^n)$.

Definition 2.21 (Correlation on $S'(\mathbb{R}^n)$)
The correlation product of a tempered distribution $F \in S'(\mathbb{R}^n)$ with $\phi \in S(\mathbb{R}^n)$ is defined by

$$F \star \phi(x) \stackrel{\text{def}}{=} \int dy\, f(y)\, \phi(y - x)\,.$$

Lemma 2.4 (Correlation on $S'(\mathbb{R}^n)$)
For all $F \in S'(\mathbb{R}^n)$, $\phi \in S(\mathbb{R}^n)$, we have $F \star \phi \in S'(\mathbb{R}^n)$.

In fact, we could have taken $\phi \in \mathcal{E}'(\mathbb{R}^n)$ instead of $\phi \in S(\mathbb{R}^n)$ in Definitions 2.20 and 2.21 as well as in Lemmas 2.3 and 2.4 above (note that $S(\mathbb{R}^n) \subset \mathcal{E}'(\mathbb{R}^n)$); the essential requirement is that ϕ has a smooth Fourier transform. However, the only thing we will ever need in this book is convolution and correlation with smooth test functions $\phi \in S(\mathbb{R}^n)$. Also, we could have formulated Lemmas 2.3 and 2.4 more tightly: the convolution and correlation products are in fact *smooth functions* in $C^\infty(\mathbb{R}^n) \cap \mathcal{P}(\mathbb{R}^n) \subset S'(\mathbb{R}^n)$. This may seem a pretty powerful result, but we don't really need it in practice; it pertains to the kind of conventional smoothness that looks nice on paper, but has no operational significance. Nevertheless, this observation does help us conceptually.

Similar to Lemma 2.2 we have

Lemma 2.5 (Generalisation of Lemma 2.2)
If $F \in S'(\mathbb{R}^n)$ and $\phi, \psi \in S(\mathbb{R}^n)$, then

$$(F \star \phi) \star \psi = F \star (\phi * \psi)\,.$$

In view of Definitions 2.20 and 2.21 one could interpret the shift invariant operators $F*$ and $F\star$ as "function-valued" tempered distributions. The mapping of a test function yields a smooth function, which can in turn be regarded as a regular tempered distribution. Note that the convolution product incorporates parity reversal: $F \star \phi = F * \tilde{\phi}$. It can be shown that every translation invariant, continuous linear mapping on $S(\mathbb{R}^n)$ has a unique representation of the form $F*$ or, equivalently, $F\star$.

A general theorem useful in the context of convolution algebras follows from the Hölder inequality (Theorem 2.1).

Theorem 2.5 (Young Inequality)
Let $1 \leq p, q \leq \infty$ with $1/p + 1/q \geq 1$, and let $f \in L^p(\Omega)$, $\phi \in L^q(\Omega)$, then

$$\|f * \phi\|_r \leq \|f\|_p \|\phi\|_q \,,$$

where $\frac{1}{r} = \frac{1}{p} + \frac{1}{q} - 1$.

See e.g. Wheeden and Zygmund [297] for a proof. For further details on distribution theory, cf. [36, 46, 59, 91, 93, 94, 253, 254, 278].

2.4 Summary

This chapter was meant to explain to the mathematician the physical notions of source fields and detector devices involved in the realization of a measurement. The physicist in turn may have gained some insight in topological function spaces and their duals, and in the theory of distributions. For the interdisciplinary image scientist it is necessary to appreciate *both* views.

We have introduced the concept of a state space Σ as the space of all possible scalar field configurations, and that of a device space Δ, the collection of all admissible detector devices. We have then formulated a paradigm that puts these two concepts into correspondence by topological duality.

In the assertoric point of view one postulates a model for Σ and prescribes the detector devices by duality: $\Delta = \Sigma'$. This case has been argued to be pathological for image analysis purposes, since there is no guarantee that the device space can actually be implemented, nor that it is well-defined or well-behaved under even the most basic operations such as Fourier transformation and differentiation.

The paradigm $\Sigma \equiv \Delta'$ on the other hand expresses the operational point of view that a scalar field manifests itself exclusively in the joint response of fully controllable and operationally realizable detector devices to which it is exposed, while it declines to investigate the hypothetical structure of metameric sources that cannot be disambiguated by those devices. In the next chapter we propose a definition of an image based on this measurement paradigm. We will continue to pursue the by now fully motivated

Suggestion 2.1
Avoid functions, use functionals.

Problems

2.1

An operator $F : V \rightarrow W$ from a vector space V to a vector space W is called ill-posed in the sense of Hadamard if it is not well-posed. It is well-posed if the following existence, uniqueness and continuity conditions are met:

- $\forall w \in W \, \exists v \in V$ such that $Fv = w$,
- If $Fv_1 = Fv_2$ then $v_1 = v_2 \; \forall v_1, v_2 \in V$, and
- $F^{\text{inv}} : W \rightarrow V$ is continuous.

a. What does it mean for differentiation on $C^{\infty}(\mathbb{R}^n) \cap L^1(\mathbb{R}^n)$ (smooth, integrable functions, equipped with the $\| \cdot \|_1$-norm topology) to be ill-posed w.r.t. this definition? Give an example.

b. Why is differentiation on $S'(\mathbb{R}^n)$ well-posed? Illustrate this using the same example as in a.

c. What about differentiation on $S(\mathbb{R}^n)$? Why not simply use this function space to model an image?

2.2

What does it mean for a functional $F \in S'(\mathbb{R}^n)$ to be continuous? Argue why this is not a strong constraint, and why it is desirable for representing physical measurements. Can you give an example of a discontinuous functional?

2.3

Define 1_Ω as the indicator function on the set Ω:

$$1_\Omega(x) = \begin{cases} 1 & \text{if } x \in \Omega \\ 0 & \text{otherwise}, \end{cases}$$

and (for compact Ω) its volumetrically normalised counterpart χ_Ω:

$$\chi_\Omega(x) = \frac{1_\Omega(x)}{\int dx \, 1_\Omega(x)} .$$

Let $\beta(a; \sigma)$ be the interior of the sphere with radius σ centred at point a: $\beta(a; \sigma) = \{x \in \mathbb{R}^n \mid \|x - a\| < \sigma\}$.

a. Show that $\lim_{\sigma \downarrow 0} \chi_{\beta(a;\sigma)} = \delta_a$.

Let $\gamma_{a;\sigma}$ be the Gaussian measure of width σ centred at a:

$$\gamma_{a;\sigma}(x) = \frac{1}{\sqrt{2\pi\sigma^2}^n} e^{-\frac{1}{2}\left(\frac{x-a}{\sigma}\right)^2} .$$

b. Show that $\gamma_{a;\sigma} \in S'(\mathbb{R}^n)$ is a probability measure, i.e. $\int dx \, \gamma_{a;\sigma}(x) = 1$.

c. Show that $\lim_{\sigma \downarrow 0} \gamma_{a;\sigma} = \delta_a$.

2.4

Let $\gamma_{a;\sigma}$ be the normalised Gaussian measure as defined in Problem 2.3.

a. Show that $\gamma_{a;\sigma}$ satisfies the isotropic linear diffusion equation $\partial_s u = \frac{1}{2}\Delta u$ for every $a \in \mathbb{R}^n$ and $s \equiv \sigma^2 > 0$.

b. Justify the notation $\gamma_{a;\sigma} = e^{\frac{1}{2}\sigma^2\triangle}\delta_a$ for $\sigma \geq 0$.

2.5
Heaviside's function is defined as $H = 1_{\mathbb{R}_0^+}$ (see Problem 2.3).

a. Show that $H \in \mathcal{S}'(\mathbb{R})$ and $H'(x) = \delta(x)$ (a prime attached to a function or distribution denotes differentiation).

b. Show that $|x| \in \mathcal{S}'(\mathbb{R})$ and that $|x|' = \text{sgn}\,(x) \equiv H(x) - H(-x)$ and $|x|'' = 2\delta(x)$.

2.6
Define
$$\phi(x) = \begin{cases} e^{-1/x} & \text{if } x > 0 \\ 0 & \text{if } x \leq 0. \end{cases}$$

a. Show that ϕ is smooth. (Hint: show that, if $x > 0$, $\phi(x) \leq m!\,x^m$ for any $m \in \mathbb{Z}_0^+$, and that the k-th order derivative $\phi^{(k)}(x) = \mathcal{O}(x^{m-2k})$ for any integer $m \geq 2k$.)

b. Construct a smooth function $\psi \in C_0^\infty(\mathbb{R}^n)$ with compact support $\text{supp}\,\psi = [a_1, b_1] \times \ldots \times [a_n, b_n]$.

c. Show that if $\alpha \in C_0^\omega(\mathbb{R}^n)$, i.e. analytical as well as compactly supported, then $\alpha \equiv 0$.

2.7
a. Show that Fourier transformation "diagonalises" linear partial differential operators with constant coefficients $P(c, \nabla) = \sum_{|\alpha|\leq m} c_\alpha \nabla_\alpha$, i.e. $P(c, \nabla) = \mathcal{F}^{-1} P(c, -i\omega)\mathcal{F}$.

b. Show that $\mathcal{F}$ "diagonalises" the convolution product: $\phi * \psi = \mathcal{F}^{-1}(\mathcal{F}\phi\mathcal{F}\psi)$.

2.8
a. Show in which precise sense Definition 2.18 is compatible with conventional differentiation if F is a regular tempered distribution corresponding to a smooth function $f \in C^\infty(\mathbb{R}^n) \cap \mathcal{P}(\mathbb{R}^n)$.

b. Explain the statement: "∇ is antisymmetric", or alternatively, "$i\nabla$ is symmetric".

c. What is the transposed operator of the linear partial differential operator defined in Result 2.1?

d. Calculate the eigenvalues and eigenvectors of $i\nabla$.

2.9
Let $P(c, \nabla) = \sum_{|\alpha|\leq m} c_\alpha \nabla_\alpha$ be a linear partial differential operator. Show that it is an operator of the type $F*$ (Definition 2.20). In other words, a linear partial differential operator can be interpreted as a source field $F \in \Sigma$.

2.10
Consider the Fourier transform $\mathcal{F}$ as defined in this chapter. Define $\widetilde{\mathcal{F}} = (2\pi)^{-n/2}\mathcal{F}$. The *scalar product* of two $\mathbb{C}$-valued functions f and g is defined as

$$f \cdot g \overset{\text{def}}{=} \int dx\, f(x)g^*(x),$$

in which g^* is the complex conjugate of g.

a. Show that $\widetilde{\mathcal{F}}^2 = P$ where P is the parity reversal operator: $P(x) = -x$.

b. An operator U on a Hilbert space $\mathcal{H}$ is called unitary if $U(\mathcal{H}) = \mathcal{H}$ and $Uf \cdot Ug = f \cdot g$ for all $f, g \in \mathcal{H}$. Show that $\widetilde{\mathcal{F}}$ is a unitary transformation on $\mathcal{H} = L^2(\mathbb{R}^n)$.

2.11
Let $v_a : \mathbb{R}^n \to \mathbb{R}^n$ be the vector field given by $v_a(x) = (x - a)/\|x - a\|^n$ for $x \neq a$. Euler's gamma function is defined as

$$\Gamma(x) \stackrel{\text{def}}{=} \int_0^\infty dz\, e^{-z} z^{x-1} .$$

a. Show that $v_a^\mu \in \mathcal{S}'(\mathbb{R}^n)$ for each component $\mu = 1, \ldots, n$.

b. Show that $\nabla \cdot v_a \equiv \sum_{\mu=1}^n \partial_\mu v_a^\mu = A_n \delta_a$, in which A_n denotes the $(n-1)$-dimensional surface area of the unit hypersphere $\{x \in \mathbb{R}^n \mid \|x - a\| = 1\}$: $A_n = 2\sqrt{\pi}^n / \Gamma(n/2)$.

2.12
a. Show that $\mathcal{F}\delta = 1$ by using the "clean" definition $\mathcal{F}F[\phi] \equiv F[\mathcal{F}\phi]$ (Definition 2.19).

b. Solve the isotropic linear diffusion equation $\partial_s u = \Delta u$ for $u \in C^\infty(\mathbb{R}^n \times \mathbb{R}^+)$ by Fourier transformation under the initial condition $u(x; s = 0) = \delta(x)$.

c. As b but with initial condition $u(x; s = 0) = f(x)$, $f \in \mathcal{S}'(\mathbb{R}^n)$.

2.13
Define the regular n-dimensional discrete spacetime grid δ_X spanned by the basis vectors $\vec{e}_\nu$, $\nu = 1, \ldots, n$: $X \equiv \{\vec{k} = \sum_{\nu=1}^n k^\nu \vec{e}_\nu \mid k^\nu \in \mathbb{Z}\}$. A dual grid δ_Ω is spanned by a dual basis of covectors $\widetilde{e}^\mu$, $\mu = 1, \ldots, n$, defined by $\widetilde{e}^\mu[\vec{e}_\nu] \equiv \delta_\nu^\mu$ (recall that covectors map vectors to numbers), i.e. unity if μ and ν coincide, and zero otherwise: $\Omega \equiv \{\vec{k} = \sum_{\mu=1}^n k_\mu \widetilde{e}^\mu \mid k_\mu \in 2\pi\mathbb{Z}\}$.

a. Show that $\mathcal{F}\delta_X = \varepsilon_\Omega \delta_\Omega$ and, vice versa, $\mathcal{F}\delta_\Omega = \varepsilon_X \delta_X$, in which $\delta_S \equiv \sum_{s \in S} \delta_s$ is a distribution confined to the grid S and ε_S is a constant.

b. Show that ε_S is the n-dimensional volume of the elementary cell of the grid S, and that $\varepsilon_X \varepsilon_\Omega = (2\pi)^n$.

2.14
Assume you have a computational routine that performs some kind of task on the basis of an input image. In this context, "source" refers to the state of the physical field of interest prior to image formation.

a. Give an operational definition of "band-limitedness" of the source. Don't specify implementation details; state the answer in terms of the performance of your routine.

b. Likewise for "noise".

2.15
Show that Definition 2.19 is a good definition. (Hint: show that, if F is a regular tempered distribution with $f \in L^1(\mathbb{R}^n)$, the definition agrees with the usual one for $L^1(\mathbb{R}^n)$, i.e. $\mathcal{F}^\dagger = \mathcal{F}$.)

> *It has appeared that, if we take any common object of the sort that is supposed to be known by the senses, what the senses immediately tell us is not the truth about the object as it is apart from us, but only the truth about certain sense-data which, so far as we can see, depend upon the relations between us and the object. Thus what we directly see and feel is merely 'appearance', which we believe to be a sign of some 'reality' behind. But if the reality is not what appears, have we any means of knowing whether there is any reality at all? And if so, have we any means of finding out what it is like?*
>
> —Bertrand Russell

CHAPTER 3

Local Samples and Images

An image is a *coherent* set of *local samples*. It is the purpose of this chapter to turn this statement into a precise, operational definition. To begin with, we shall take it for granted that spacetime is a *continuum*, not because it *is*, but simply because a flotilla of useful mathematical tools is based on this hypothesis. See e.g. Herman Weyl [296]:

> "Vom Wesen des Raumes bleibt dem Mathematiker bei solcher Abstraktion nur die eine Wahrheit in Händen: daß er ein dreidimensionales Kontinuum ist."

It should be appreciated that there is really no compelling reason for assuming this (nor the opposite!) to be true, and that promising approaches essentially based on a *discrete* topology, e.g. in information theory [66, 107, 133], do exist. This fundamental issue—continuum versus lattice representation—clearly affects image analysis; matters like these are often to be found at the branching points of methodological main streams.

Coordinates like $x, y, z \in \mathbb{R}^n$ will be used invariably to label points $\overline{x}, \overline{y}, \overline{z} \in \mathbf{M}$ in spacetime. The existence of a global coordinate chart in classical spacetime allows us to be sloppy and talk about a point $x \in \mathbb{R}^n$ even if we mean $\overline{x} \in \mathbf{M}$. However, one should always keep in mind that coordinates can be randomly chosen and altered; we will return to this in Section 3.2.

We shall sometimes distinguish between spatial, temporal, and spatiotemporal images, depending upon whether these are intended to capture a single instantaneous snapshot (a "picture"), a sequence of successive snapshots viewed at a single point or non-localised (a "signal"), or a video sequence of successive spatial "slices" (a "movie"). Depending upon the level of abstraction, the distinction of spacetime into space and time may or may not be relevant. Whenever needed we shall assume that the first coordinate of x refers to time, and the other $d = n-1$ coordinates to space: $x = (x^0 = t; \vec{x})$. We sometimes use the self-explanatory factorisation $\mathbf{M} = \mathbf{T} \times \mathbf{X}$ in order to distinguish the respective manifolds.

3.1 Local Samples

Let us start with a definition of a local sample. From the previous chapter, in particular Assumption 2.1, Page 27, the following definition presents itself.

Definition 3.1 (Local Sample)
A local sample is a contraction of source and detector instances:

$$\lambda : \Sigma \times \Delta \to \mathbb{R} : (F, \phi) \mapsto F[\phi].$$

Abstract notation: $\lambda = \{F, \phi\}$.

In sloppy form: $\lambda = \int dx\, f(x)\, \phi(x)$. Samples are *always finite*, as opposed to point values of the underlying source field. Apart from this they may be positive, zero, or negative. A positive sample does not imply positivity of the underlying source field; a source is positive if *all* samples obtained by positive definite detectors turn out positive. If we want to compare different samples, we have to gauge our detectors by a conventional normalisation. For the moment we will require neither positivity nor normalisation.

Since Definition 3.1 claims to define a *local* sample, we have to be able to tell what its *base point* is. In order to do so we need an explicit definition of a *projection map* $\pi : \Delta \to \mathbf{M}$ which associates each detector element $\phi \in \Delta$ with its corresponding base point $x = \pi[\phi]$. This in turn assumes that we can perceive of Δ as a "bundle" of *local device spaces* Δ_x, comprising one "fibre" for each base point: $\Delta = \cup_{x \in \mathbf{M}} \Delta_x$. The inverse image $\pi^{-1}(x)$ of a base point x is, by definition, the entire local device space Δ_x at that point. A *local state space* Σ_x is then established as the physical degrees of freedom probed by a local device space Δ_x, i.e. "what we are looking at" with a localised detector. We will return to a precise definition of π in Section 3.9. For the moment it suffices to think of the base point as a the "centre of gravity" for the filter ϕ. The base point we would like to attribute to a sample is of course the one corresponding to the detector, but note that there is no way of telling from the *value* of a local sample "where it's at"; the geometric notion of a base point is established as an extrinsic detector property (a *label*).

Obviously, local samples are obtained at finite resolution. Again, being a spatiotemporal property, resolution cannot be inferred from a sample's value, only from its underlying aperture. A precise definition of the resolution of a local detector requires us to define a notion of *extent* or *inner scale* for that detector. A definition of inner scale will also be postponed until Section 3.9; think of it for the moment as the width of a central region, containing the filter's base point, where most of the filter's weight is concentrated (it is clearly not very useful to relate inner scale to detector support, since by construction this may be all of spacetime). See also Problem 3.2.

We can consider the transformation (*push forward*) of a detector under an arbitrary spacetime automorphism, i.e. a "warping", or a smooth transformation of spacetime with smooth inverse.

Definition 3.2 (Push Forward)
Let $\theta : \mathbf{M} \to \mathbf{M} : x \mapsto \theta(x)$ be a smooth spacetime automorphism. The push forward of a filter is then defined as the mapping

$$\theta_* : \Delta_x \to \Delta_{\theta(x)} : \phi \mapsto \theta_* \phi \stackrel{\text{def}}{=} J_{\theta^{\text{inv}}} \, \phi \circ \theta^{\text{inv}} \, ,$$

with Jacobian determinant $J_\chi \equiv | \det \nabla \chi |$.

This induces a natural, so-called *pull back* (also called "reciprocal image") of the source.

Definition 3.3 (Pull Back)
With the automorphism θ and its push forward θ_* as defined in Definition 3.2, the pull back of the source is defined as the mapping

$$\theta^* : \Sigma_{\theta(x)} \to \Sigma_x : F \mapsto \theta^* F \quad \text{defined by} \quad \theta^* F[\phi] \stackrel{\text{def}}{=} F[\theta_* \phi] \, .$$

In sloppy form this states that $\theta^* f \equiv f \circ \theta$, which physicists tend to refer to as "scalar field transformation" (Problem 3.3). Note that if ϕ lives at base point x, then its push forward $\theta_* \phi$ is associated with the *mapped* point $\theta(x)$, which explains its name. Naturally, pull back works the other way around.

Push forward and pull back are instances of a so-called "carry along" principle. If we have two communicating objects—i.c. sources plus detectors producing a response—then a change of either will in general be reflected in the output. Reversely, a given change in output can be explained as being caused by a change in either object. For example, shifting a patient underneath a scanner will have the same effect as moving the scanner in opposite sense over a stationary patient. This principle generalises to arbitrary deformations beyond rigid transformations (at least conceptually: one of the options is not necessarily in the interest of the patient). The idea is that at least one of these dual views is practicable and legitimate (e.g. processing scanner output).

It would be formally more correct to attach base points to sources and detectors matching the labels of Σ and Δ in Definitions 3.2 and 3.3, but that would yield rather cumbersome notations. There ought to be no confusion if we simply keep in mind the following commutative diagram:

$$
\begin{array}{ccc}
\Delta_x & \xrightarrow{\ \theta_*\ } & \Delta_y \\
\pi \downarrow & & \downarrow \pi \qquad \pi \circ \theta_* = \theta \circ \pi \\
x & \xrightarrow{\ \theta\ } & y \\
\pi \uparrow & & \uparrow \pi \qquad \pi \circ \theta^* = \theta^{\mathrm{inv}} \circ \pi \\
\Sigma_x & \xleftarrow{\ \theta^*\ } & \Sigma_y
\end{array}
\tag{3.1}
$$

In the literature on distribution theory one often pulls back the test functions rather than the distributions. Although it is strictly a matter of convention, it is common practice in image analysis to normalise the filters by volumetric integration, thereby implying the transformation law of Definition 3.2; the Jacobian takes care that normalisation is preserved despite transformations. This implies that ϕ is a *scalar density*, or *relative scalar*, with the dimension of an inverse volume, as opposed to the source f, which is an *absolute scalar* having an independent dimension (it may of course be dimensionless). The advantage of our convention, exploited later, is that

- images carry the same dimension as the underlying sources, and can thus in turn be regarded as source fields (image processing), and

- unlike densities—which the image may be *intended* to represent—, scalars can be represented in terms of plain numbers (image representation).

3.2   Covariance versus Invariance

A proper image definition should not depend upon an arbitrary system of coordinates. In this section we distinguish between a spacetime point $\bar{x} \in \mathrm{M}$, and its coordinate representation $x \in \mathbb{R}^n$, in order to reveal coordinate independence. For the sake of simplicity, this will be the only notational distinction made between intrinsic (manifold proper) and extrinsic (coordinatised) objects. It will be clear from the context whether a mapping (filter or source function, automorphism, etc.) maps points or coordinates.

Definition 3.1 is *generally covariant*. The meaning of this can be explained in two equivalent ways. If we refrain from using coordinates, it boils down to the following mathematical triviality.

Triviality 3.1 (General Covariance: Active View)
Let $\theta : \mathrm{M} \to \mathrm{M} : \bar{x} \mapsto \bar{y} \equiv \theta(\bar{x})$ be a smooth spacetime automorphism, with inverse $\eta : \mathrm{M} \to \mathrm{M} : \bar{y} \mapsto \bar{x} \equiv \eta(\bar{y})$. If $F \in \Sigma_{\bar{x}}$, $\phi \in \Delta_{\bar{x}}$, so that $\eta^* F \in \Sigma_{\bar{y}}$ and $\theta_* \phi \in \Delta_{\bar{y}}$, then trivially

$$
\eta^* F[\theta_* \phi] = F[\phi] .
$$

This is an unconventional way of stating general covariance, because one does not usually state trivialities[1] at all! It looks less trivial though when expressed in terms of coordinates. The "principle of general covariance" then dictates certain "coordinate transformation laws" for the parametrised objects of interest, making it look less trivial than it really is. The crux is that, on a coordinate level, we may "creatively mis-interpret" the manifold automorphism in Triviality 3.1: once having stated it in terms of a *given* coordinate system (*active* view), we can re-interpret the coordinate expression as a *change of coordinates* (*passive* view). That is, from the coordinate expression *per se* we cannot tell whether it is intended to represent an actual *dragging* of points along the manifold, expressed in terms of a *fixed* coordinate system, or merely a *relabelling* of points, i.e. a *change* of coordinates, leaving the manifold unaffected.

Triviality 3.2 (General Covariance: Passive View)
Let $\theta : \mathbb{R}^n \to \mathbb{R}^n : x \mapsto y \equiv \theta(x)$ *be the coordinate representation of the automorphism defined in Triviality 3.1, with inverse* $\eta : \mathbb{R}^n \to \mathbb{R}^n : y \mapsto x \equiv \eta(y)$. *Let* $f(x)$ *and* $\phi(x)$ *be parametrisations of F and* ϕ *in terms of a fiducial coordinate system* x, *and assume that* $f'(y)$ *and* $\phi'(y)$ *are the reparametrisations of* $f(x)$ *and* $\phi(x)$ *corresponding to* $\eta^* F$ *and* $\theta_* \phi$, *respectively, but with* θ *re-interpreted as a coordinate transformation. Then we have*

$$f'(y) = f(x) \quad \text{and} \quad \phi'(y) = |\det \frac{\partial x}{\partial y}| \phi(x).$$

The active view on covariance boils down to the ability to warp spacetime and everything living in it in a consistent manner, such that all *interactions* (the coincidence of events) remain unaltered. The effort of doing so is purely *mathematical*; the physical spacetime manifold derives its intrinsic existence only indirectly via such interactions, not as a "container" filled with objects. Thus if everything is dragged along with the point mapping, then nothing has happened *physically*, an observation already pointed out by Albert Einstein:

> "There is no such thing as an empty space, i.e. a space without field[2]. Spacetime does not claim existence on its own, but only as a structural quality of the field."

This view is shared by Moritz Schlick, who argues that space and time lack physical objectivity:

> "Space and time are not measurable in themselves: they only form a framework into which we arrange physical events."

The passive view on covariance is nothing but the appreciation that coordinates are mere dummies, the choice of which should not affect physically meaningful expressions. Thus if we do introduce coordinates, then either

- we adopt the transformation rules in Triviality 3.2 *on principle*, or

---

[1] From a *physical* viewpoint it is not *quite* trivial; the objects on the l.h.s. live at $\bar{y} \in M$, the ones on the r.h.s. at $\bar{x} \in M$; apparently one can deform "empty spacetime" *arbitrarily*!

[2] I.e. metric field in Einstein's context.

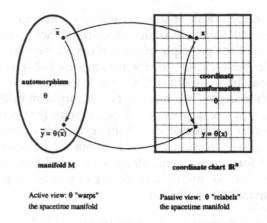

Figure 3.1: Active versus passive point of view on general covariance: the mapping $x \mapsto y = \theta(x)$ can be interpreted either as a coordinate transformation, so that x and y correspond to the *same point* $\bar{x}$ in *different coordinate systems* (passive view), or as a manifold transformation, so that x and y refer to *different points* $\bar{x}$ and $\bar{y}$ in the *same coordinate system* (active view).

- we decide on an unambiguous coordinate convention

in order to guarantee *manifest* covariance. Note that a restriction to Cartesian coordinates, frequently employed in the literature, is *not* unambiguous, and requires us to account for "Cartesian transformation rules" (a proper subgroup of general automorphisms). Figure 3.1 illustrates the active and passive points of view. See also Problems 3.5 and 3.6.

Clearly, the "principle of general covariance" is *not* a physical principle; *any* spacetime theory can be stated in generally covariant form[3] [92, Section II.2] [211, Section 12.5]. In particular any image model—if it does not aim to describe grid details—can be stated in such a form. Whereas coordinate-free notation is conceptually convenient, the use of coordinates has obvious computational merits. We will therefore use both.

A nontrivial notion is that of *invariance*, or *symmetry* under the action of an arbitrary group of spacetime transformations. Invariance is often confused with covariance, creating intricate paradoxes. Here is a typical one: say, we want to design a linear, spatial filter for a particular image processing task, using Cartesian coordinates. The semantics of the problem dictates isotropy and homogeneity (say), so we require our filter to be invariant under rotations and translations, i.e. "coordinate transformations" of the type $x \mapsto y = Rx + a$, in which R is a rotation matrix. This constrains our solutions to convolution type filters depending upon a single, radial coordinate $r = \|x\|$. On the other hand, however, we have just argued that coordinate transformations cannot produce constraints! The paradox resolves itself once we appreciate that invariance is not the same as covariance.

---

[3]This has even led people to reject Einstein's theory of general relativity, because it was conjectured that this theory was *deduced* from the physically void principle of general covariance!

Definition 3.4 (Invariance)
Notation as in Triviality 3.1. A source $F \in \Sigma$ is called invariant if

$$\eta^* F = F.$$

Likewise a detector $\phi \in \Delta$ is called invariant if

$$\theta_* \phi = \phi.$$

It is understood that left and right hand sides are to be evaluated at the same base point. In terms of coordinates, using the same notation as in Triviality 3.2, this implies:

$$f'(y) \overset{\text{cov}}{=} f(x) \overset{\text{inv}}{=} f(y),$$
$$\phi'(y) \overset{\text{cov}}{=} |\det \tfrac{\partial x}{\partial y}| \phi(x) \overset{\text{inv}}{=} \phi(y).$$

In other words, f' and ϕ' are *functionally identical* to f and ϕ under a symmetry transformation (or a coordinate transformation that formally corresponds to that). Invariance is meant to be interpreted from an *active* point of view. Confusion, such as sketched in the above paradox, arises due to mis-interpretation at the coordinate level and abuse of terminology; in the case at hand we did not mean to shift and rotate our coordinate frame (a physically void action), but our actual filter object. The abstract part of Definition 3.4 avoids coordinates and is consequently transparent and unambiguous.

As opposed to covariance, invariance does pose a physical constraint; the larger the group, the stronger the constraint. Clearly a very strong (and rather trivial) constraint would result if we would require the general group of spacetime automorphisms to be a symmetry group for our detectors (or sources, although there is in general very little to require here; but think of synthetically created *test images*, or *statistical symmetry assumptions* about sources in the world). An object invariant under this most general group may be called a *topological invariant*. A constant number is a trivial, but important example: counting events in spacetime does not depend upon how we deform it (diffeomorphically).

Ensembles of spacetime objects may exhibit stronger symmetries than individual members. Indeed, by our choice of $\Delta = S(\mathbb{R}^n)$, we observe that under automorphisms θ we have the "closure" property $\theta_* \Delta = \Delta$, a fact that will in general cease to hold if we replace Δ by a proper subset. This shows that for $\Delta = S(\mathbb{R}^n)$ Definition 3.1 reflects the topological structure of spacetime. But, in anticipation of things to come, it also entails a warning that we should not arbitrarily constrain Δ if we want to maintain consistency with our conception of spacetime: in a way, Δ *is* our spacetime!

3.3 Linearity

The premise underlying Definition 3.1 is of course *linearity*. This is natural from a geometrical point of view if a local sample is indeed intended to represent a density (or n-form): [158, 211, 264, 265].

One might nevertheless conjecture that we should not exclude nonlinear representations. This is certainly legitimate if one knows something about the physics of image formation and the details of reconstruction, and wants to bring grey-value samples into linear correspondence with the physical quantity of interest (quantitative reconstruction). But if we lack such *a priori* knowledge, or if we assume that quantitative reconstruction has already been accomplished, then there is no reason for introducing nonlinearities, at least not for the purpose of a structural representation. It is important to appreciate that this is not a limitation, for even retrospective compensation for reconstruction artifacts is a semantical task which, like any other task in image processing, requires a syntactical representation to start out from. Without loss of generality we may take that propriospective representation to be linear.

In the following section we consider linearity in somewhat more detail.

3.3.1 ✳ Linearisation from an Abstract Viewpoint

It is convenient to simplify expressions by the so-called *condensed index notation*: see Box 3.1. Using this convention, let us parametrise an arbitrary source $g \in \mathbb{R}^l$ by g_α, and detector $\psi \in \mathbb{R}^m$ by ψ^i, with $\alpha = 1, \ldots, l$ and $i = 1, \ldots, m$. We may assume that $l \geq m$ (otherwise we have a redundant set of detectors and we may skip some). Since we do not assume equality, samples will generally represent metameric classes of source configurations (note that both l and m may be infinite).

Generally a (not necessarily linear) sample, $\lambda \in \mathbb{R}$ say, will depend nontrivially upon source as well as detector: $\lambda = \lambda(g, \psi)$. Assuming stable null response in the absence of a source, and in the limit of vanishing detector strength, we can formally expand $\lambda = g_\alpha R_i^\alpha(g)\psi^i + \mathcal{O}(g\psi^2)$. We may assume that the matrix $R_i^\alpha(g)$ has full rank m (otherwise, cancel redundant combinations of ψ and redefine m accordingly). Define the mapping from $\mathbb{R}^l$ onto $\mathbb{R}^m$ by $f_i \equiv g_\alpha R_i^\alpha(g)$; all g that are thus mapped to a fixed f are encapsulated in a metameric class. Finally we can invoke a suitable diffeomorphism, mapping $\psi \mapsto \phi$ so as to get rid of the $\mathcal{O}(g\psi^2)$-residual, and we obtain a canonical, linear source field representation $\lambda = f_i\phi^i$. Of course these are merely formal manipulations; the actual reconstruction stage should ideally produce such a canonical representation in the form of a "raw image".

Linear representations are in some sense least committed, since all grey-values carry equal weight. However, it depends upon our goals and priors whether or not we may want to bias the representation *a posteriori* by semantically inspired nonlinearities. Retrospective correction for "reconstruction nonlinearities" is an example. Also, in image renderings one often emphasizes certain grey-values by a nonlinear transformation of the data $\lambda \mapsto F(\lambda)$ (histogram equalisation, gamma-correction, contrast and brightness adjustment, etc.). As long as F is invertible, we obtain conceptually equivalent representations, in the sense that there is no loss of information *prior to quantisation* and in the *absence of noise*.

Non-invertible nonlinearities are *the* way to create new metameric classes from old ones. This is the core of semantical operations in image processing, image analysis and pattern recognition. After all, if we manage to *explicitly* encapsulate all task-irrelevant image details into a single class object—i.e. if we know the right metamers—we have in fact disclosed the solution space!

3.4 Images

Let us now turn to the definition of an image. An image is a *coherent* set of local samples. According to Koenderink [149] (recall Footnote 1 on Page 1):

> "[...] pictures are pictures relative to some input device (format), in them-selves they are but records. A record may be a different picture to different input devices."

Coherence can be obtained in many ways, but the essence is that it exists only with respect to the conventional format of an input device, or with respect to an external observer[4]. We may want to impose coherence on Δ in a way that reflects our *a priori* spacetime model. We are particularly interested in the New-tonian model, but one can think of other models as well, e.g. that of a Galilean spacetime, which incorporates a symmetry under constant relative motion (so-called Galilean boosts [79]), a Minkowski spacetime (Lorentz invariance) or even a curved spacetime (invariance under all spacetime automorphisms). Often we want to emphasize only a particular symmetry aspect. In that case we account only for the relevant subgroup. In principle, however, we do not admit incom-plete spacetime models; otherwise we should have constrained our symmetry group right from the start.

For the present discussion let us de-emphasize the details of the model, and consider an arbitrary, smoothly parametrised symmetry group Θ—a so-called real-analytical *Lie group*: see Box 3.2—assumed to act on the spacetime manifold.

The idea is to construct parametrised ensembles of detectors that reflect the structure of the symmetry group and thus that of our spacetime model. A method to accomplish this is to build in the symmetry group of that model, as follows. Start with an arbitrary base point, the origin say, or rather with a filter ϕ assumed to be attached to that point (recall Page 40: $\pi[\phi] = 0$). This is of course in no way a preferred one, but once we have such a filter, we can push it forward under the symmetry group Θ, the parametrisation of which requires a set of (discrete or continuous) parameters[5], collectively denoted by ℓ, say. If Θ acts *transitively* on spacetime, meaning that we can reach any point x from our point of departure by a suitable choice of group parameters, then "copies" of ϕ may be distributed all over spacetime in this way. Transitivity is essential to avoid that our spacetime domain looks like a swiss cheese, filled with gaps or other topological artifacts. Transitivity will certainly hold if the translation group $T(n)$ is a subgroup of Θ.

---

[4]If, in turn, we want to explain the observer's own visual percept without infinite recursion, we cannot use pictures. Pictures may exist in the brain, but are not looked at by little men inside; recall Box 2.2.

[5]Θ may be an infinite dimensional Lie group, meaning that we need infinitely many parameters. This is for instance the case with the group of all spacetime automorphisms.

We will henceforth assume that this is the case, in other words, we commit our-selves from now on to the assumption that spacetime is *homogeneous* (a complete definition of Θ will be given in Section 3.7).

Parametrising an arbitrary transformation as $\theta \equiv \theta_\ell \in \Theta$, the push forward $\theta_*\phi$ will be an ℓ-parametrised filter given by Definition 3.2. If we do this for every tag ℓ, the result is an *orbit* of detectors within Δ. We can make transitivity explicit by splitting ℓ into a spatiotemporal *scanning* parameter $x \in T(n) \sim \mathbb{R}^n$ (it could be a Fourier coordinate as well, or in general any Lie parameter corresponding to a so-called *representation* of the translation group) and a *tuning* parameter (com-prising a discrete and/or continuous spectrum of values; it may also be void), s say: $\ell = (x; s)$. By transitivity, starting at the origin, the result of the group action will be an s-parametrised ensemble of filter instances at every base point x. We may repeat this for any other filter living at the origin.

At this point the push forward procedure merely yields a stratification into invariant orbits of the filters we already assumed to exist all over the place, but as soon as we select a proper subset of $\Delta = S(\mathbb{R}^n)$, then this particular strat-ification will become significant. Given a more restrictive class of filters (such as the ensemble of "point operators" to be introduced in Section 3.9), the pro-cedure tells us how to glue these together. For each $\phi \in \Delta$ the invariant orbits $\Theta_*\phi \equiv \{\theta_*\phi \,|\, \theta \in \Theta\} \subset \Delta$ reflect the nature of our spacetime model; by construc-tion, they project down to the spacetime manifold and have all its characterising symmetries built in. The spacetime symmetry group embodied in the defini-tion thus provides the desired spatiotemporal coherence. The least committed representation of an image should aim to incorporate no more than the *a priori* spacetime model.

Definition 3.5 (Image)
See Definitions 3.1, 3.2 and 3.3. Let Θ be a transitive spacetime symmetry group. Furthermore, let $F \in \Sigma$ be a source field, $\phi \in \Delta$ a local detector with base point $\pi[\phi] = 0$, and $\lambda = F[\phi]$ a corresponding local sample. Then an image L is defined as the Θ-orbit of local samples:

$$L = \{\theta^*\lambda \,|\, \theta \in \Theta\} \quad \text{with} \quad \theta^*\lambda \stackrel{\text{def}}{=} \theta^*F[\phi] = F[\theta_*\phi].$$

Abstract notation: $L \equiv \{F, \phi, \Theta\}$.

We will indicate the orbit of source and detector by Θ^*F and $\Theta_*\phi$, respectively, so $\Theta^*F = \{\theta^*F \,|\, \theta \in \Theta\}$, $\Theta_*\phi = \{\theta_*\phi \,|\, \theta \in \Theta\}$, etc. Note that, if $\Delta = S(\mathbb{R}^n)$, the Θ-orbits $\Theta^*F \subset \Sigma$ and $\Theta_*\phi \subset \Delta$ are invariant subspaces of Σ and Δ, respectively, for any choice of $F \in \Sigma$ and $\phi \in \Delta$. Of course, in practice one considers only finite, connected regions of the orbits (finite field of view, finite time span, finite parameter range, etc.).

At this level of abstraction the essence of our definition stands out most clearly. The essence is that an image entails three things [73, 75, 78]:

- a given source $F \in \Sigma$ (obviously),

- a conventional detector $\phi \in \Delta$ to contract the source into a numeric sample, and

- a symmetry group Θ that reflects the *a priori* spacetime model.

In particular, it does *not* entail coordinates or pixel indices. Note that a derivative $\nabla_\alpha L \equiv \{F, \nabla_\alpha^\uparrow \phi, \Theta\}$ is itself a well-defined image.

Definition 3.5 embodies the entire spacetime symmetry group. This guarantees full symmetry without any symmetry requirements whatsoever on the underlying filter. At this point we still lack a defining paradigm for this filter; any $\phi \in \mathcal{S}(\mathbb{R}^n)$ is in principle a feasible "prototype" for the family $\Theta_* \phi$.

If Θ represents a connected Lie group, then L is a function of its Lie parameters, but the definition does not exclude discrete symmetries such as reflections. In particular, if we parametrise the group Θ by $\ell = (x; s)$ and freeze the tuning parameter s, then, by spatiotemporal transitivity, we may interpret an image as a globally defined source field $L_s \in \Sigma$, with function representation $L(x; s)$. Definition 3.5 can then be regarded as a parametrised ensemble of spatiotemporal "image sections" (slices parametrised by s), i.e. images in the previously defined sense in which one only incorporates spacetime homogeneity.

Lemma 3.1 (Image Section)
See Definitions 2.21, 3.1, and Lemma 2.4 for notation. If $\Theta = \mathrm{T}(n)$, the spacetime translation group, then Definition 3.5 reduces to

$$\Lambda : \Sigma \times \Delta \to \Sigma : (F, \phi) \mapsto F \star \phi.$$

Abstract notation: $\Lambda = \{F \star, \phi\}$.

Proof 3.1 (Lemma 3.1)
Problem 3.10.

It is important to appreciate that an image section generally embodies an *incomplete* spacetime model. Definition 3.5 takes the full model into account and is therefore preferred. However, if we keep the tuning parameter s in the back of our mind we can use the representation of Lemma 3.1 instead. Of course, Lemma 3.1 would be an instance of Definition 3.5 if the *full* spacetime symmetry group happens to be $\mathrm{T}(n)$, i.e. if s is void. However, we will see later that this is *inconsistent* with basic image processing requirements! It will be shown that consistent models will have to account for spacetime *scaling*.

By the assumption of transitivity, the procedure of following an orbit described in Definition 3.5 provides a dense spatiotemporal arrangement—in theory a continuum, in practice of course some discrete grid—of (ensembles of) local detectors. When using a pixel grid we should not allow for "pixel-uncorrelated" samples, since that would mess up our topology; it would give us a mere *set* of local samples, which does not reflect the *coherent* structure of the underlying field configuration. Oversampling enforces coherence. In the next sections we will be more specific about source (Section 3.5), group (Section 3.7) and detector paradigms (Section 3.9).

3.5 Raw Images

It is clear that we cannot store a "naked" source field, since we are always dealing with an image of it[6]. The image acquisition device in combination with the reconstruction procedure has resulted in a "raw image" $F = \{F_0, \phi_0, \Theta_0\}$, in which the source of interest F_0 is confounded with a specific point spread function ϕ_0 subjected to the action of some group Θ_0 (typically the incomplete group of discrete translations on a lattice), the details of which are reconstruction and modality specific.

Raw images are typically represented as rastered data and stored in a disk file. This is however a "bad" representation, since all modality limitations and grid artifacts are *explicit* in a raw image; after all, reconstruction aims to capture the source of interest by means of a (discretised) scalar function, the kind of thing we already rejected in Section 2.1 as being an undesirable representation for most image analysis purposes. By definition, *relevant* information is independent of (although likely affected by) the details of the grid and the limitations of the acquisition instrument; a medical doctor is interested in the pathology of his patient, not of his CT scanner.

If, however, we perceive of an image as yet another source field, then to some extent we can "overrule", or rather *hide* the measurement pathology inherent to a raw image. Since it is generally the only source at our disposal, we will henceforth take the raw image to be (a representation of) our initial source field F, and model it as a general point measure (recall the remark on rastered images on Page 33).

Definition 3.6 (Raw Image)
A raw image is a linear combination of point distributions on a grid X:

$$F = \sum_{x \in X} f(x)\, \delta_x \,.$$

This means that we hide the physical image formation details of $F = \{F_0, \phi_0, \Theta_0\}$, which belong to the domain of image reconstruction. For syntactical purposes this is legitimate as long as we can verify that the effect of doing so is negligible in our computations (if not, we may want to improve the reconstruction method or plead for better equipment). In semantical routines we may need to account for the image formation details (*"a priori* knowledge"). The source model of Definition 3.6 is quite practical in image analysis, since it directly maps to the typical image file formats comprising header information (grid definition: X) and pixel data (the numbers $f(x)$, $x \in X$), but usually not all details of the physics of image formation.

Of course we can think of the distributional derivative of a raw image as a linear combination of derivatives of the Dirac distribution.

---

[6]Cf. Kant's notion of *"das Ding an sich"*; recall the quotation on Page 13, and Footnote 4 on Page 16.

3.6   Static versus Dynamic Representations

The detector choice is one of convention, and thus subject to debate. An unambiguous definition must therefore either entail a clear motivation of selection criteria, or otherwise treat the detector as a dummy variable. In the latter case one essentially defines an image as the coherent set of sources $\Theta^* F$, rather than samples. This amounts to identifying an image with an unspecified linear image processing routine with given input data F and spacetime model Θ. This is indeed a rather uncommitted representation, but the only feasible machine representation is to store the raw data of F. It has already been argued why the raw image of Definition 3.6 is itself not a good representation, but rather a limiting case of Definition 3.5, in which grid artifacts are predominant. A complete machine representation of local samples according to Definition 3.5 would be pretty straightforward, were it not for the prohibitive factor that $\Delta = \mathcal{S}(\mathbb{R}^n)$ constitutes an infinite dimensional function space. It is obviously impossible to implement a discrete set of filters that is sufficiently dense in such a space. Whereas it has been argued that this is not a problem for image processing, in which one typically implements, or dynamically allocates a small subset of filters suitably adapted to a specific task (*dynamic representation*), it *does* pose a problem for a *static representation* if the details of such a task cannot be anticipated. However, we can exploit the fact that, by construction, images form a linear space; the superposition principle allows us to single out any subspace $\Gamma \subset \Delta$ as long as it spans Δ. If $\mathrm{span}\,\Gamma = \Delta$, an arbitrary element of Δ can always be *effectively* realized by any particular image processing application based on the representation Γ. We may call such a static representation *uncommitted*.

Clearly one should not take "uncommitted" too literally, in the sense of being "void of *a priori* assumptions". In particular we did commit ourselves to a particular device space $\Delta = \mathcal{S}(\mathbb{R}^n)$, albeit a very flexible one, and we are about to commit ourselves to a particular spacetime model, even though it may not seem a very shocking one. The aptness of such commitments must always be evaluated in a practical situation. It is nevertheless plausible that most image analysis tasks are well served by the *a priori* assumptions underlying our image representation (the "proto-semantics" of coupling syntax to a physical environment [161]). Making these explicit does not only serve the model's transparency, but also its susceptibility for future extensions and modifications.

If we want the subspace Γ to be compatible with our spacetime model, it has to be an invariant subspace of Δ under the action of the spacetime symmetry group: $\Theta_* \Gamma = \Gamma$ (again, notation is self-explanatory: $\Theta_* \Gamma = \{\theta_* \phi \mid \theta \in \Theta, \phi \in \Gamma\}$ etc.). Manifest invariance is guaranteed when considering a Θ-orbit $\Gamma_0 \equiv \Theta_* \phi$ of a paradigmatic, "zeroth order" filter $\phi \in \Delta$. Such a filter could be used to define a *point* in operational sense: Box 3.3. The coherent spacetime manifold then arises by projection of smoothly connected copies of this prototypical point: $\mathbf{M} = \pi[\Theta_* \phi]$. A representation Γ_0 of zeroth order image structure induces a representation of differential structure up to arbitrary order k, Γ_k say, viz. by considering the orbit of the transposed derivatives of ϕ up to order k: $\Gamma_k \equiv \Theta_* \nabla_\alpha^\dagger \phi$ with $|\alpha| \leq k$ (recall Definition 2.18, Page 31). In this way we obtain $\Gamma \equiv \Gamma_\infty$ endowed with an explicit hierarchy according to differential order.

This leaves us with two unsolved problems: an explicit spacetime model Θ and a filter paradigm that unambiguously establishes the form of the basic point operator ϕ. The rest of this chapter is meant to fill in the missing details needed for an actual implementation of Γ. The line of approach will be as follows.

- We decide on an explicit spacetime model, i.e. on an explicit spacetime symmetry group Θ, the *Newtonian symmetry group*: Section 3.7.

- We establish a basic, zeroth order point operator ϕ on the basis of a consistency requirement: Sections 3.8 and 3.9.

- We construct its orbit under the action of the group Θ: $\Gamma_0 = \Theta_* \phi$. By differentiation, this induces a representation Γ_k up to arbitrary order k: Section 3.10.

- We verify whether the resulting family $\Gamma \equiv \Gamma_\infty$ is complete and indeed spans $\Delta = \mathcal{S}(\mathbb{R}^n)$: Section 3.11.

For such an outstandingly interdisciplinary field as image analysis, it is important to *understand* results beyond any formally correct construction. For this reason we will discuss the resulting image representation from several points of view in the next chapter.

3.7 The Newtonian Spacetime Model

In this section we will elaborate on the details of the classical, Newtonian spacetime model. Although this is the kind of phenomenological spacetime we are all familiar with, it is geometrically much more intricate than Einstein's relativistic spacetime model. It was the remarkable mathematician Élie Cartan who explained its geometry only *after* the relativistic model had been established [32].

Nevertheless, the traditional Newtonian model is believed to be most appropriate for almost all imaging situations that arise in practice. According to this model, (empty) spacetime is a flat manifold stratified into "planes of simultaneity" over absolute time [92, Section III.1] [211, Sections 12.2, 12.3 and 21.4]. If we disregard time we could call it the Euclidean model—flat space has a Euclidean metric—but in all cases we explicitly include *scaling*, something that is often either taken for granted or overlooked when people use this terminology. See Figure 3.2 and Box 3.4.

We will not scrutinise all geometrical details of the classical spacetime model here. For our purpose it suffices to consider it from the invariance point of view. For later use we will define the slightly more general, so-called Galilean model, which contains the Newtonian model of interest as a stationary limit.

Definition 3.7 (Galilean/Newtonian Spacetime Model)
The Galilean spacetime model is defined by the symmetry group Θ *acting on "empty spacetime" by affine*[7] *transformations of the type*

$$\theta_{A,x}(z) = A\,z + x\,,$$

in which $x = (x^0 = t; \vec{x})$ *parametrises a spacetime translation, and* A *a spatial rotation, spatially isotropic scaling, and uniform motion, which can be uniquely decomposed as* $A = e^\Lambda\,X = X\,e^\Lambda$, *with*

$$\Lambda = \left[\begin{array}{c|c} \lambda^0 & \emptyset \\ \hline \emptyset & \lambda I \end{array}\right] \quad \text{and} \quad X = \left[\begin{array}{c|c} 1 & \emptyset \\ \hline \vec{v} & R \end{array}\right].$$

Here, $R \in \mathrm{SO}(n-1)$ *is a spatial rotation matrix:* $RR^\mathsf{T} = R^\mathsf{T}R = I$, $\det R = 1$, $\vec{v}$ *is a constant spatial velocity, and* $\lambda^0, \lambda \in \mathbb{R}$ *are real numbers. The Newtonian spacetime model is defined likewise, but without the velocity degree of freedom:* $\vec{v} \equiv \vec{0}$.

The "transvection" X that results if we put $R = I$ is called a Galilean boost [211, Section 12.3]. See also Problem 3.7.

It is often natural to extend Θ by admitting discrete parity transformations (spatial reflections and time reversal). This is certainly the case for a mere syntactical description; only semantical routines are potentially able to distinguish an image from a mirrored instance (think of character recognition), or a causally displayed movie from a reversed one (not an easy task). However, the inclusion of parity is not crucially important to what follows; whenever it is relevant, it will be pointed out.

Definition 3.7 defines classical spacetime via the action of a classical invariance group on "empty spacetime". That is rather peculiar if you come to think of it; how can something act on a void? The operational significance of this is of course in terms of the induced action on real objects, i.c. source fields and detector devices. It is important to appreciate that the action of a given group *on a physical system* depends upon *system specific parameters*. One refers to specific realizations of such physical actions as "group representations". In a way it is better to say that the collection of all group representations affecting physical objects induces the action on "empty spacetime", rather than the other way around. This was of course the essence of Einstein's remark on Page 43, but it also touches the very core of the local sign problem (recall Box 2.2, Page 24): one needs to "embody" empty spacetime in terms of physical machines.

Examples of system specific parameters are the *present moment* and the *foveal point* in active vision; we return to this in the next chapter. Here we consider only the simplest case (no nontrivial system parameters), appropriate for off-line analysis. Putting Definition 3.2 in the specific context of Definition 3.7 we then arrive at the following

---

[7]Only affine transformations are compatible with a flat spacetime, or with a so-called "flat connection" [92].

Result 3.1 (Galilean/Newtonian Spacetime Compatible Filters)
See Definition 3.7. Given a prototypical point operator $\phi \in \Delta$, a family of point operators compatible with the Newtonian or Galilean spacetime model is given by the orbit $\Gamma_0 = \Theta_*\phi$, consisting of parametrised filters of the type

$$\phi_{A,x}(z) = |\det A^{\text{inv}}|\, \phi\left(A^{\text{inv}}(z - x)\right) ,$$

in which we have $|\det A^{\text{inv}}| = e^{-\text{tr}\,\Lambda}$ and $A^{\text{inv}} = e^{-\Lambda} X^{\text{inv}}$, with

$$X^{\text{inv}} = \left[\begin{array}{c|c} 1 & \emptyset \\ \hline -R^T \vec{v} & R^T \end{array} \right] .$$

The filter $\phi_{A,x}$ lives at scan point x and is tuned to spatiotemporal scale e^Λ, spatial velocity $\vec{v}$ ($\vec{v} \equiv \vec{0}$ respectively), and spatial orientation R relative to its default values. Using the entire filter ensemble $\Gamma_0 = \Theta_*\phi$ ensures that, regardless the choice of operator ϕ, all points in the image domain will be treated on equal foot (homogeneity), all orientations are *a priori* equivalent (isotropy, of course only within spatial slices), and any size and duration is as good as any other (scale invariance). Especially the significance of the last requirement is not always appreciated in image analysis, but it is absolutely crucial if we want to free image models from pixel artifacts. Scales may be expressed in proportion to grid constants ("pixel scale"), but any *fixed* relationship would imply an *ad hoc* scale bias, and an undesirable confounding with the details of the grid.

It frequently happens in practice, for example in 3D medical imaging, that the image source is specified on a regular, but non-cubic spatial grid (recall Definition 3.6 and Problem 2.13 on Page 38). Since scales must be expressed as multiples of grid units, we must take care to compensate for this. This simply boils down to a replacement of the 2-parameter matrix Λ by a general diagonal matrix compensating for this grid anisotropy. Note that there is still only 1 independent spatial parameter!

3.8 Image Processing

In this section we concentrate on necessary conditions for selecting candidate point operators $\phi \in \Delta$. The actual construction is deferred until the next section. With spacetime given by Definition 3.7, Result 3.1 yields the complete zeroth order filter space $\Gamma_0 = \Theta_*\phi$. An *ad hoc* choice is precluded by a *consistency requirement*.

Recall that in the previous chapter we expressed our desire to perceive of a sampled image as yet another source field. This implies the possibility to sample this source field in turn (i.e. by carrying out an observation). Indeed, the *crux of image processing* is the ability to cascade observations: Figure 3.3.

Crux 3.1 (Image Processing)
An image is itself a source field that can be sampled in a cascade fashion.

This calls for additional structure on Δ. The mathematical notion closest to a cascade structure is an *algebra*. An algebra $\mathcal{A}$ (over the field $\mathbb{R}$) is a group endowed with two internal operations and one external operation, usually called "addition", "multiplication" (or "concatenation") and "scalar multiplication", respectively. Addition and scalar multiplication is enabled by the vector space structure already imposed on our device space, so the new thing is the ability to "multiply" elements. The requirements for this are as follows. Denoting the infix multiplication operator by $\circ$, we have, for all $a, b, c \in \mathcal{A}$:

$$(a \circ b) \circ c = a \circ (b \circ c), \tag{3.2}$$
$$a \circ (b + c) = a \circ b + a \circ c, \tag{3.3}$$
$$(a + b) \circ c = a \circ c + b \circ c. \tag{3.4}$$

In words, $\circ$ is associative, and distributive with respect to addition. Moreover, scalar multiplication must be such that for all $a, b \in \mathcal{A}$ and $\lambda \in \mathbb{R}$,

$$\lambda(a \circ b) = (\lambda a) \circ b = a \circ (\lambda b). \tag{3.5}$$

Note that an algebra does not necessarily have an identity element $e \in \mathcal{A}$ such that $e \circ a = a \circ e = a$ for all $a \in \mathcal{A}$. If it does, it is called an "algebra with identity". If in addition every element $a \in \mathcal{A}$ has an inverse $a^{-1} \in \mathcal{A}$, meaning $a \circ a^{-1} = a^{-1} \circ a = e$, one calls it a "regular algebra". A "singular algebra" is one in which we cannot invert the elements. Also, multiplication is generally not required to commute. If $a \circ b = b \circ a$ for all $a, b \in \mathcal{A}$, we have a "commutative algebra". We will see now how device space naturally acquires an algebraic structure by the consistency requirement of Crux 3.1. First some definitions.

Definition 3.8 (Tensor Product)
The tensor product of two $\mathbb{R}$-valued operators X and Y is the operator $X \otimes Y :$ Dom $X \times$ Dom $Y \rightarrow \mathbb{R}$ defined by

$$X \otimes Y(x; y) \overset{\text{def}}{=} X(x) \cdot Y(y).$$

A k-fold tensor product is straightforwardly defined.

Definition 3.9 (Scan Operator)
The k-fold scan operator $\overset{k}{\mathsf{S}}$ is defined for every $k \in \mathbb{Z}_0^+$ as the vector summation operator

$$\overset{k}{\mathsf{S}}: \underbrace{\mathbb{R}^n \times \ldots \times \mathbb{R}^n}_{k} \rightarrow \mathbb{R}^n : (x_1, \ldots, x_k) \mapsto \overset{k}{\mathsf{S}}(x_1, \ldots, x_k) \overset{\text{def}}{=} \sum_{i=1}^{k} x_i.$$

By definition, $\overset{0}{\mathsf{S}} \equiv 0$. If clear from the context we will omit the superscript k.

Starting at the origin, the scan operator follows the route indicated by the vector arguments inserted into its slots. A k-fold local sample can now formally be defined by the pull back of the source under the scan operator, or, operationally, by the push forward of a k-fold tensor product of detectors.

Definition 3.10 (k-Fold Local Sample)

See Definitions 3.1, 3.9 and 3.8. A k-fold local sample $\lambda_k \in \mathbb{R}$ is a contraction of source and k consecutive detector instances:

$$\lambda_k : \Sigma \times \underbrace{\Delta \times \ldots \times \Delta}_{k} \to \mathbb{R} : (F, \phi_1, \ldots, \phi_k) \mapsto S^* F[\phi_1 \otimes \ldots \otimes \phi_k] \overset{\text{def}}{=} F[S_*(\phi_1 \otimes \ldots \otimes \phi_k)].$$

Schematically:

$$
\begin{array}{ccc}
\phi_1 \otimes \ldots \otimes \phi_k \in \Delta_{x_1} \times \ldots \times \Delta_{x_k} & \overset{\overset{k}{S_*}}{\longrightarrow} & \overset{k}{S_*}(\phi_1 \otimes \ldots \otimes \phi_k) \in \Delta_y \\
\downarrow & & \downarrow \\
(x_1, \ldots, x_k) \in \mathbb{R}^n \times \ldots \times \mathbb{R}^n & \overset{\overset{k}{S}}{\longrightarrow} & y = \overset{k}{S}(x_1, \ldots, x_k) \\
\uparrow & & \uparrow \\
\overset{k}{S}{}^* F \in \Sigma_{x_1} \times \ldots \times \Sigma_{x_k} & \overset{\overset{k}{S^*}}{\longleftarrow} & F \in \Sigma_y
\end{array}
\qquad (3.6)
$$

Note that Definition 3.1 corresponds to the case $k = 1$. A k-fold filtered image can be defined as usual through the action of the spacetime symmetry group (Definition 3.5).

The dual interpretations of Definition 3.10 are as follows. Instead of F we now have a generalised type of source field $S^* F$ capable of swallowing any tensor product of independent detectors so as to produce a numeric sample in a multi-linear continuous way. Such an object is also called a *cotensor of rank k*. In an abstract sense it is just our original source field F; the pull back under S merely tells it how to interact with multiple detectors. This is clear from the dual point of view: $S_*(\phi_1 \otimes \ldots \otimes \phi_k)$ is the *effective* filter associated with the k independent filters $\phi_1, \ldots, \phi_k$ probing the original source field F. In fact we have

Lemma 3.2 (Effective Detector for k-Fold Local Sample)

See Definition 3.8 and Definition 3.10.

$$S_*(\phi_1 \otimes \ldots \otimes \phi_k) = \phi_1 * \ldots * \phi_k.$$

Proof 3.2 (Lemma 3.2)
Problem 3.11.

In order for Definition 3.10 to be a correct definition we must show that it indeed corresponds to a k-fold consecutive sampling of the initial source field F.

Proof 3.3 (Consistency of Definition 3.10)
Assume $\Theta = \mathrm{T}(n)$, so that $\Theta^* F[\phi] = F * \phi$. The induction hypothesis is that the effective detector corresponding to a k-fold local sample, obtained by applying k independent filters $\phi_1, \ldots, \phi_k$ in succession, exists and is given by Lemma 3.2. Denoting this sample by $\lambda_k = F * \Phi_k$, in which $\Phi_k \equiv \phi_1 * \ldots * \phi_k$, we find, after sampling it once more with another filter ϕ_{k+1}, using Lemma 2.5 on Page 34:

$$\lambda_{k+1} \overset{\text{def}}{=} (F * \Phi_k) * \phi_{k+1} = F * (\Phi_k * \phi_{k+1}) \overset{\text{def}}{=} F * \Phi_{k+1}.$$

The general case, including filter tuning, is left to the reader. This completes the consistency proof.

Because of the convolution-algebraic structure of device space, we do not have to keep track of filtering stage and processing history; local samples, and hence images, are identically and unambiguously defined at all levels of linear processing.

After these considerations, notably Definition 3.10 and Lemma 3.2, it is clear that the consistency requirement of Crux 3.1 can be stated as follows.

Observation 3.1 (Consistency Requirement for Image Processing)
Device space Δ must form a linear convolution algebra.

In other words we must satisfy the closure property $\phi, \psi \in \Delta \Rightarrow \phi * \psi \in \Delta$ (i.e. the only nontrivial algebraic requirement: Problem 3.13). In that case we can indeed interpret any k-fold local sample as just another sample in the original sense of Definition 3.1. This is necessary if we want image processing to be consistent with the duality principle of Paradigm 2.1, Page 21; by virtue of algebraic closure, the infinite chain in Figure 3.3 can be equivalently represented by a single piece as in Figure 2.2 on Page 22.

Observation 3.1 holds in the spacetime domain. In the Fourier domain convolution is replaced by the usual kind of multiplication, but the algebraic closure property is maintained as an abstract structure independent of the filter domain we choose to work with.

From Lemma 2.1 on Page 29 it follows that the consistency requirement of Observation 3.1 is indeed satisfied if $\Delta = S(\mathbb{R}^n)$. However, it should also be clear that for an arbitrary choice of device space this requirement is quite likely to be violated! We conclude that Assumption 2.1 on Page 27 was indeed justified, and postulate the following *linear image processing paradigm*.

Paradigm 3.1 (Linear Image Processing)
Recall Paradigm 2.1, Page 21 and Definition 2.4, Page 27:

$$\Delta \stackrel{\text{def}}{=} S(\mathbb{R}^n) \qquad \text{whence} \qquad \Sigma \stackrel{\text{def}}{=} S'(\mathbb{R}^n).$$

3.9 The Point Operator

The aim of this section is to establish the explicit form of the point operator. We need a few conventions and definitions.

Definition 3.11 ($S^+(\mathbb{R}^n)$)
Recall Definition 2.4 on Page 27. The class of positive smooth test functions is defined as

$$S^+(\mathbb{R}^n) \stackrel{\text{def}}{=} \{\phi \in S(\mathbb{R}^n) \mid \phi(x) \geq 0 \,\forall x \in \mathbb{R}^n \text{ and } \phi \not\equiv 0\}.$$

Assumption 3.1 (Positivity)
The basic point operator is a positive filter: $\phi \in S^+(\mathbb{R}^n)$.

This guarantees that whenever exposed to a positive source (or a positive measure: recall Definition 2.17 on Page 31; see also Problem 3.14) the point operator will always return a positive sample. The positivity constraint reflects the "zeroth order" nature of a point operator.

Definition 3.12 (Linear Momenta)
The linear k-th order momentum of a positive filter $\phi \in S^+(\mathbb{R}^n)$ is defined as the contravariant tensor of rank $k \in \mathbb{Z}_0^+$, the components of which are given in Cartesian coordinates by

$$m_k{}^{\mu_1 \cdots \mu_k}[\phi] \overset{\text{def}}{=} \int dz\, z^{\mu_1} \ldots z^{\mu_k}\, \phi(z)\,.$$

(The definition in arbitrary coordinates follows by suitable transformation; Box 3.5 explains the so-called *tensor index notation*.) In other words, a linear filter momentum is obtained as the filter's response when subjected to a hypothetical source field $x^{\mu_1} \ldots x^{\mu_k}$ (a monomial, regular tempered distribution). It is often natural to normalise linear momenta through division by $m_0[\phi]$, the "geometrical" result of which is of course nonlinear. In particular, a straightforward way to define the base point of a point operator is by evaluating its "centre of gravity", as follows.

Definition 3.13 (Projection Map)
The projection map π is defined on $S^+(\mathbb{R}^n)$ as the normalised first order momentum, the coordinates of which are given by

$$\pi^\mu[\phi] = \frac{m_1{}^\mu[\phi]}{m_0[\phi]}\,.$$

We have defined the projection map only for positive filters $\phi \in S^+(\mathbb{R}^n)$ (Problem 3.15). For $\phi \equiv 0$ we might, if desired, define $\pi[0] \equiv 0$—the only sensible definite choice in view of symmetry—although it is better to think of the null filter as non-localised, or (still better) everywhere-localised.

Besides being localised, nontrivial filters are also characterised by the fact that they have a certain extent; although not necessarily of compact support, they are always "essentially compact" by virtue of rapid decay. In fact, a way to study filter characteristics in all generality is by evaluating momenta relative to the filter's base point.

Definition 3.14 (Central Momenta)
Recall Definition 3.13. The central k-th order momentum $\sigma_k{}^{\mu_1 \cdots \mu_k}[\phi]$ is similar to the corresponding linear momentum of Definition 3.12, but with monomial $(z - x)^{\mu_1} \ldots (z - x)^{\mu_k}$ centred at $x = \pi[\phi]$ rather than the origin.

Unlike linear momenta, central momenta are shift invariant. Thus they are indeed characteristics of the filter's profile independent of its base point. Again one may normalise central momenta by $\sigma_0[\phi] = m_0[\phi]$; the lowest order nontrivial instance will then be of second order. In contrast to, say, the extent of its support region, a properly normalised second order central momentum provides a suitable measure for filter extent.

Definition 3.15 (Inner Scale Operator)
See Definition 3.14. The inner scale operator $\sigma[\phi]$ is defined on $S^+(\mathbb{R}^n)$ as the square root of the normalised second order central momentum $s[\phi]$, the coordinates of which are given by

$$s^{\mu\nu}[\phi] = \frac{\sigma_2^{\mu\nu}[\phi]}{\sigma_0[\phi]}.$$

Again we could set $\sigma[0] \equiv 0$ for definiteness, although any scale is equally feasible for the null filter. The variance matrix $s[\phi]$ is symmetric positive definite, hence can be put on diagonal form in a Cartesian coordinate frame by a mere rotation, and moreover, has a well-defined square root. Each eigenvalue expresses the extent of the filter in the direction of its corresponding eigenvector. Degeneracies apart[8], these eigenvectors are mutually perpendicular.

Corollary 3.1 (Inner Scale Operator)
See Definition 3.15. For every $\phi \in S^+(\mathbb{R}^n)$ there exists a Cartesian coordinate frame, and an n-tuple of positive numbers σ_μ ($\mu = 0, \ldots, d$), such that

$$s[\phi] = \sigma^2[\phi],$$

with positive diagonal inner scale matrix $\sigma[\phi] = \mathrm{diag}\,\{\sigma_0, \ldots, \sigma_d\}$.

The σ_μ will be called the inner scale parameters of ϕ (in the direction of the corresponding eigenvectors). The proof of Corollary 3.1 is left as an exercise to the reader.

Lemma 3.3 (Base Point and Inner Scale of Convolution Products)
Recall Definitions 3.9, 3.13 and 3.15. Define $S^*\pi \equiv \pi \circ S_*$ and similarly $S^*s \equiv s \circ S_*$. Then we have

$$S^*\pi[\phi_1 \otimes \ldots \otimes \phi_k] = \pi[\phi_1] + \ldots + \pi[\phi_k],$$
$$S^*s[\phi_1 \otimes \ldots \otimes \phi_k] = s[\phi_1] + \ldots + s[\phi_k].$$

Proof 3.4 (Lemma 3.3)
Problem 3.11.

Note that $S^*\pi$ and S^*s are essentially the projection map and the variance, or squared inner scale operator, respectively, but dressed up to produce a single base point and variance matrix from a specified tensor product of independent filters. According to Lemma 3.2, Lemma 3.3 says that the base point of a convolution product of filters equals the sum of base points of the filters:

$$\pi[\phi_1 * \ldots * \phi_k] = \pi[\phi_1] + \ldots + \pi[\phi_k], \tag{3.7}$$

and similarly, that the variance matrix of a convolution product of filters equals the sum of variance matrices of the filters:

$$s[\phi_1 * \ldots * \phi_k] = s[\phi_1] + \ldots + s[\phi_k]. \tag{3.8}$$

---

[8]I.e. in the anisotropic case; in the actual isotropic case we have a 1-dimensional temporal eigenspace perpendicular to a d-fold degenerate spatial eigenspace.

Thus one should add the *squares* of the inner scales in order to arrive at the effective inner scale of a convolution product of filters (recall Corollary 3.1). Convolution always increases filter extent (the terms on the r.h.s. of Equation 3.8 are always strictly positive), a phenomenon known as "blur". It is natural to consider filters with isotropic spatial scale, but recall that typical grids, especially in 3D medical imaging, are not necessarily isotropic, so that we may need to adjust the spatial scale parameters separately relative to the grid in order to enforce physical isotropy. Of course it would be better if image acquisition and reconstruction would take care of isotropic resolution, but this is not always possible due to technical limitations or because of other considerations. We will henceforth assume that grid anisotropies have been taken care of, so that $\sigma_1 = \ldots = \sigma_d$. The isotropic inner scale parameter for the spatial domain will be indicated by σ. The independent temporal inner scale parameter σ_0 will henceforth be indicated by the more mnemonic symbol τ.

Applying Definitions 3.13 and 3.15 to the Newtonian case of Result 3.1 will indeed yield base point and filter extent as expected (i.e. $\pi[\phi] = x$ and $\sigma[\phi] = e^\Lambda$, respectively), provided the basic filter has been carefully prototyped so as to correspond to $x = 0$ and $\Lambda = 0$. The following gauge is always realizable by suitable spacetime symmetry transformation and superposition (Problem 3.18).

Definition 3.16 (Prototypical Point)
A prototypical point is a point operator $\phi \in S^+(\mathbb{R}^n)$ which has even parity and is gauged as follows:

$$m_0[\phi] = 1, \qquad m_1[\phi] = \vec{0}, \qquad m_2[\phi] = I.$$

Alternatively,

$$1[\phi] = 1, \qquad \pi[\phi] = \vec{0}, \qquad \sigma[\phi] = I.$$

Put differently, a prototypical point corresponds to the origin of an arbitrary coordinate system, has unit extent in all directions and is normalised to unit weight. It is an *ad hoc* member of an invariant family of generic point operators that results if we subject it to the spacetime symmetry group of Definition 3.7; recall Result 3.1. Only the first constraint is an invariant normalisation for the entire zeroth order family (with this normalisation, a positive filter has a formal interpretation as a *probability measure*). Even parity is a natural, but rather weak requirement for a point operator in view of the vector space structure of device space, for if a filter is not even, we can always symmetrise it by adding a parity reversed instance. Note that odd point operators do not exist (Problem 3.17).

The algebraic closure property of Observation 3.1 can be easily satisfied if we take device space "large enough". The existence of a closed algebra based on a *single* point operator is far less obvious. From Lemma 3.3 it follows that autoconvolutions of prototypical point operators must be *renormalised* by a rescaling in order to restore the conditions of Definition 3.16. In particular it follows that the prototypical point operator inducing an autoconvolution algebra must be *self-similar*.

Corollary 3.2 (Self-Similarity of Prototypical Point)
Recall Assumption 3.1, Definition 3.15 and Definition 3.16. Let $\phi \in \mathcal{S}^+(\mathbb{R}^n)$ be a prototypical point generating an autoconvolution algebra, and let $\phi_s = \theta_* \phi$ be the point operator obtained by the scaling transformation $\theta : z \mapsto \sigma^{\text{inv}} z$ corresponding to a variance matrix $s = \sigma^2$, with positive diagonal scale matrix

$$\sigma = \text{diag}\{\tau; \underbrace{\sigma, \ldots, \sigma}_{d}\},$$

then ϕ satisfies the following self-similarity or semigroup property:

$$\phi_r * \phi_s(z) = \phi_{r+s}(z).$$

Proof 3.5 (Corollary 3.2)
The only not entirely trivial element in the corollary is to verify that the matrix σ as defined by the scaling transformation θ is indeed the inner scale operator defined in Definition 3.15. It is easily verified that we indeed have $s = s[\phi_s]$. The rest follows from Lemma 3.3, or alternatively, from Equation 3.8.

This finally leads us to the main result of this section.

Result 3.2 (The Point Operator)
See Definition 3.7, Result 3.1 and Corollary 3.2. The unique self-similar prototypical point operator $\phi \in \mathcal{S}^+(\mathbb{R}^n)$ is given by the isotropic, separable, normalised Gaussian of unit width:

$$\phi(z) = \frac{1}{\sqrt{2\pi}^n} e^{-\frac{1}{2}z^2}.$$

In Newtonian spacetime, an arbitrary point operator $\phi \in \mathcal{S}^+(\mathbb{R}^n)$ is fully characterised by its base point and its spatial and temporal inner scales, and is given by

$$\phi(z; x, \sigma) = \det \rho \, \phi(\rho(z - x)),$$

in which $\rho \equiv \sigma^{\text{inv}}$, the resolution or inverse scale matrix.

The class of point operators will be denoted by $\mathcal{G}^+(\mathbb{R}^n) \subset \mathcal{S}^+(\mathbb{R}^n)$.

Proof 3.6 (Result 3.2)
Although there are quite a few self-similar functions, there is only one that is compatible with the positivity constraint and the smoothness requirement underlying our definition of a prototypical point operator, in other words, the solution in $\mathcal{S}^+(\mathbb{R}^n)$ is unique. All other potential candidates fail because they do not belong to this space (in fact, most are not even in $\mathcal{S}(\mathbb{R}^n)$). For a detailed proof the reader is referred to Pauwels *et al.* [235]. In this paper the authors solve the self-similarity equation for the less restrictive case $\phi \in C^0(\mathbb{R}) \cap L^1(\mathbb{R}^n)$ and without the positivity constraint. In that case the solution space is determined by a family of filters, of which Result 3.2 gives the unique solution in $\mathcal{S}^+(\mathbb{R}^n)$. (More details can also be found in Problem 3.19.)

The following lemmas are useful in practice.

Lemma 3.4 (Fourier Transform of Prototypical Point Operator)
Recall Definition 3.16 and Result 3.2. The Fourier transform of the prototypical point operator is given by

$$\widehat{\phi}(\omega) = e^{-\frac{1}{2}\omega^2} \,.$$

Lemma 3.5 (Fourier Transform and Affine Transformation)
Recall Definition 3.7, Result 3.1 and Lemma 3.4. The Fourier transform of $\phi_{\boldsymbol{A},x}(z)$ is given by

$$\widehat{\phi}_{\boldsymbol{A},x}(\omega) = e^{i\omega x}\, \widehat{\phi}(\boldsymbol{A}\,\omega) \,.$$

Note that the last lemma holds for arbitrary affine transformations beyond the limitations of classical spacetime.

Proof 3.7 (Lemmas 3.4 and 3.5)
Problem 3.21.

Before we continue, let us adopt a new coordinate convention that turns out useful in those cases in which an explicit distinction into space and time is formally immaterial (the convention used thus far was explained on Page 40).

Definition 3.17 (The ct-Convention)
Instead of $x = (x^0 = t; \vec{x})$ we henceforth use "pseudo-isotropic" spacetime coordinates $x = (x^0 = ct; \vec{x})$, in which $c > 0$ is a free parameter, indicating the ratio of spatial and temporal inner scale parameters:

$$c \stackrel{\text{def}}{=} \frac{\sigma}{\tau} \,.$$

This convention applies to all spacetime coordinate labels. Note that c has the dimension of a velocity. It is similar to, but should not be confused with the convention used for Minkowski spacetime in the theory of special relativity, in which c figures as a physical *constant* (the speed of light). In the present context c is a *free* parameter introduced to relate the independent spatial and temporal inner scales. Its significance is merely notational; in the ct-convention we have only one explicit, *pseudo-isotropic scale*, viz. σ, the other one is absorbed into the coordinatisation. The trick basically boils down to the public fact that 2 equals $1 + 1$.

There ought to be no confusion about this; we must always keep in mind that spacetime is *not* isotropic, so if notation in this book suggests otherwise, then we are apparently dealing with the ct-convention. Time scale can always be made explicit again, if desired, by suitable choice of c: $\tau = \sigma/c$. In the ct-convention we can maintain all previous results, notably Corollary 3.2 and Result 3.2, except that now we have

$$\boldsymbol{\sigma} = \sigma\boldsymbol{I} \,. \tag{3.9}$$

Similarly we have $\boldsymbol{s} = s\boldsymbol{I}$ (Definition 3.1). With only one explicit scale parameter, we will often write the point operator $\phi(z; x, \boldsymbol{\sigma})$ as $\phi(z; x, \sigma)$. If it is origin-based ($x = 0$) we will simply write $\phi(z; \sigma)$, and if, in addition, $\sigma = 1$, we are back at the standard form $\phi(z)$. Thus different operators with the same name tag can be distinguished from the way they are "prototyped", i.e. by checking the number and type of arguments. Although this is a usual habit in physics, it is good to point it out anyway as it tends to utterly confuse mathematicians. It serves intuition and has the additional advantage that we are less likely to run out of mnemonic symbols.

3.10 Differential Operators

The unique zeroth order point operator of Result 3.2 induces a complete family of higher order differential operators known as the *Gaussian family* [167].

Definition 3.18 (Prototypical Derivative)
Recall Definition 2.18 on Page 31 and Result 3.2. A prototypical derivative of order k is defined in Cartesian coordinates as a k-th order partial derivative of the prototypical point operator $\phi \in \mathcal{G}^+(\mathbb{R}^n)$:

$$\phi_{\mu_1\ldots\mu_k}(z) = \partial_{\mu_1\ldots\mu_k}\phi(z),$$

with $k \in \mathbb{Z}_0^+$ and $\mu_1,\ldots,\mu_k = 0,\ldots,d$.

The notation on the r.h.s. is a convenient shorthand for a k-th order partial derivative w.r.t. the spatiotemporal coordinates $x^{\mu_1},\ldots,x^{\mu_k}$, but holds *only in a Cartesian frame*. The l.h.s. can, however, be defined in *any* coordinate system by suitable transformation of the r.h.s. (most easily described in terms of so-called *covariant derivatives*: this will be explained in Chapter 5). One can alternatively consider the Fourier representation with the help of the following lemma.

Lemma 3.6 (Fourier Transform and Differentiation)
See Definition 3.18 and Lemma 3.4. The Fourier transform of $\phi_{\mu_1\ldots\mu_k}(z)$ is given by

$$\hat{\phi}_{\mu_1\ldots\mu_k}(\omega) = (-i\omega_{\mu_1})\ldots(-i\omega_{\mu_k})\,\hat{\phi}(\omega).$$

Proof 3.8 (Lemma 3.6)
Problem 3.21.

Derivatives are automatically normalised by virtue of zeroth order normalisation.

Corollary 3.3 (Induced Normalisation)
If $\phi \in \mathcal{S}^+(\mathbb{R}^n)$ is normalised to unit weight, $m_0[\phi] = 1$, then

$$m_k{}^{\mu_1\ldots\mu_k}[\phi_{\nu_1\ldots\nu_k}] = (-1)^k k!\, S\left\{\delta_{\nu_1}^{\mu_1}\ldots\delta_{\nu_k}^{\mu_k}\right\}.$$

Here, δ_ν^μ is the Kronecker symbol[9] ($\delta_\nu^\mu = 1$ if and only if $\mu = \nu$, otherwise $\delta_\nu^\mu = 0$), and S denotes index symmetrisation. More explicitly:

$$S\left\{\delta_{\nu_1}^{\mu_1}\ldots\delta_{\nu_k}^{\mu_k}\right\} \stackrel{\text{def}}{=} \begin{cases} 1 & \text{if upper and lower indices match pairwise (in any order),} \\ 0 & \text{otherwise.} \end{cases}$$

---

[9]Actually a tensor; see Chapter 5.

Proof 3.9 (Corollary 3.3)
Straightforward from induction using partial integration. Note that smoothness and rapid decay are essential conditions.

The Gaussian family now arises from the action of the Newtonian group on the prototypical derivatives.

Result 3.3 (The Gaussian Family)
See Definition 3.18. In Newtonian spacetime, an arbitrary derivative operator is fully characterised by its order (in each independent direction), its base point and its spatial and temporal inner scales (again we write $\rho \equiv \sigma^{inv}$ for notational convenience), and is given by

$$\phi_{\mu_1 \ldots \mu_k}(z; x, \sigma) = \det \rho \, \phi_{\mu_1 \ldots \mu_k}\left(\rho\left(z - x\right)\right) \, .$$

The induced family of derivatives will be collectively denoted by $\mathcal{G}(\mathbb{R}^n)$ and will be referred to as the Gaussian family.

Note that we have computed the push forward of the prototypical derivatives of Definition 3.18, *not* the derivatives of the push forward of the prototypical point operator (first formula in Result 3.2). The latter would have brought in a resolution-dependent amplitude σ^{-1} or τ^{-1} for each order of differentiation, depending upon whether the corresponding partial derivative is in spatial or temporal direction. Put differently, we have actually taken derivatives of the general point operator (second formula in Result 3.2) w.r.t. *naturally scaled* coordinates $\tilde{x} \equiv \rho \, x$ rather than w.r.t. the dimensionful x. This implies that derivatives of different orders are dimensionally comparable, and that they are of the same order of magnitude for a "typical image". It becomes relevant as soon as we compare different orders or mix space and time derivatives. Also, when using different levels of scale, it is important to take properly normalised derivatives.

Example 3.1 (Natural Scaling)
- The Helmholtz operator $\Delta - m^2$, in which Δ is the spatial Laplacian and m is a constant, has an operational counterpart in the form of a 1-parameter family of linear filters $\left(\sigma^2\Delta - m^2\right)\phi(z; x, \sigma)$; if we leave out the scaling factor it has no meaning.

- The d'Alembertian $\square \equiv \Delta - c^{-2}\partial_{tt}$, or wave operator—no ct-convention here—corresponds to the 2-parameter filter family $\left(\sigma^2\Delta - c^{-2}\tau^2\partial_{tt}\right)\phi(z; x, \sigma)$.

- Suppose we are looking for an extremum of a differential quantity over scale. Then we had better use dimensionless derivatives in the comparison of local extrema, otherwise we are likely to end up with a biased global extremum somewhere near the coarsest or finest level of resolution.

Lindeberg proposes a steerable, deliberate scale bias for certain scale selection purposes in which immediate localisation is a major aim [192, 193, 194]. To this end he introduces an adaptable scale exponent γ via $\sigma\nabla \to \sigma^\gamma\nabla$, and shows that the resulting family of normalisation methods spans the class of scale selection methods for which local maxima over scales of normalised differential entities are preserved under rescalings of the original data. Choices of γ other than unity must, however, be semantically motivated, and adapted to the specific task at hand.

Apart from the natural scaling phenomenon induced by differentiation, it is of interest to study the effect of scaling and differentiation on (external, i.e. source-induced) noise. For additive, pixel-uncorrelated as well as Gaussian-correlated noise with a Gaussian distribution, this has been done by Blom [20]. Note that noise is a semantical concept; it requires a model of a *noise source* $\delta f \in \Sigma$ that perturbs the measurements. The same is *a fortiori* the case with any definition of a *signal-to-noise-ratio*.

The Gaussian family allows us to take image derivatives by straightforward linear filtering according to Definition 2.18 on Page 31.

Notation 3.1 (Image Derivatives)
We shall henceforth abbreviate

$$L_{\mu_1 \dots \mu_k} \stackrel{\text{def}}{=} (-1)^k F[\phi_{\mu_1 \dots \mu_k}].$$

The sample's base point and scale will be clear from the context.

3.11 Completeness

The Gaussian family of Result 3.3 is a *complete* family. This is clear from the observation that, if expressed in Cartesian coordinates, its members correspond to partial differential operators. Indeed, according to Taylor's theorem we can in principle reconstruct any smooth function on a full neighbourhood from its derivatives in one interior point (recall that images are, in an well-defined sense, smooth functions of space and time). Completeness is a conceptually important, yet merely formal requirement. Clearly nobody will ever use the entire family in practice, but completeness confirms that we are not overlooking anything in our model, in other words, that there is no *a priori* limitation to structural representations based on the Gaussian filter family. Restrictions to finite order are most naturally studied in the framework of so-called *local jets*. We will return to this in Chapter 4.

Another way of looking at the issue of completeness may provide alternative insight, viz. by studying the so-called *self-similarity solutions* of the isotropic diffusion equation [167, 231].

Lemma 3.7 (Gaussian Family and Diffusion Equation)
Recall Definition 2.13 on Page 30, Result 2.2 on Page 32, and Definition 3.17. The Gaussian filter family of Result 3.3 corresponds to the solution space of the following initial value problem on the scale-spacetime domain $\mathbb{R}^n \times \mathbb{R}^+$ with parameters $x \in \mathbb{R}^n$ (base point), and $\mu_1, \dots, \mu_k = 0, \dots, d, k \in \mathbb{Z}_0^+$ (order):

$$\begin{cases} \partial_s u & = \tfrac{1}{2} \Delta u \\ \lim_{s \downarrow 0} u(\,.\,; s) & = \partial_{\mu_1 \dots \mu_k} \delta_x \,, \end{cases}$$

in which the initial conditions are derivatives of the Dirac point distribution at point x, and in which Δ is the spatiotemporal Laplacian in ct-convention.

That is, $u(z; x, s) = \phi_{\mu_1 \dots \mu_k}(z; x, \sigma)$, in which the evolution parameter relates to inner scale according to

$$s = \sigma^2 = c^2 \tau^2,\tag{3.10}$$

so it is indeed, as notation suggests, just the variance parameter.

Proof 3.10 (Lemma 3.7)
This was basically Problem 2.4 (hint: Fourier transformation).

Declining from the ct-convention the initial value problem of Lemma 3.7 may be split into two, one for isotropic space and one for the time axis. However, neither use of ct-convention nor explicit separation of variables takes into account the natural *coupling* of spatial and temporal scales. It seems reasonable to expect that the *a priori* arbitrary velocity parameter c is in some sense "optimal" if its value is (locally) adapted to the image's *kinematic* structure. The underlying heuristics is based on conservation: image structure is typically not spatiotemporally uncorrelated, but determined by spatial entities that may arbitrarily deform, but are nevertheless preserved over time. This idea is formalised by the concept of *multiscale optic flow*, which forms the subject of Chapter 6.

The interpretation of Lemma 3.7 is that we can perceive of a Gaussian derivative filter as a "diffused" differential operator rather than a "punctal" one confined to an infinitesimal neighbourhood (recall Problem 2.9 and see also Problem 3.22). Alternatively we can think of it as the image of a point source that corresponds to an exact derivative operator. Classical differentiation thus corresponds to the hypothetical "classical limit" of vanishing scale.

In the theory of differential equations the Gaussian point operator is known as the *Green's function* or *propagator* of the diffusion equation. The *scale-space image* of an arbitrary source field $F \in \mathcal{G}'(\mathbb{R}^n)$ is obtained by exploiting the linear superposition principle; it is the solution to the diffusion equation with the initial condition given by the raw image (Definition 3.6). Image derivatives can be likewise explained; either we insert distributionally defined derivatives of the raw image as the initial condition, or, more practically, we take the raw image itself and transpose the derivatives to the Green's function.

In view of scaling (recall the Pi theorem, Box 3.4) it is natural to substitute variables $(x; s) \mapsto (\xi; u)$:

$$\begin{cases} \xi(x; s) &= \dfrac{x}{\sqrt{2s}} \\ \eta(x; s) &= s, \end{cases}\tag{3.11}$$

i.e. we adopt the natural units appropriate for each level of resolution. In fact, as Koenderink observed [167], we can get rid of explicit scale dependencies *altogether* if we decouple the solution from a multiplicative "envelope" or window of width $\sqrt{2}$ times the inner scale σ.

Definition 3.19 (Gaussian Envelope)
See Result 3.2 and the remarks on Page 62 concerning the notation. The Gaussian envelope at scale level s is defined as

$$\mathcal{A}(\xi; \eta) \stackrel{\text{def}}{=} \phi(x; 2s).$$

This separable envelope accounts for the scaling phenomenon of a filter due to its nature as a *density*. By dimensional analysis of higher order members of the Gaussian family we may expect to find interesting solutions if we pull out a scaling factor of integral degree, corresponding to differential order. Therefore the following decomposition appears natural.

Definition 3.20 (Decomposition Gaussian Family)
Recall Result 3.3 and Definition 3.19. The functions $\psi_{\mu_1 \ldots \mu_k}(\xi)$ are defined by

$$\phi_{\mu_1 \ldots \mu_k}(x; s) \stackrel{\text{def}}{=} \frac{N_k}{\sqrt{4s}^k} \, \mathcal{A}(\xi; \eta) \, \psi_{\mu_1 \ldots \mu_k}(\xi) \, .$$

The normalisation constant N_k is provided for later convenience, but is clearly inessential. Combining Definition 3.20 with the diffusion equation of Lemma 3.7 yields the following result, which at the same time shows that the decomposition is consistent.

Result 3.4 (Sturm-Liouville Eigenvalue Problem)
See Lemma 3.7 and Definition 3.20.

$$H \, \psi_{\mu_1 \ldots \mu_k} = \mathrm{E}_k \, \psi_{\mu_1 \ldots \mu_k} \, ,$$

with "Hamiltonian"

$$H \stackrel{\text{def}}{=} N + \frac{n}{2} \, I \, ,$$

in which the so-called creation and annihilation operators $A^\dagger$ and A are defined, respectively, as the first order operators $A^\dagger \equiv \frac{1}{\sqrt{2}} (\xi - \nabla)$ and $A \equiv \frac{1}{\sqrt{2}} (\xi + \nabla)$, and $N \equiv A^\dagger A$.

A Sturm-Liouville problem possesses a *complete, orthonormal set of eigenfunctions*, i.e. in Cartesian coordinates—in which the eigenvalue problem is separable— n-fold products of 1-dimensional *Hermite functions*. Result 3.4 will appeal to the physicist: it is identical to the stationary Schrödinger equation for the n-dimensional harmonic oscillator in quantum mechanics. The eigenvalue spectrum corresponds to the set of differential orders $k \in \mathbb{Z}_0^+$, and is degenerate if $n > 1$, except for $k = 0$. Without proof we present its Cartesian solution. For further details the reader is referred to Koenderink [167] (e.g. different coordinate representations, significance for vision) and to Problem 3.23 (an abstract proof that requires no coordinates, nor an explicit choice of spatiotemporal or Fourier representation).

Claim 3.1 (Eigenfunctions)
The $L^2(\mathbb{R}^n)$-normalised solution of Result 3.4 in a Cartesian coordinate frame is given by

$$\psi_{\mu_1 \ldots \mu_k}(\xi) = \frac{1}{\sqrt{2^k k! \sqrt{\pi}}} \, e^{-\frac{1}{2}\xi^2} \, H_{\mu_1 \ldots \mu_k}(\xi) \, ,$$

with multidimensional Hermite polynomials

$$H_{\mu_1 \ldots \mu_k}(\xi) \stackrel{\text{def}}{=} (-1)^k e^{\xi^2} \partial_{\mu_1 \ldots \mu_k} e^{-\xi^2} \, .$$

3.12 Discretisation Schemes

When turning concepts into computations, apparent trivialities may become significant issues. *Robustness* has been anticipated by admitting only *continuous* functionals and function transforms in our theory. This is why we had to give up point mappings (functions of spacetime), classical differentiation, etc., at least in their naive forms. These are compelling matters one has to deal with before one can even start implementing. Clearly, there are also optional design issues. A typical example is the trade-off that usually exists between computational speed and accuracy; we can only find a reasonable compromise if we balance *pro's* and *con's* using a task-oriented criterion.

Here we address one such issue, viz. that of *discretisation*. More specifically, we pose the question: how should we discretise a continuously parametrised model? The question is one of *rate of progression* along a parametrised curve that allows us to select a discrete parameter value at regular "time[10]" intervals (*modulo* affine transformations). Since we do not want the sampling to depend upon the parametrisation, we need to find the *natural* rate, corresponding to a so-called *canonical parametrisation*. Note that the size of the sampling interval is an entirely different matter, and is logically of lower priority (it presumes canonical parametrisation as well as semantical input).

When implementing our image model, we need a prescription for selecting discrete values for the Lie parameters attached to our filters. It is not prudent to simply take a constant sampling rate along the orbits of the group, for the effect of this depends upon the parametrisation. What we need is a kind of "speed-control" for traversing each orbit that takes us from one filter instance—scan point and tuning parameter—to the next, such that we can take samples at constant "time" intervals without introducing an *a priori* bias. Put differently, we need a *canonical parametrisation* of the spacetime group [231, 255]. The question of canonical parametrisation is thus the analogue counterpart of the discretisation problem.

When we reparametrise the spacetime group (or any coordinatised manifold), we affect the coordinate representation of all objects within it, i.c. the flow field of Lie generators ("scanning" and "tuning" operators). Although geometrical objects themselves are parametrisation independent (recall the discussion on covariance in Section 3.2), computations are typically performed in parameter space, which is why one has to think about the problem.

In general, Equation A.103 and Diagram A.104 of Appendix A comes into play. Let $f : \mathbb{R}^N \to \mathbb{R}^N$ be the desired reparametrisation (N being the number of parameters, or the *dimension of the Lie group*). The linear operators f_* and f^* are the manifestations of f on vectors and covectors, respectively. Since Lie generators are vectors (or equivalently, derivations), we can apply f_* to obtain their corresponding expressions after reparametrisation. The question is how f_* relates to f, and which f is the natural one to choose. Once we know this, duality determines f^*.

---

[10]"Time" in the sense of sequential processing.

Definition 3.21 (Derivative Map)
Let $f : \mathbb{R}^N \to \mathbb{R}^N : p \mapsto q \equiv f(p)$ be any reparametrisation. Then we define

$$f_* : \mathbb{R}^N_p \to \mathbb{R}^N_q : v_p \mapsto v'_q \equiv Df(p)(v_p)\Big|_q ,$$

in which $Df(p)$ denotes the derivative map of f at p.

(For reasons of rigour and clarity base points have been attached to vector arguments and tangent spaces; we usually refrain from this to keep the notation simple. For a thorough explication, see e.g. Spivak [265, Volume I].) An example may clarify this definition.

Example 3.2
See Definition 3.21. Let us label parameters by $a, b = 1, \ldots, N$. The linear derivative map has matrix representation

$$[Df(p)]^a_b = \frac{\partial f^a}{\partial p^b}(p).$$

Hence, if we map a coordinate vector in the p-domain, we find

$$f_*\left(\frac{\partial}{\partial p^a}\right) = \sum_{b=1}^{N} \frac{\partial f^b}{\partial p^a}(p) \frac{\partial}{\partial q^b}\Big|_q ,$$

which is a vector in the q-domain. If this is so, then a form in the q-domain must have a "reciprocal image"

$$f^*\left(dq^a\right) = \sum_{b=1}^{N} \frac{\partial f^a}{\partial p^b}(p) \, dp^b\Big|_p ,$$

a form in the p-domain. This follows from duality: Problem 3.24.

Definition 3.22 (Canonical Group Parametrisation)
Let Θ be a spacetime group of dimension $N = \dim \Theta$, with generators

$$\mathcal{L}_a \stackrel{\text{def}}{=} \sum_{b=1}^{N} v^b_a(\ell) \frac{\partial}{\partial \ell^b} \quad (a = 1, \ldots, N),$$

when expressed in terms of group parameters $\ell \in \mathbb{R}^N$. Let $f : \mathbb{R}^N \to \mathbb{R}^N : \bar{\ell} \mapsto \ell \equiv f(\bar{\ell})$ be a reparametrisation, and $f_* \equiv Df$ its derivative map (to be evaluated at correspondingly mapped base point $\ell = f(\bar{\ell})$). Then $\bar{\ell}$ defines a canonical group parametrisation if f is such that

$$f_* \bar{\mathcal{L}}_a = \mathcal{L}_a ,$$

for all $a = 1, \ldots, N$, in which $\bar{\mathcal{L}}_a$ is similar to $\mathcal{L}_a$, but with $\bar{v}^b_a(\bar{\ell}) = \delta^b_a$ instead of $v^b_a(\ell)$.

(Of course this definition has a dual counterpart: Problem 3.25.) The generators $\mathcal{L}_a$ are also called *Killing vector fields* on the group manifold relative to the action of the spacetime group [144, 211]. In canonical coordinates the group looks like a mere translation, or unit-velocity flow. Consequently, canonical parameters admit sampling on a regular grid without *a priori* bias.

Example 3.3 (Canonical Parametrisation of Newtonian Spacetime)
To illustrate canonical parametrisation, let us take the Newtonian spacetime group of Definition 3.7, Page 52.

- Spacetime translation is canonical as it stands.

- Since the point operator is rotationally symmetric, we do not need an actual realization of the canonical parameters of the rotation group.

- Finally, canonical scale parameters are determined as follows. In ct-convention we have one scale parameter σ, which we reparametrise: $\sigma = f(\lambda)$. The Lie generator for a rescaling is $\mathcal{L} = \sigma\, d/d\sigma$, with canonical counterpart $\bar{\mathcal{L}} = d/d\lambda$. According to Definition 3.22 we have $f_* \bar{\mathcal{L}} = \mathcal{L}$, which, using Definition 3.21, yields a first order differential equation for f, viz. $f'(\lambda) = f(\lambda)$. This determines the reparametrisation up to an integration constant $\sigma_0 > 0$:

$$\sigma = \sigma_0\, e^\lambda .$$

Note that f, not f_*, accounts for the "hidden scale" σ_0. If $\sigma = \sigma I$ and $\Lambda = \lambda I$, we get $\sigma = \sigma_0 \exp \Lambda$. In other words, using $\det \exp \Lambda = \exp \operatorname{tr} \Lambda$ (which holds for *any* matrix Λ),

$$\phi(z; x, \sigma \equiv \sigma_0 \exp \Lambda) = \exp(-\operatorname{tr} \Lambda)\, \phi\left(\exp(-\Lambda)\,(z - x)\right) .$$

We may now discretise into L equally weighted scale levels, say $\lambda = k\delta\lambda$ with constant step size $\delta\lambda$, and $k = 0, \ldots, L - 1$. Declining from ct-convention we get

$$\sigma = \sigma_0\, e^\lambda \quad \text{and} \quad \tau = \tau_0\, e^\mu ,$$

for isotropic space and time[11], respectively. See also Problem 3.26. Figure 3.4 shows an MRI image at eight equidistant levels of canonical scale.

Of course, the recipe for canonical parametrisation does not tell us anything about physical constraints on the canonical domain, such as limitations of scope and graininess. Neither does it fix the actual *size* of the sampling interval; the canonical domain addressed by the canonical parameters $\bar{\ell}$ is a continuum still to be replaced by a (regular) grid. This requires independent considerations. For certain applications a fairly coarse sampling may suffice, while others may require a very fine one. Perhaps one may even want to define the sampling interval *dynamically*, e.g. when zooming in on a region of interest [189]. Considerations of this sort fall in the category "semantics".

Nevertheless, one can say something about the structure that results from a given discretisation constant. For example, we might wonder whether we should take sampling rate to depend upon the dimension of spacetime. We consider again an example for the nontrivial case of scale discretisation.

Example 3.4 (Scale Discretisation: Structural Considerations)
Consider a d-dimensional image in isotropic space. Problem 3.27 makes it plausible that

- the *a priori* probability for a *given* critical point (i.e. a point at which the gradient vanishes) to be an extremum (maximum or minimum) equals $p_e = 2^{-(d-1)}$, and

- the probability for it to be a hypersaddle is complementary: $p_s = 1 - p_e$.

---

[11] Inner scale will henceforth be denoted by σ; the symbol τ will be reserved for other purposes.

	$\delta\alpha = 10\%$	$\delta\alpha = 30\%$	$\delta\alpha = 100\%$
$d = 1$	0.100	0.300	1.000
$d = 2$	0.050	0.150	0.500
$d = 3$	0.033	0.100	0.333

Table 3.1: Logarithmic scale steps $\delta\lambda$ for several values of *a priori* relative decrease $\delta\alpha = -\delta \log N$ in spatial dimensions $d = 1, 2, 3$.

This tells us something about *relative* numbers of critical points of various types, but nothing about *absolute* counts. We may expect the number of critical points to be proportional to the number of "natural volume elements", or "structural degrees of freedom", i.e. σ^{-d}. This is the best guess when lacking prior knowledge about image structure (think of pixel uncorrelated noise, or otherwise of statistics on large ensembles of unrelated images), and is the only one consistent with scale invariance.

Denote the number of critical points as a function of canonical scale by $N(\lambda)$, and its decay rate by γ, i.e. $N' = -\gamma N$ (the "half-time" thus equals $\lambda_{1/2} = \gamma^{-1} \log 2$). We have just argued that the *a priori* expectation—disregarding effects induced by coherences—is that γ equals dimension d. The same result holds if we consider only extrema.

This observation gives us an *interpretation* of a given sampling rate in terms of the *a priori* expected decrease of structural features, which may be exploited to estimate scale steps in advance. If $\sigma_k = \sigma_0 \exp(k\delta\lambda)$ for $k = 0, \ldots, L - 1$ (with σ_0 equal to the spatial grid constant, say), then the relative decrease of critical points equals $\delta\alpha \equiv -\delta \log N = \gamma\delta\lambda = d\delta\lambda$. Therefore, if we wish to control inter-slice decay $\delta\alpha$ (a convenient quantitative measure of structural change), we choose a desirable value for it, and subsequently discretise accordingly: $\delta\lambda \equiv \delta\alpha/d$. In particular we see that *higher dimensional images* then require *finer scale sampling*, which is unfortunate, as these are already computationally more expensive. Practical example: say we want roughly $\delta\alpha = 30\%$ decay per scale level, then we take $\delta\lambda = 0.30$ for 1-dimensional signals, $\delta\lambda = 0.15$ for 2-dimensional images, and $\delta\lambda = 0.10$ in the case of 3-dimensional volume data: Table 3.1.

Knowing how to interpret scale discretisation in terms of structural rate of change, we can, to some extent, *reverse* the line of reasoning in the previous example. The presence of coherent structures is likely to cause the expected relation between scale sampling and decay rate to be violated in any *particular* situation. When computing a given scale-space image, we may adapt scale steps so as to enforce a *fixed* decay rate (of course this cannot be a feed-forward procedure). The result will be a fine-tuned sampling rate $\delta\lambda_k$ that varies with the level of scale $k = 0, \ldots, L - 1$. The number $d_k \equiv \delta\alpha/\delta\lambda_k$ indicates a kind of "effective dimension" of the image at the k-th level of resolution. It will typically be close to d, but will deviate from it depending upon the degree of coherence or symmetry (cf. the notion of "fractal dimension" [108]): Figure 3.5. An example of a degenerate case is the bar-code on the cover of this book, for which the effective dimension will be close to one (in the absence of noise). A similar verification for the 1-dimensional case has been reported by Lindeberg [191, Chapter 7].

It might not seem very useful to fine-tune scale discretisation steps by the retrospective constant-decay method, but there are cases in which the method turns out useful. In particular, if a canonical parametrisation is too difficult to obtain in closed-form, then it seems a plausible heuristic strategy for determining step-sizes. Vincken used this idea in an extension of his "hyperstack segmentation" algorithm [285, Chapter 6], in which links are established between consecutive layers of a scale-space representation, resulting in a tree-like structure that branches off in the high-resolution direction (see also Koenderink [150] for the original idea of scale-space linking, and Koster [177] for heuristic linking models). The core idea of multiscale segmentation is that following all links downward in scale, starting from an appropriately selected seed point, or "root", will project to a "meaningful" segment at the finest level of scale. Each such segment thus has an extremely simplified "single-degree-of-freedom" representation at a sufficiently coarse level of resolution. Vincken showed experimentally that linear scale-space can be replaced by other, nonlinear, one-parameter families. Similar segmentation results are then obtained if the evolution parameter is discretised according to the heuristic method of constant-decay as explained above. Figure 3.6 illustrates the method of multiscale segmentation. A different multiscale segmentation strategy is based on the idea of *watersheds*. Figure 3.7 illustrates the method as implemented by Fogh [88].

3.13 Summary and Discussion

In this chapter two basically equivalent image definitions have been proposed. Both are based on the duality paradigm developed in the previous chapter, and on an explicit spacetime model.

The first definition is based on the general device space $\Delta = \mathcal{S}(\mathbb{R}^n)$, comprising all smooth functions of rapid decay, and is operational insofar that it can be *dynamically* implemented. As such it is suitable for image processing purposes, in which one is beforehand with an explicitly formulated task, so that it is feasible to implement a small, *suitably adapted* subset of detectors. Many—but not all—linear filters proposed in the literature for various image processing purposes fit into this detector paradigm as special instances.

The second definition proposed in this chapter is based on the Gaussian family $\Delta = \mathcal{G}(\mathbb{R}^n)$, and is equivalent to the former by completeness. Unlike the former, this definition is suited for a *static* image representation, known as *Gaussian scale-space*, in the sense that the implementation of filters *precedes* a precise specification of a task. This is feasible by virtue of the facts that the Gaussian family is *finite dimensional* and yet *complete*. The crucial step from $\mathcal{S}(\mathbb{R}^n)$ to $\mathcal{G}(\mathbb{R}^n)$ is the choice of a *point operator*. It has been shown that there is actually no choice; only the Gaussian fits consistently in a general image processing framework.

The filter classes $\mathcal{S}(\mathbb{R}^n)$ and $\mathcal{G}(\mathbb{R}^n)$ have the "right topology", enabling us to take derivatives and to carry out convolutions and Fourier transforms in a decent manner, which are very basic, low-level operations in image processing. Many alternative filters proposed in the literature fail to provide a nice differential, or even a consistent algebraic structure. Although one may not need all this in any *specific* case, endowing raw image data with such a nice and yet uncommitted structure encourages the use of powerful mathematical machinery that would otherwise not be applicable. As such it provides a general foundation for a rigorous, systematic and progressive approach to image analysis.

Both $S(\mathbb{R}^n)$ as well as $\mathcal{G}(\mathbb{R}^n)$ form an algebra based on the convolution product. Apart from the source fields they are intended to represent, the image definitions incorporate the details of the sampling apertures underlying the realization of a grey-value sample, and have consequently infinitely many degrees of freedom even at a single location in spacetime: the "test functions" used to model the aperture profiles. These are also referred to as "detector devices", "filters","templates", "feature detectors", "local neighbourhood operators", "sensors", "point spread functions", or "receptive fields", depending upon context and scientific inclination.

Images have been defined in terms of local samples based on such measurement paradigms by incorporating a spacetime symmetry group in the form of an analytical Lie group. The action of this group on any paradigmatic filter accounts for the kind of spatiotemporal coherence that distinguishes an image from a mere set of local samples, and pertains to our intuition about the nature of spacetime as a homogeneous and pseudo-isotropic "void". In particular, scale provides topology, which is the skeleton of any geometry. Quite a strong topology in fact; by virtue of the scaling degree of freedom *any* two points in an image will overlap at suitably coarse resolution. Moreover, the proposed image definitions are *generally covariant*, which implies that they do not rely on an *ad hoc* choice of coordinates.

To model biological vision (optically guided behaviour) one may want to extend the image model so as to embody the spacetime symmetry group in an *intrinsic* fashion, rather than by way of spacetime labelling. This amounts to solving the local sign problem. Although the proposed definition does not account for this, it may nevertheless be partly satisfactory in the context of "front-end" vision.

Any selection of detectors amounts to a formation of metameric source field configurations that cannot be disambiguated. One might be tempted to consider this undesirable, but the (hypothetical) alternative would be even less fortunate. Metamerism allows us to encapsulate all but the *essential* degrees of freedom. "Essence" requires the specification of a task (semantics); for our purpose of an uncommitted image representation this task could be specified as the embodiment of a coherent spacetime model (syntax or proto-semantics).

Images are *infinitely differentiable* in a *linear, well-posed* and *operationel* sense. This enables a robust, numerically stable implementation of differential operators up to some finite, but *a priori* unrestricted order, based on straightforward linear filtering[12]. In the literature nonlinear methods have been proposed to extract image derivatives in a robust way. This is not only more cumbersome, but also less natural. In fact it will be very hard[13] to convince a mathematician of anything else than a linear scheme: differentiation is, *by definition*, linear!

---

[12]Limitations to finite order no longer find their cause in a conceptual weakness (ill-posedness), but it should be obvious that discretisation and noise always impose computational limitations.

[13]Probably impossible, but everyone has a weak moment...

☞ Box 3.1 (Condensed Index Notation)

Condensed indices were introduced by Bryce DeWitt in the context of quantum field theory [54], who used "pseudo discrete" indices to label both discrete as well as continuous degrees of freedom of a field. The continuous part usually labels the quantity's spatiotemporal or Fourier coordinates. Thus for example ϕ^i, using a condensed index $i \equiv (x; m)$ with $x \in \mathbb{R}^n$ and $m \in \mathbb{Z}^k$, corresponds to a k-dimensional function $\phi_m(x)$ on an n-dimensional continuum. It is basically a notational convention, *not* a discretisation scheme, which helps to considerably simplify notations—and thus theoretical manipulations—without loss of generality. For example, the summation convention, when applied to a condensed index, entails a summation over its discrete part as well as an integration over its continuous part. Furthermore, functions of ϕ^i correspond to functionals of $\phi_m(x)$, partial derivatives with respect to ϕ^i to functional derivatives with respect to $\phi_m(x)$, etc. To summarise:

Let $i \equiv (x; m)$ and $j \equiv (y; p)$ denote condensed indices, each comprising a continuous label $x, y \in \mathbb{R}^n$ as well as a discrete label $m, p \in \mathbb{Z}^k$, then the following conventions apply:

$$i \overset{\text{def}}{=} (x; m)$$

$$\phi^i \overset{\text{def}}{=} \phi_m(x)$$

$$\sum_i \overset{\text{def}}{=} \sum_{m \in \mathbb{Z}^k} \int_{\mathbb{R}^n} dx$$

$$J(\phi) \overset{\text{def}}{=} J[\phi]$$

$$\frac{\partial}{\partial \phi^i} \overset{\text{def}}{=} \frac{\delta}{\delta \phi_m(x)}$$

$$\delta_i^j \overset{\text{def}}{=} \delta_m^p \delta(x - y)$$

Example: make the following identifications: $i \equiv y \in \mathbb{R}$, $\alpha \equiv (x; m) \in \mathbb{R} \times \mathbb{Z}_0^+$, $\phi^i \equiv \phi(y) \in \mathcal{A}$, $R_{\alpha i} \equiv \frac{d^m}{dx^m} \delta(x - y) \in L(\mathcal{A}, \mathcal{B})$. Here, the indices i and α refer to field components with respect to function spaces $\mathcal{A}$ and $\mathcal{B}$, respectively, and $L(\mathcal{A}, \mathcal{B})$ denotes the vector space of linear transformations $\mathcal{A} \to \mathcal{B}$. Then the linear mapping

$$F_\alpha(\phi) = R_{\alpha i} \phi^i$$

corresponds to the m-th order derivative

$$F_m(x)[\phi] = \int dy \, \frac{d^m}{dx^m} \delta(x-y)\phi(y) = \frac{d^m \phi(x)}{dx^m} \, .$$

The simplicity of the condensed index notation may be deceptive. The thing to keep in mind is that things may not be "as finite as they look". For example, if $\alpha \equiv (x; m) \in \mathbb{R} \times \mathbb{Z}$, a scalar product $f_\alpha g^\alpha$ corresponds to a series (over $m \in \mathbb{Z}$) of integrals (over $x \in \mathbb{R}$), hence requires a restriction to proper function spaces and summable sequences in order to be well-defined. Apart from this, the operator norm of a linear operator, such as F_α above, or the corresponding matrix norm of $R_{\alpha i}$, may be ill-defined; contrary to the finite dimensional case, linear operators are not automatically continuous!

☞ **Box 3.2 (Lie Groups)**

The theory of Lie groups was developed by Sophus Lie around 1870 (the term is due to Élie Cartan). Together with F. Engel he wrote a comprehensive account of his work in three volumes [184]. Lie mainly focused on the parametrisation aspect; he observed that many groups can be parametrised by a finite number of continuous parameters in an analytic fashion. He investigated their fundamental role in the context of differential equations [231] and geometry. Lie's pioneering work led Felix Klein to the formulation of his "Erlanger Programm", devoted to the study of geometries and their corresponding automorphisms [148]. Hermann Weyl studied Lie groups as topological manifolds, and contributed to the theory of *Lie algebras*. A modern compilation of the results of Sophus Lie has been written by Ibragimov [127]. Nowadays Lie group theory has become an indispensable tool in physics and many areas of applied mathematics. An account in the context of machine vision has been given by Van Gool *et al.* [99], and by Olver *et al.* [230]. In this book it suffices to appreciate the main idea, viz. that a Lie group is a group which at the same time is a finite-dimensional differentiable manifold. It "acts" on a manifold (spacetime in our case), and as such affects all objects living on that manifold (sources and detectors in our case). This is why it is useful in the context of image analysis as well; we can "address" spacetime via the operationally defined action of a Lie group (the "spacetime group[a]").

Typical examples, and the only ones we shall need, are the *general linear group* (and more specifically some of its subgroups) in n-dimensional spacetime, $GL(n)$, and the spacetime *translation group* $T(n)$.

---

[a]Recall that "spacetime group" refers to the *invariance* group of spatial and temporal metrics, *not* to the *covariance* group of spacetimeautomorphisms.

☞ Box 3.3 (Points and Infinitesimals)

Philosophical accounts on the "point" concept are numerous. Attempts to define a point date back at least to the Pythagoreans, who defined it as a "monad having position" (μονας προσλαβουσα θεσιν). This conception of a point as something *indivisible* (monad) having a *position* is also shared by Aristotle. However, Aristotle clearly had in mind a kind of "infinitesimal", and more or less identified a point—he used the term στιγμη —with its position tag (τοπος). Indeed, he considered it impossible that points could ever be combined so as to make up a continuum, say a line.

According to Euclid [109], a point—σημειον—is "that which has no part" (σημειον εστιν, ου μερος ουθεν). This declines from the assumption that indivisibility entails "no weight", or "size". Thus it seems an adequate description in the context of this book, in which we conceive of a point as an integral operator designed to merge polymerous input into a monad (the formation of metamerism!). Geometrically, our point operator also complies with Aristotle's concept to some extent, insofar that it has position (its base point). However it is clearly not "infinitesimal".

It is interesting to note that Plato seemed to have disagreed with Aristotle's definition, calling it "geometrical fiction" (γεωμετρικον δογμα), which—from an operational point of view—indeed it is! It should be noted that the matter is by no means of mere historical interest. In *non-standard analysis*, a relatively new branch of mathematics, infinitesimals are introduced in a rigorous way, [142, 143, 244], and according to Kurt Gödel,

> "there are good reasons to believe that non-standard analysis, in one version or other, will be the analysis of the future."

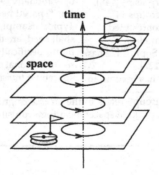

Figure 3.2: Newtonian spacetime model: a symmetry transformation is an affine transformation, consisting of a spacetime shift (zeroth order), a spatial rotation, a temporal and an isotropic spatial scaling (first order). A discrete spatial or temporal parity reversal may be taken into account as well (not indicated). The transformation can be visualised by its effect on a "gauge figure": a cylinder in temporal direction of given thickness and circular cross-section. The zeroth order part brings the spatiotemporal centres of the gauge figures into alignment, while the first order part adjusts phase (spatial rotation), radius (spatial scaling), and thickness (temporal scaling).

☞ Box 3.4 (Scale Invariance)

Page 52: "[...], but in all cases we explicitly include *scaling*, something that is often either taken for granted or overlooked when people use this terminology."

It is a remarkable, not at all self-evident fact that all physical laws are subject to scale invariance. It was Joseph Fourier who pointed out in 1822 that every physical quantity has an "exponent of dimension", an observation he made as a general remark in his treatise "Théorie Analytique de la Chaleur" [90, pp. 126–130] (the same work in which he initiated the development of Fourier analysis). The significance of this idea has been investigated more systematically by Lord Rayleigh [200].

Scale invariance expresses the freedom to reparametrise the (typically very few) independent "hidden scales" involved in a physical quantity. Scale invariance poses a very strong *a priori* constraint: any relation between physical quantities must remain valid after an arbitrary rescaling of the independent hidden scales.

An equivalent but more familiar way of enforcing scale invariance is by associating a dimensional unit with each physical quantity, corresponding to its hidden scale, and demanding consistent use of units. So simple dimensional analysis will reveal scale invariance.

A rigorous way of formulating dimensional analysis is through the *Pi theorem*: if we take into account the rescaling of all independent hidden scales, then we may reformulate the dimensionality constraint as follows (see Olver [231, pp. 218–221] for a proof and some examples):

Pi Theorem: *Let $z_\mu, \mu = 1 \ldots N$, be independent fundamental physical quantities, which scale according to*

$$z_\mu \mapsto \lambda_\mu z_\mu \qquad \mu = 1 \ldots N,$$

and let $x_i, i = 1 \ldots M$, be derived quantities, scaling according to

$$x_i \mapsto \prod_{\mu=1}^{N} \lambda_\mu^{\alpha_{\mu i}}\, x_i \qquad i = 1 \ldots M,$$

for some $N \times M$ matrix of constants $A = (\alpha_{\mu i})$, prescribed by the physical dependence of the quantities x_i on the fundamental units labelled by the index μ. Let R be the rank of this matrix A. Then there exist $M - R$ independent dimensionless monomials

$$\xi_a = \prod_{i=1}^{M} x_i^{\beta_{ia}} \qquad a = 1 \ldots M - R,$$

with the property that any other dimensionless quantity can be written as a function of the ξ_a. The $M \times (M - R)$ matrix $B = (\beta_{ia})$ satisfies the linear system:

$$AB = 0.$$

So if $F(x_1, \ldots, x_M) = 0$ expresses a relationship between physical quantities, then there exists an equivalent relation of the form $\mathcal{F}(\xi_1, \ldots, \xi_{M-R}) = 0$.

It is an obvious advantage to redefine the variables in a theory so as to render them dimensionless, because one never has to worry about any scaling constraints and usually ends up with fewer parameters. In the context of images the relevant hidden scales are the spatiotemporal grid constants (pixel scale and frame interval) and the grey-value sampling quantum ("bin").

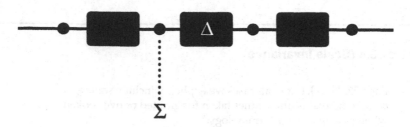

Figure 3.3: Image processing calls for an algebraic structure on device space.

☞ **Box 3.5 (Tensor Index Notation)**

Tensor indices are frequently employed in classical geometry and tensor calculus [1, 141, 181, 185, 211, 265]. A tensor index can assume as many different values as there are spacetime dimensions. We shall use Greek indices from the middle of the alphabet ($\mu, \nu, \rho, \ldots$) to denote a *spatiotemporal index*, i.e. a tensor index which may refer to either space (values in the range $1, \ldots, d$) or time (value 0). If time is excluded we will call it a *spatial index*, for which we will reserve the middle part of the Latin alphabet ($i, j, k, \ldots$). The distinction is important in view of the anisotropic transformations of classical spacetime models, in which time and space do not "mix".

The essence of tensor indices—as opposed to, say, multi-indices—is that they are "sensitive" to spacetime transformations, because they explicitly relate to a *coordinate basis* that transforms in a particular way. Details of this classical "tensor transformation law" will be discussed in Chapter 5, and can also be found in the above-mentioned literature. For the moment it suffices to point out the formal role of tensor indices as "unspecified" component labels. The following example illustrates this.

Whenever we write a thing like $T^{\mu\nu}$, we are referring to one of the $n \times n$ ordered components of some geometrical object, $\boldsymbol{T}$ say, relative to a coordinate basis $\{\vec{e}_0, \ldots, \vec{e}_d\}$, viz. the component indexed by the pair (μ, ν). Often, however, it is meant to be read as *the object itself*, i.e., as

$$\boldsymbol{T} \stackrel{\text{def}}{=} \sum_{\mu,\nu=0,\ldots,d} T^{\mu\nu}\, \vec{e}_\mu \otimes \vec{e}_\nu \,.$$

The classical convention of writing $T^{\mu\nu}$ is admittedly sloppy, but it *does* tell us more about the object than the form $\boldsymbol{T}$ does, because it is a self-contained shorthand for the r.h.s. expression.

A useful convention in tensor index manipulations is the so-called *Einstein summation convention*. Whenever a tensor index occurs twice, once as an upper and once as a lower index, one is supposed to sum over the entire index range. Thus for example $T^\mu_\mu \equiv T^0_0 + \ldots + T^d_d$.

Tensor calculus will be more systematically explored and applied in Chapter 5. Appendix A contains a summary.

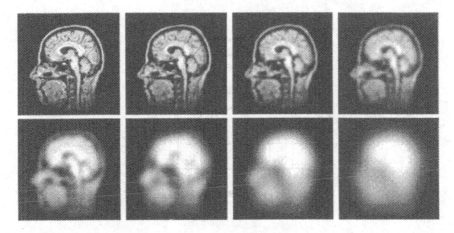

Figure 3.4: Illustration of a Gaussian scale-space representation: Depicted is a typical 2D slice (comprising 256×256 pixels) from a 3D MRI brain image at exponentially increasing inner scales $\sigma = 1.00, 1.54, 2.36, 3.62, 5.55, 8.52, 13.1, 20.1$ pixels, respectively. (Blurring was carried out within the 2D slices.)

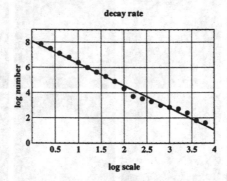

Figure 3.5: The top row shows the "elliptic regions", calculated for a typical 256×256 sagittal slice of an MR brain scan (first image) at three levels of scale. An elliptic region is defined as the splodge formed by a connected set of pixels for which the determinant of the Hessian is positive: $F[\phi_{xx}]F[\phi_{yy}] - F[\phi_{xy}]^2 > 0$. The (interpolated) plot below shows the logarithm of the number of such regions as a function of logarithmic scale (scale is measured in pixel units). The graph is almost linear over a significant range of scales: $\lambda \in [0.20, 3.80]$, i.e. $\sigma \in [1.22, 44.7]$. A linear fit yields a slope of -1.77 ± 0.29, in agreement with the *a priori* expectation value of -2. Adapted, with permission, from Vincken [285, Figures 6.4 and 6.5].

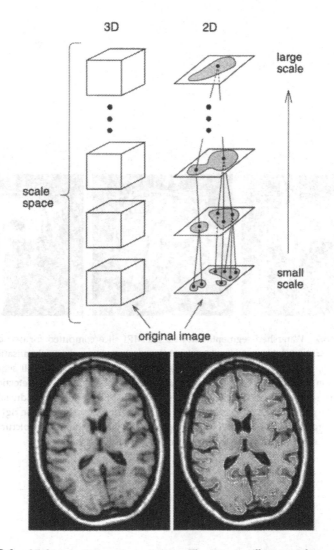

Figure 3.6: Multiscale image segmentation. The top row illustrates the multiscale linking phase. The bottom row shows a practical example: on the left we have one slice taken from an MR scan, the right shows the projection of a root point, apparently capturing to the global cortex structure. With permission from Vincken [285].

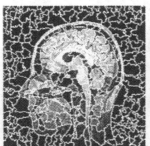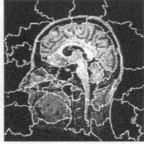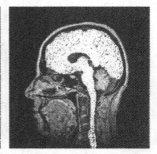

Figure 3.7: Watershed segmentation of a 2D MRI slice computed for two detection scales, $\sigma = 4.59$ (left), and $\sigma = 11.9$ (middle), projected to a chosen localisation scale, $\sigma = 0.75$, and superimposed on the image. The right image shows all segments at localisation scale that link to a few manually selected segments at the detection scales (± 5 mouse clicks for each of the two scales). Note that this linking to a fiducial level of inner scale generates a nested hierarchy of segments; the watersheds in the right picture can be merged in a hierarchical fashion, reproducing the left and middle pictures at two successive stages. With permission from Fogh [88].

Problems

3.1

Let $\gamma_{a;\sigma}$ be the normalised Gaussian as defined in Problem 2.3, and let f_a be the function defined for $x \neq a$ by $f_a(x) = -\int_0^\infty d\sigma\, \sigma \gamma_{a;\sigma}(x)$.

a. Show that f_a is locally integrable on $\mathbb{R}^n$ and can thus be regarded as a distribution.

b. Show that

$$f_a(x) = \begin{cases} A_n^{-1}\frac{1}{2-n}\|x - a\|^{2-n} & \text{if } n \neq 2 \\ \frac{1}{2\pi}\log\|x - a\| & \text{if } n = 2, \end{cases}$$

in which the constant A_n is defined as in Problem 2.11.

c. Show that $f_0* = \Delta^{-1}$, the inverse Laplacian, or $f_0 * \Delta = \Delta f_0* = \mathrm{id}$.

3.2

See Definitions 3.12 and 3.14. Let $\phi_{a;\Lambda}$ be the Gaussian filter given by

$$\phi_{a;\Lambda}(x) = \frac{1}{\sqrt{2\pi}^n}\frac{1}{\sqrt{\det\Lambda}}e^{-\frac{1}{2}(x-a)^\mu\Lambda_{\mu\nu}^{-1}(x-a)^\nu},$$

in which Λ is a symmetric, positive definite matrix with inverse Λ^{-1}.

a. Show that $\sigma_k{}^{\mu_1\cdots\mu_k}[\phi_{a;\Lambda}]$ does not depend on the filter's centre location a.

b. Calculate the zeroth, first and second order central momenta of $\phi_{a;\Lambda}$ in a suitable coordinate system. Show also that in a general coordinate system $\sigma_2^{\mu\nu}[\phi_{a;\Lambda}] = \Lambda^{\mu\nu}$.

c. Show that, in the 1-dimensional case (with $\Lambda \equiv \sigma^2$), $\sigma_{2k+1}[\phi_{a;\sigma^2}] = 0$ and

$$\sigma_{2k}[\phi_{a;\sigma^2}] = (-2\sigma^2)^k \left\{ \frac{d^k}{d\lambda^k} \int \frac{dx}{\sqrt{2\pi\sigma^2}} e^{-\lambda x^2/2\sigma^2} \right\}_{\lambda=1}.$$

d. Calculate $\sigma_k{}^{\mu_1\cdots\mu_k}[\phi_{a;\Lambda}]$ in a coordinate system that diagonalises Λ.

3.3

Take Definition 3.3 as your point of departure and show that Definition 3.2 is consistent with it.

3.4

Show that the Gaussian of Problem 3.2 corresponds to the push forward of a normalised, standard Gaussian $\phi(x) = \frac{1}{\sqrt{2\pi}^n}e^{-\frac{1}{2}x^2}$ by an affine transformation.

3.5

Let η and θ be spacetime automorphisms.

a. Show that $(\eta \circ \theta)_* = \eta_* \circ \theta_*$ and $(\eta \circ \theta)^* = \theta^* \circ \eta^*$.

b. Verify Triviality 3.1 using coordinate free notation.

We are now going to express the principle of covariance Triviality 3.1 in terms of coordinates. Let $F \in \Sigma_x$, $\phi \in \Delta_x$ be instances of a source and a detector, parametrised in a coordinate system x by a scalar "function" $f(x)$ and a scalar density $\phi(x)$, respectively. Let η be the inverse of θ, mapping a point labelled by x to $y \equiv \theta(x)$, and vice versa $x \equiv \eta(y)$ (*active* point of view: we are not changing coordinate systems here, but carrying out a manifold transformation). Alternatively, this can be interpreted from a *passive* point of view as a mere *coordinate reparametrisation* $x \equiv \eta(y)$, attributing a new coordinates $y \equiv \theta(x)$ to the point initially labelled by x. Furthermore, take $f'(y)$ and $\phi'(y)$ to be the parametrisations corresponding to $\eta^* F \in \Sigma_y$ and $\theta_* \phi \in \Delta_y$ in the mapped (or relabelled) point (primes indicate transformation, not differentiation).

c. Show that $f'(y) = f(x)$ ("scalar transformation"). Give an "active" and a "passive" interpretation.

d. Verify also that $\phi'(y) = |\det \frac{\partial x}{\partial y}| \phi(x)$ ("scalar density transformation"), and explain likewise.

e. Express the principle of covariance for the sample $\lambda = F[\phi]$ in terms of its coordinate parametrisation.

3.6
Relate general covariance (Triviality 3.1) to the tracking of a physical object, and discuss the appropriateness of this relationship.

3.7
See Definition 3.7.

a. Show that in n-dimensional spacetime, $\dim \Theta = \frac{1}{2}n^2 + \frac{1}{2}n + 2$ (Galilean spacetime model).

b. Show that if $\vec{v} = \vec{0}$ in n-dimensional spacetime, $\dim \Theta = \frac{1}{2}n^2 - \frac{1}{2}n + 3$ (Newtonian spacetime model).

c. Disregarding time, the relevant subgroups in d-dimensional space have $\dim \Theta = \frac{1}{2}d^2 + \frac{3}{2}d + 1$ and $\dim \Theta = \frac{1}{2}d^2 + \frac{1}{2}d + 1$, respectively (Euclidean space model with or without Galilean boosts).

3.8
See Definition 3.7. Calculate the eigenvalues and eigenvectors of a Galilean boost in n dimensions,

$$X = \left[\begin{array}{c|c} 1 & \emptyset \\ \hline \vec{v} & I \end{array} \right],$$

and sketch the result for the $(1+1)$-dimensional case.

3.9
See Box 3.4. Derive a system of first order partial differential equations for $F(x_1, \ldots, x_M)$ expressing the scale invariance constraint.

3.10
See Definitions 2.20, 2.21 and 3.5, and Lemma 2.4. Suppose $\Theta = T(n)$, the spacetime translation group.

a. Show that $\Theta^* F = F\star$.

b. Show that, for every multi-index α, $\nabla_\alpha F \star \phi = F \star \nabla_\alpha^\dagger \phi$ and $\nabla_\alpha F \star \phi = F \star \nabla_\alpha \phi$.

3.11
See Definitions 3.9, 3.13 and 3.15 and Lemmas 3.2 and 3.3. Assume $\phi \in \mathcal{S}^+(\mathbb{R}^n)$.

a. Show that the push forward of a tensor product of filters under the scan operator S equals their convolution product (Lemma 3.2).

b. Show that the base point of a convolution product of filters equals the sum of base points of the filters: $\pi[\phi_1 * \ldots * \phi_k] = \pi[\phi_1] + \ldots + \pi[\phi_k]$.

c. Show that the scale matrix of a convolution product of filters equals the sum of scale matrices of the filters: $s[\phi_1 * \ldots * \phi_k] = s[\phi_1] + \ldots + s[\phi_k]$.

d. Define $S^*\pi \equiv \pi \circ S_*$ and show that $S^*\pi[\phi_1 \otimes \ldots \otimes \phi_k] = \pi[\phi_1] + \ldots + \pi[\phi_k]$. Note that $S^*\pi$ is essentially the projection map itself, made fit to produce a single base point from a specified tensor product of independent filters.

e. Define $S^*s \equiv s \circ S_*$ and show that $S^*s[\phi_1 \otimes \ldots \otimes \phi_k] = s[\phi_1] + \ldots + s[\phi_k]$. Note that S^*s is essentially the inner scale operator itself, made fit to produce a single scale matrix from a specified tensor product of independent filters.

f. Does additivity still hold if we replace $\pi[\phi]$ and $s[\phi]$ by the unnormalised first and second order momenta $m_1[\phi]$ and $m_2[\phi]$?

3.12 (Frequency Tuning)
Define $f_k(x) = f(x)\,e^{ikx}$ for any $f \in S'(\mathbb{R}^n)$ and likewise $\phi_k(x) = \phi(x)\,e^{ikx}$ for any $\phi \in S(\mathbb{R}^n)$, and let $F_k[\phi] = \int dz\, f_k(z)\,\phi(z)$, $F[\phi_k] = \int dz\, f(z)\,\phi_k(z)$, so that by trivial duality $F_k[\phi] = F[\phi_k]$ (*Gabor filtering*).

a. Argue why duality does not trivially carry over to higher order derivatives.

b. Show that $\nabla F[\phi_k] = (\nabla - ik)\,F_k[\phi]$, and give an interpretation relating differentiation and Gabor filtering.

3.13
See Observation 3.1.

a. Show that if $\phi, \psi \in \Delta \Rightarrow \phi * \psi \in \Delta$ holds for all detectors in Δ, then Δ is indeed a linear convolution algebra.

b. Verify that $\Delta = S(\mathbb{R}^n)$ meets the closure property stated under a.

c. Does $\Delta = S(\mathbb{R}^n)$ have an identity element? Is it a regular algebra (i.e. can one invert its elements)? Is it commutative?

3.14
See Definition 2.17 and Assumption 3.1.

a. Show that if $f \in C^0(\mathbb{R}^n) \cap P(\mathbb{R}^n) \subset S'(\mathbb{R}^n)$, then $f \geq 0$ as a function implies that $f \geq 0$ as a regular tempered distribution, *vice versa*.

b. Prove: every positive distribution is a positive measure.

3.15
See Definitions 3.13 and 3.15.

a. Argue (or show by example) that these definitions would produce counter-intuitive results if the positivity constraint were removed.

b. How should one define base point, or projection map, and inner scale for an arbitrary *derivative* of a positive filter?

3.16
Relate Definitions 3.12, 3.13 and 3.14 to the *generating function*

$$M[\phi](x) \stackrel{\text{def}}{=} \int dz\, \exp(z \cdot x)\,\phi(z).$$

3.17

See Assumption 3.1 and Definition 3.16. Show that there exists no point operator with odd parity.

3.18

Page 60: "The following gauge is always realizable by suitable spacetime symmetry transformation and superposition..." Prove this statement, assuming only $\phi \in S^+(\mathbb{R}^n)$. (Hint: integrate over the rotation group $SO(n-1)$ and apply a suitable spacetime translation and scaling.)

3.19

Recall Corollary 3.2 and Proof 3.6. Pauwels *et al.* [235] have shown that if a basic filter $\chi \in C^0(\mathbb{R}) \cap L^1(\mathbb{R})$, satisfying $1[\chi] = 1$ and $\chi(x) = \chi(-x)$, is pushed forward by a rescaling (scale invariance!), yielding a scale-parametrised filter

$$\chi_\sigma(x) \stackrel{\text{def}}{=} \frac{1}{\sigma}\chi(\frac{x}{\sigma}),$$

and if there exists a sufficiently smooth reparametrisation $\psi(t) = \sigma$, $\phi_t(x) \equiv \chi_\sigma(x)$, such that this filter satisfies the semigroup property

$$\phi_s * \phi_t = \phi_{s+t},$$

then its Fourier transform must be of the form

$$\widehat{\phi_t}(\omega) = e^{-\alpha|\omega|^\alpha t},$$

in which $\alpha > 0$ is a constant (and $\sigma = \psi(t) = t^{1/\alpha}$). Show that if in addition $\phi \in S^+(\mathbb{R})$, then $\alpha = 2$. (Hint: require regularity and compute the second order momentum $m_2[\phi_t]$ of ϕ_t.)

3.20

Recall Corollary 3.2.

a. Show that a prototypical point $\phi \in S^+(\mathbb{R}^n)$ satisfying the self-similarity condition is separable:

$$\phi(x) = \prod_{\mu=0}^{d} \psi(x^\mu).$$

b. Show that the 1-dimensional function $\psi \in S^+(\mathbb{R})$ satisfies

$$\psi * \psi(x) = \frac{1}{\sqrt{2}}\psi(\frac{x}{\sqrt{2}}).$$

3.21

Fill in the details of Proofs 3.7 and 3.8.

3.22

Define the operators $\chi_+^m \in S'(\mathbb{R})$ for $m \in \mathbb{Z}$ as follows (see Problem 2.5 for definition of $H(x)$):

$$\chi_+^m(x) = \begin{cases} \frac{1}{(m-1)!}x^{m-1}H(x) & \text{if } m > 0 \\ \delta^{(-m)}(x) & \text{if } m \leq 0, \end{cases}$$

in which a parenthesised superscript denotes order of differentiation.

a. Show that $I_+^m \equiv \chi_+^m * : S(\mathbb{R}) \to S(\mathbb{R})$ is an m-th order integration if m is positive, the identity operator if m is zero, and a $(-m)$-th order differentiation if m is negative.

b. Show that

- $\frac{d}{dx}\chi_+^{m+1} = \chi_+^m$,
- $x\chi_+^m = m\chi_+^{m+1}$, and
- $x\frac{d}{dx}\chi_+^m = (m-1)\chi_+^m$.

c. Show that $I_+^m \circ I_+^n = I_+^{m+n}$.

(This idea of coupling integration and differentiation has been extended to a complex analytical family of distributions χ_+^z for all "complex orders" $z \in \mathbb{C}$ by Marcel Riesz [57, 243].)

3.23
This problem is meant to fill in some details of Result 3.4. For simplicity consider the one-dimensional case in items **a-h**.

a. Show that N is Hermitian, and that all its eigenvalues are real and nonnegative.

b. Show that the creation and annihilation operator do not commute, and that their commutator is given by $[A, A^\dagger] = 1$. (Notation: $[A, B] \equiv AB - BA$.)

c. Prove by induction that $[A, A^{\dagger k}] = kA^{\dagger k-1}$ and from there, that for an analytical function f, $[A, f(A^\dagger)] = f'(A^\dagger)$.

d. Show that the smallest eigenvalue of N is in fact zero, and that the corresponding eigenfunction is nondegenerate, and is uniquely defined by:

$$\left\{ \begin{array}{ll} A\psi_0 & = 0 \\ \|\psi_0\|_2 & = 1. \end{array} \right.$$

e. Show that $\psi_k = \frac{1}{\sqrt{k!}}A^{\dagger k}\psi_0$ is a normalised eigenfunction of N with integer eigenvalue k.

f. Show that the eigenvalue spectrum of H is given by $E_k = k + \frac{1}{2}$, and that the eigenfunctions are the same as for N.

g. Show that $A^\dagger\psi_k = \sqrt{k+1}\psi_{k+1}$ and $A\psi_k = \sqrt{k}\psi_{k-1}$. (This of course explains the terminology.)

h. Prove that the set $\{\psi_k \mid k \in \mathbb{Z}_0^+\}$ forms a complete and orthonormal basis of $L^2(\mathbb{R})$.
Hint: show that $\left\{ \begin{array}{ll} \psi_k \cdot \psi_l & = \delta_{kl} \\ \sum_{k \in \mathbb{Z}_0^+} \psi_k(x)\psi_k(y) & = \delta(x-y). \end{array} \right.$

i. Complete the proof of Result 3.4.

3.24 (Duality Claim Example 3.2)
Recall Example 3.2, and the definition of duality, see e.g. Equation A.103 in Appendix A.

a. Show that if $g \equiv f^{\text{inv}}$, then

$$\omega(v) = g^*\omega(f_*v).$$

b. Let dp^a and $\partial/\partial p^b$ be a dual basis of forms and vectors $(a, b = 1, \ldots, N)$. By definition this means that

$$dp^a(\partial/\partial p^b) \stackrel{\text{def}}{=} \delta^a_b \,,$$

(cf. also Problem 2.13). Prove the last statement in Example 3.2:

$$g^* : \mathbb{R}^N_p \to \mathbb{R}^N_q : \omega_p \mapsto \omega'_q \equiv Dg(q)(\omega_p)\Big|_q \,.$$

(Hints: [brute force method:] write out the equality under **a** in terms of coordinates relative to the dual basis, or [smart method:] exploit the symmetrical nature of duality.)

3.25 (Dual of Definition 3.22)

Page 69: "...Of course this definition [i.e. Definition 3.22] has a dual counterpart..." Recall the definition of duality, see e.g. Equation A.103 in Appendix A. Let

$$\Omega^a \stackrel{\text{def}}{=} \sum_{b=1}^{N} \omega^a_b(\ell)\, d\ell^b \quad (a = 1, \ldots, N),$$

and

$$f^* \bar{\Omega}^a \stackrel{\text{def}}{=} \Omega^a \,.$$

(Recall that one must pull back *transformed* covectors to obtain originals.) Show that

$$\sum_{c=1}^{N} \omega^a_c(\ell)\, v^c_b(\ell) = \delta^a_b \,.$$

(Hint: consider measures $\bar{\Omega}^a$ in $\bar{\ell}$-coordinates, dual to the vectors $\bar{\mathcal{L}}_a$: $\bar{\Omega}^a[\bar{\mathcal{L}}_b] \equiv \delta^a_b$.)

3.26

Work out the argument of canonical scale parametrisation in Example 3.3 via the dual route of Problem 3.25.

3.27 (Classification of Critical Points)

Consider an isotropic spatial image in d dimensions. If we take a one-dimensional section through a critical point somewhere in the image, we will find either a (local) minimum or a maximum. Odds are that both occur with equal probabilities, and that we won't find a single indifferent critical point where both first as well as second order directional derivatives vanish.

a. Why? (Hint: use an argument of *genericity*.)

Using self-explanatory notation we thus have generically $p_+ = p_- = 1/2$, and $p_0 = 0$. Now we may consider another scanline through the critical point of interest, perpendicular to the first one, and repeat the inspection. Again we find either a maximum or a minimum with equal probabilities, yet independent of those of the previous scanline. Continuing in this way we can classify critical points into maxima $(+, \ldots, +)$, minima $(-, \ldots, -)$, as well as various sorts of "hypersaddles" characterised by toggling signs one by one.

b. Argue that this line of reasoning makes it plausible that

- the probability for a *given* critical point to be an extremum (maximum or minimum) equals $p_e = 2^{-(d-1)}$, and
- the probability for it to be a hypersaddle is complementary: $p_s = 1 - p_e$.

CHAPTER **4**

The Scale-Space Paradigm

The conclusion reached at the end of Chapter 3 can be summarised by the following *scale-space paradigm*.

Paradigm 4.1 (Scale-Space)
Recall Paradigm 2.1 on Page 21, and Result 3.3 on Page 64. As an alternative to Paradigm 3.1, Page 57, we may define, without loss of generality,

$$\Delta \overset{\text{def}}{=} \mathcal{G}(\mathbb{R}^n) \quad \text{whence} \quad \Sigma \overset{\text{def}}{=} \mathcal{G}'(\mathbb{R}^n).$$

In this chapter we will study it in somewhat more detail.

4.1 The Concept of Scale and Some Analogies

The Gaussian point operator, which constitutes the pivot in Paradigm 4.1, has many nice properties. Not surprisingly, there are many different approaches arriving at this very same result. The paradigm was first introduced—in a rather nice axiomatic way—by Taizo Iijima in 1962 in the context of optical character recognition [128], and was followed by a series of other papers in the seventies. Another early axiomatic approach, in the same context, has been described by Nobuyuki Otsu in his 1981 PhD thesis [233]. These early accounts remained essentially unnoticed until they were recently translated by Weickert *et al.* [293].

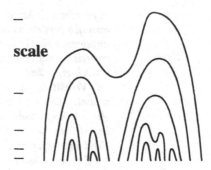

Figure 4.1: Prohibition of spurious resolution entails that iso-grey-level contours always close with a local convexity towards the direction of increasing scales.

Figure 4.2: Illustration of spurious resolution: **Left:** original image. **Middle:** unsharply focused image; the filter in this case roughly corresponds to an indicator function on a disk, the size of which is a measure of inner scale. **Right:** histogram equalisation applied to a centre portion of the unsharply focused image. Spurious detail is most clearly visible at frequencies where contrast is inverted. (Original photographs by courtesy of Koenderink and Van Doorn.)

The paradigm was first introduced in the English literature by Witkin [303], who also proposed the term "scale-space image". Witkin decided to take the Gaussian as the basic scale-space filter for its "well-behavedness".

Koenderink introduced a rigorous notion of *spurious resolution* and showed that Gaussian scale-space theory arises by necessity from the requirement that no spurious detail should be generated when resolution is diminished [150]. Thus coarse scales provide truly simpler pictures than fine scales, much like in cartographic generalisation [215, 245], cf. Figures 4.1–4.2.

Both Koenderink and Witkin already pointed out in their early papers that the actual challenge is to understand image structure at all levels of resolution *simultaneously*. This "deep structure"— the unfolding of structure over scale—is one of the outstanding challenges any multi-resolution theory[1] will have to cope with.

---

[1] The problem affects vision research as well; visual systems often embody a sensorium comprising receptive fields of various sizes.

 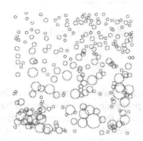 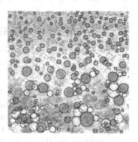

Figure 4.3: Lindeberg proposes a method for probing "deep structure" using an automatic scale selection criterion, and applies this to "blob detection" [187, 191, 194]. The left image shows a sunflower field. "Dominating scales" at various locations are most easily revealed by measuring blob responses in scale-space (only the 200 most salient ones have been displayed). The middle image shows the selected scales after back-projection into the image plane. The right image is an overlay illustrating that the scales thus obtained indeed correspond to the ones that are leaping to the eye. Note that it is nontrivial to obtain similar results in a context in which the scale degree of freedom has not been made explicit. Illustration by courtesy of Lindeberg.

Koenderink points out that the scale-space representation defines a versal family of spatial images, to which Thom's classification theorem can be applied [96, 237, 270]; see also [158, Chapter 9], [41, 42, 84, 102, 101, 134, 135, 137, 138, 152, 156, 162, 170, 187, 188, 191], and Boxes 4.2 and 4.3. He also proposes a "projection" rule to link images at successive scales [150, 156], which has formed the basis for many multiscale strategies in computer vision and image analysis applications, e.g. *edge focusing* [17, 202], *segmentation* [285, 286], *shape from texture* [197], *scale selection* [189, 190, 195], etc. [103, 105, 191, 266]; see Figure 4.3.

Since it is actually the *a priori* freedom of scale that allows us to free ourselves from grid artifacts, multiscale strategies appear inescapable. Indeed, one can observe an increasing appreciation of resolution as an essential control parameter in image analysis over the last decade or so. It is the pivotal parameter not only in scale-space theory (further references on basics: [11, 20, 83, 186]), but also in *wavelet theory* [34, 37, 38, 44, 45, 56, 113, 132, 176, 209, 210, 247, 251, 268, 269, 275], and in *mathematical morphology* [21, 23, 24, 56, 110, 111, 130, 256, 257]; see also Box 4.1. In several other standard techniques resolution comes in more implicitly, e.g. in the disguise of a multiplier of a regularising term in "functional minimisation" [74, 221]; this will be illustrated in Section 4.1.3.

Figures 4.4 and 4.5 illustrate the Gaussian scale-space paradigm. Two fundamentally different representations of "the same image" are compared, one binary rendering, with pixel values either 0 or 1, and one grey-tone rendering, with pixel values in the range $\{0, \ldots, 255\}$ (two common cases). At pixel scales the representations are totally incomparable, but at sufficiently coarse resolution they convey the same information. It is a good test case to run image analysis routines designed for grey-scale images on such binary, "atomic" simulations, or on any

☞ **Box 4.1 (Mathematical Morphology)**

Mathematical morphology for grey-scale images can be put in an image processing framework completely analogous to the one proposed in this book, the essential difference being the details of duality. This expresses a difference in the concept of measurement. Linear, topological duality expresses the desire to interpret detectors as *volumetric* probing devices (the sources are "transparent"). In mathematical morphology measurements are to be taken as *tactile* probes, e.g. "depth"; it is most naturally conceived of as a theory of *surface contact* (the sources are "opaque" but confined). At a sufficiently high level of abstraction, linear and morphological models boil down to identical algebraic structures.

It is of some interest to reveal this abstract "super-structure" encompassing both theories as disguises of one and the same model. For one thing, this would allow us to translate all results obtained for the linear theory, as far as these retain their validity in the generalised framework, to the domain of mathematical morphology,

vice versa. Indeed, Dorst and Van den Boomgaard have pointed out that scale-space theory has properties completely analogous to mathematical morphology based on a *quadratic structuring element*, the morphological counterpart of a Gaussian point operator [22, 23, 56] (cf. also [130]). They have also shown that convolution becomes *tangential dilation* (a local operation as opposed to ordinary dilation), that Fourier transformation translates to the so-called *slope transform*, and so forth.

Another reason for disclosing the common structure is that it reveals the appropriateness of each model for a specific image modality and functional goal in a transparent way. The morphological model seems ideally suited for a syntactical definition of *range images*. Ultimately, *every* raw image arises from the integration of some density. The question is which syntactical convention applied to structure the raw data subserves semantics best.

other visually reasonable halftoning [236, 280, 281]. A well-designed routine is robust and should only produce small-scale artifacts related to "insufficient resolution" (quantisation and discretisation noise). If the scale of interest is well beyond grain limits, performance should be comparable. For example, certain visual tasks can certainly be carried out on either rendering.

4.1.1 ✳ Scale and Brownian Motion: Einstein's Argument

We can also view a scale-space image as the continuum limit of a discrete probabilistic process, viz. a random walk of quanta during a time span proportional to inner scale. We could following Einstein's treatise of random walk [61, Chapter 1] in the context of the irregular motion of particles suspended in a liquid (Brownian motion).

Einstein's argument is intuitive, though a bit shaky (Problem 4.2). Assume we have a discrete grid, at each voxel (unit volume element) of which we have a number of quanta (e.g. an integer-valued raw image). Let us assume the situation

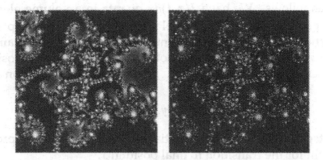

Figure 4.4: **Left:** Same as Figure 2.1, shown here for ease of comparison. This illustration is a device specific black-and-white rendering of the original 8-bit per pixel grey-values, sampled on a regular 512×512 pixel grid. **Right:** A rendering of a binary simulation of the same image after applying Ulichney's so-called "blue-noise" halftoning algorithm [280, 281] to the original grey-scale data. The idea behind halftoning, or "dithering", is to trade-off grey-scale resolution against spatial resolution for the purpose of display on binary devices, such as printers and black-and-white monitors, with minimal visual artifacts. The loss of spatial resolution is usually acceptable if the pitch is sufficiently small. In this binary simulation, the original grid size has been maintained. The algorithm is not entirely deterministic (in addition to a definite local neighbourhood operation, a random number generator is invoked at each binary pixel calculation to vary a threshold), and the result *at pixel scale* is essentially stochastic.

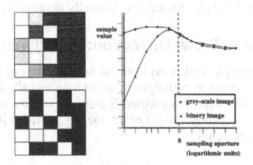

Figure 4.5: **Left:** A zoom-in on a 5×5-neighbourhood of pixel $x = (238, 255)$ of the original, 8 bit-per-pixel grey-scale image of the fractal, and of the 1 bit-per-pixel blue-noise simulation, as shown in Figure 4.4. **Right:** A comparison of the scale traces at the pixel of interest for the grey-scale and binary image, using logarithmic scales within the range $\sigma \in [0.3, 3.0]$. Grey-scale units have been chosen in such a way that the images have identical grey-value averages in order to allow mutual comparison. The dotted line indicates the transient scale $\sigma = \sqrt{3}/6$ separating sub-pixel and supra-pixel domains; it is the scale of a Gaussian filter that yields the same amount of blur as a unit square pixel in the absence of neighbourhood correlations. Note that despite the dramatic effect of halftoning at pixel scale, the effect becomes negligible already at scales slightly larger than unity. Neither human beings nor image applications require grey-scale images if the scales of interest are sufficiently large relative to pixel scale!

at time t is as follows. We have $f(x,t)dx$ quanta in a volume element dx that contains the point with coordinate label $x \in \mathbb{R}^d$. After a time step s quanta will have experienced a spatial displacement $x \to y$ following a certain probability density $\psi_s(y; x, t)$. If probabilities do not depend upon current position and are stationary, then $\psi_s(y; x, t) = \phi_s(y - x)$. Investigating the situation at x we then find

$$f(x, t + s) = \int dy \, f(y, t) \, \phi_s(x - y) \,, \qquad (4.1)$$

where we integrate over all possible initial positions y with the appropriate probabilistic weight for the transition to final position x.

Under the condition that s is very small, and that corresponding step sizes are typically small as well (meaning $\phi_s(z)$ vanishes rapidly with increasing $\|z\|$), we can replace left and right hand sides by their lowest order Taylor terms. Assuming isotropic probabilities, the result of this is

$$\frac{\partial f}{\partial t} = D \triangle f \,, \qquad (4.2)$$

in which the diffusion coefficient D is essentially isotropic inner scale (recall Definition 3.15, Page 59):

$$D \overset{\text{def}}{=} \frac{1}{2s}\sigma^2 \,, \qquad (4.3)$$

in which σ denotes the width of the probability density function ϕ_s. Identifying s with σ^2, which means defining the evolution parameter of the random walk process as the measure of inner scale, one formally obtains scale-space theory.

4.1.2 ✳ Scale and Brownian Motion: Functional Intergration

A relatively new method, which exploits the *full* structure of a functional as opposed the so-called variation principle, is based on a so-called *functional integral*, or *path integral* [35, 74, 203]. This method is inspired on Feynman's postulate for the quantum-mechanical description of interacting particles [69]. To get the gist of it, consider the functional

$$F[q(t)] = \exp\left\{ -\frac{1}{2} \int_{t_0}^{T} dt \, \|\dot{q}(t)\|^2 \right\} \,, \qquad (4.4)$$

defined for all one-dimensional paths $q(t)$ on the interval $t \in [t_0, T]$, with fixed boundary conditions $q(t_0) \equiv x_0$, $q(T) \equiv x$. One can think of $q(t)$ as the trajectory of a hypothetical particle as a function of pseudo-time (an evolution parameter which, with a modest amount of foresight, can be related to scale), and of (minus the logarithm of) $F[q(t)]$ as its "action functional", under the assumption of free motion.

The "classical solution" derived using the variation principle is a *deterministic* path, which can be solved from the Euler-Lagrange equation

$$\frac{\delta F}{\delta q}[q(t)] = 0 \,, \qquad (4.5)$$

subject to the boundary conditions, i.e.

$$\ddot{q}(t) = 0 \quad \text{with} \quad q(t_0) = x_0 \quad \text{and} \quad q(T) = x. \tag{4.6}$$

Thus the dynamics of the functional dictates "uniform motion" in the classical sense:

$$q_c(t) = x_0 + \frac{x - x_0}{T - t_0} (t - t_0). \tag{4.7}$$

Note that the exponential mapping in Equation 4.4 is inessential; monotonic mappings do not affect stationary points.

In order to describe stochastic random walk, consider the heuristic expression

$$Z = \int \mathcal{D}q(t) \, F[q(t)], \tag{4.8}$$

which is meant to be read as an integration over all possible functions $q(t)$ subject to fixed boundary conditions. Clearly this does rule out form ambiguities of the type mentioned above; in particular the exponential mapping matters.

In order to define the functional integral (Equation 4.8) rigorously, the functional $F[q(t)]$ is replaced by an $N-1$ parameter function $F(q_1, \ldots, q_{N-1})$, as follows. Divide the time interval into N small intervals of width δt, say $t_i = t_0 + i\delta t$, $N\delta t = T - t_0$. The continuous path can now be approximated by the set of $N-1$ intermediate points $q_1, \ldots, q_{N-1}$ and the two fixed boundary conditions at t_0 and $t_N \equiv T$, i.e. x_0 and x, respectively:

$$F[q(t)] = \exp\left\{ -\frac{1}{2} \int_{t_0}^{T} dt \, \dot{q}^2(t) \right\} \approx \exp\left\{ -\frac{1}{2}\delta t \sum_{i=1}^{N} \dot{q}_i^2 \right\} = F(q_1, \ldots, q_{N-1}),$$
$$\tag{4.9}$$

in which we have defined

$$\dot{q}_i \stackrel{\text{def}}{=} \frac{q_i - q_{i-1}}{\delta t}, \tag{4.10}$$

the mean velocity on the i-th interval. See Figure 4.6.

If at all sensibly definable as the limit of a finite-dimensional integral over the control points of the discretised path $q(t) \sim \{q_i = q(t_i)\}_{i=0,\ldots,N}$ between the fixed end-points $q_0 \equiv x_0$ and $q_N \equiv x$, functional integration apparently entails an infinite number of integration dummies: the $N-1$ intermediate control points pushed to the limit $N \to \infty$. Even if one refrains from taking the limit one needs to be cautious, for the result may and typically will depend upon the discretisation details. Therefore, let us define Z by some limiting procedure for the approximate expression

$$Z \approx Z_N \equiv \frac{1}{A_N} \int dq_1 \ldots dq_{N-1} \, F(q_1, \ldots, q_{N-1}), \tag{4.11}$$

in which the normalising factor A_N must be such as to make the limit $N \to \infty$ possible. In general one has to decide on the form of the *measure*; the shift invariant Lebesgue measures dq_i express homogeneity of configuration space. Other

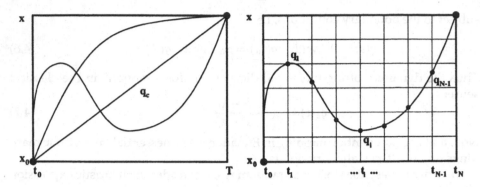

Figure 4.6: Left: some relevant paths; q_c is the classical one singled out by the variation principle and the boundary conditions. **Right**: discrete picture; the control points $q_i = q(t_i)$, $i = 1, \ldots, N - 1$, are used as integration dummies, while boundary values are kept fixed.

symmetry considerations will require appropriate group invariant measures instead. An alternative way of putting it is that one needs to think about a canonical parametrisation of $F[q(t)]$.

Consider the following change of integration variables:

$$
\begin{cases}
y_1 & = \quad q_1 - q_0 \\
& \vdots \\
y_{N-1} & = \quad q_{N-1} - q_{N-2} \, .
\end{cases}
\tag{4.12}
$$

The transformation $\vec{q} \to \vec{y}$ is orthogonal, hence has unit Jacobian. Let us introduce an N-th dummy, $y_N = q_N - q_{N-1}$, which is not independent, since we have $\sum_{i=1}^{N} y_i = x - x_0$ constant. Nevertheless, we may integrate over y_N after incorporating the constraint with the help of a δ-function, as follows:

$$
Z_N(x, T \,|\, x_0, t_0) = \frac{1}{A_N} \int dy_1 \ldots dy_N \, \delta \left(\sum_{i=1}^{N} y_i - (x - x_0) \right) \prod_{i=1}^{N} \exp \left\{ -\frac{1}{2\delta t} y_i^2 \right\} .
\tag{4.13}
$$

After exponentiating the δ-function,

$$
\delta \left(\sum_{i=1}^{N} y_i - (x - x_0) \right) = \int \frac{d\omega}{2\pi} \exp \left\{ i\omega \left(x - x_0 - \sum_{i=1}^{N} y_i \right) \right\} ,
\tag{4.14}
$$

we obtain a Gaussian integral over $\vec{y}$ and ω, which depends upon A_N. A consistent way to fix A_N, yielding the so-called "Wiener measure", is obtained by imposing the normalisation

$$
\int dx \, Z_N(x, T \,|\, x_0, t_0) \equiv 1 ,
\tag{4.15}
$$

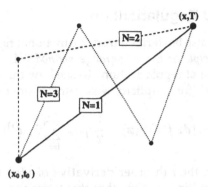

Figure 4.7: Any path corresponding to $(N-1)$-fold intermediate "scattering" is *a priori* equally probable: $\mathcal{Z}(x, T \,|\, x_0, t_0) = \mathcal{Z}_N(x, T \,|\, x_0, t_0)$. It is understood that one integrates over all control point locations.

which allows us to think of the functional integral as a *transition probability density*. With this measure the functional integral becomes

$$\mathcal{Z}(x, T \,|\, x_0, t_0) = \frac{1}{\sqrt{2\pi(T - t_0)}} \exp\left\{ -\frac{1}{2} \frac{(x - x_0)^2}{T - t_0} \right\}. \tag{4.16}$$

Note that in this simple case the number of control points N happens to be immaterial and the limit becomes trivial: $\mathcal{Z}(x, T \,|\, x_0, t_0) = \mathcal{Z}_N(x, T \,|\, x_0, t_0)$, for all $N \geq 1$. The interpretation of this is illustrated in Figure 4.7. Also,

$$\mathcal{Z}(x, T \,|\, x_0, t_0) = \mathcal{Z}(x - x_0 \,;\, T - t_0), \tag{4.17}$$

a two-fold shift invariance property already manifest in the form of the functional $F[q(t)]$ and in the definition of the measure $\mathcal{D}q(t)$. Finally, it is not so difficult to show that the probability density satisfies the semigroup property (indeed, already suggested by the form of Equation 4.8)

$$\mathcal{Z}(x, T \,|\, x_0, t_0) = \int dy \, \mathcal{Z}(x, T \,|\, y, t) \, \mathcal{Z}(y, t \,|\, x_0, t_0), \tag{4.18}$$

for any $t \in (t_0, T)$. Interpreting total evolution time as scale, $T - t_0 \equiv s \equiv \sigma^2$, we have a formal interpretation of the linear scale-space kernel as a kind of "vacuum-to-vacuum" transition probability density in the absence of a potential field. (From an affine reparametrisation $t \mapsto m(t - t_0)$ it shows that the "mass parameter" $m > 0$ in the reparametrised Lagrangian amounts to a rescaling of inner scale.)

This example gives a fair idea of how to proceed in order to carry out functional integration in general, and how to arrive at traditional scale-space theory starting from an extremely simple functional in particular.

4.1.3 ✳ Scale and Regularisation

Scale-space theory can also be derived from a functional representation, using the standard *variation principle* or *Euler-Lagrange method* [74, 221], a technique often employed in the context of regularisation. To see how this comes about, consider the following functional (for simplicity we again consider the 1D case):

$$S[\,u(x)\,;\,f(x)\,] = \int dx\, f(x)\, u(x) - \frac{1}{2} \int dx \sum_{i=0}^{\infty} \lambda_i\, u^{(i)}(x)\, u^{(i)}(x)\,, \qquad (4.19)$$

in which $u^{(i)}(x)$ means the i-th order derivative of $u(x)$ with respect to x, and in which each λ_i is a multiplier. Note that this functional is the extension of the usual source term $F[u]$ with a data independent quadratic functional, and that $f \equiv 0$ implies the trivial stationary solution $u \equiv 0$.

The form of the functional is not unique, but *any* functional containing only quadratic terms of the form $u^{(i)}(x)u^{(j)}(x)$, can be written as in Equation 4.19, at least with suitable boundary conditions in effect (Problem 4.3). The generalisation to n dimensions is straightforward. In an isotropic n-dimensional space one does not need additional multipliers; the rule is one for each order (Problem 4.4).

The Euler-Lagrange equation is

$$\sum_{i=0}^{\infty} (-1)^i \lambda_i u^{(2i)}(x) = f(x)\,. \qquad (4.20)$$

It is now obvious how one should define the multipliers in order to obtain scale-space theory; set

$$\lambda_i \stackrel{\text{def}}{=} \frac{t^i}{i!}\,, \qquad (4.21)$$

for some positive t (the dimension of which is apparently squared spatial scale[2]), and we recognise the series expansion of the exponential map:

$$\exp\left\{ -t\frac{d^2}{dx^2} \right\} u(x;t) = f(x)\,. \qquad (4.22)$$

The formal solution is

$$u(x;t) = \exp\left\{ t\frac{d^2}{dx^2} \right\} f(x)\,. \qquad (4.23)$$

To see that this is indeed just linear scale-space theory, differentiate with respect to t, and consider the initial condition at $t = 0$:

$$\begin{cases} u_t & = & u_{xx} \\ u(x;t=0) & = & f(x)\,. \end{cases} \qquad (4.24)$$

For a point source $f(x) = \delta(x)$ this yields the Green's function of the diffusion equation, which is indeed just the Gaussian point operator: $u(x) = \phi(x)$. In

---

[2]Note that if one admits only one dimensional unit, then—without loss of generality—$\lambda_i \propto t^i$ by necessity, leaving only a combinatorial freedom.

general $f(x)$ may represent any raw image, in which case $u(x; t)$ is its scale-space extension.

It should be noted that the traditional approach (Equation 4.5) is qualitatively different, and requires a significantly more complicated functional, involving derivatives to *all* orders in an infinite series of terms, each of which requires an (in principle independent) multiplier. In practice, however, one usually accounts for only a finite, typically small number of terms in functional minimisation approaches. It is apparent that in doing so one is likely to miss interesting cases when using the standard variational principle, even something as straightforward as Gaussian scale-space theory.

4.1.4   ✳ Scale and Entropy

It is clear that resolution controls the amount of spatiotemporal information in an image. This statement can be given a precise information theoretic basis using the concept of *entropy*[3], introduced by Shannon [258] and Wiener [302].

For a positive scalar source field we may define its entropy *relative to the measure m > 0* as

$$S[f; m] \overset{\text{def}}{=} -\int dx\, m(x)\, f(x) \log f(x)\,, \tag{4.25}$$

assuming the source to be normalised relative to the measure: $F[m] = 1$ (otherwise, rescale m). We need the measure for obvious dimensional reasons; it is the standard relative to which we can express "amount of information" in terms of mere numbers (cf. Jaynes' discussion on prior probabilities [133]).

Detectors are densities, and a mechanism for converting them to scalars is needed before one can apply the same definition. For reasons of consistency we therefore apply the same definition to $f \to \phi/m$ (again assuming positivity):

$$S[\phi; m] \overset{\text{def}}{=} -\int dx\, \phi(x) \log \frac{\phi(x)}{m(x)}\,. \tag{4.26}$$

Note that normalisation now boils down to the usual one: $1[\phi] = 1$. A trivial case is obtained if $m = \phi$, i.e. if we measure the information content of ϕ relative to itself: $S[\phi; \phi] = 0$.

In general we may define the *entropy of a local sample* $\lambda = \{F, \phi\}$ by the nonlinear functional

$$S[F, \phi; m] \overset{\text{def}}{=} -\int dx\, f(x)\phi(x) \log \frac{f(x)\phi(x)}{m(x)}\,. \tag{4.27}$$

This kind of entropy pertains to the sample's intrinsic structure, the profile of sampling aperture times source field "under the integral". Note that entropy is well-defined only on a suitable subspace of $\Sigma \times \Delta$ and given an adequate measure.

What about the measure? It must have the dimension of an inverse spatiotemporal volume, so let us write

$$m(\ell) = \ell^{-n}\,, \tag{4.28}$$

---

[3]Alfréd Rényi proposed a generalisation of entropy that conveys full information content *modulo* localisation [240, 241, 267] akin to the local sign problem.

in which ℓ denotes the length of a fiducial yardstick. In the homogeneous case (ℓ independent of x) we can distinguish at least two relevant cases, depending upon whether we want to evaluate "amount of structure"

- relative to inner scale, $\ell \propto \sigma$, or

- relative to the grid constant, $\ell \propto \varepsilon$.

The actual constant of proportionality is inessential (although a suitably chosen value may turn out convenient); the essence is that we measure lengths and intervals in terms of dimensionless coordinates $\xi^\mu = x^\mu/\ell$, with ℓ adapted to intrinsic or extrinsic resolution ("blur" or "graininess", respectively).

For the Gaussian point operator one can do the computations analytically, and the result is intuitive.

Result 4.1 (Entropy of Gaussian Point Operator)
Recall Result 3.2, Page 61, and Equations 4.26 and 4.28. For $\ell = \varepsilon\sqrt{2\pi e}$ we have $S[\phi; m] = n\lambda$, in which the canonical scale parameter[4] λ is given by the reparametrisation $\sigma(\lambda) \equiv \varepsilon\, e^\lambda$. For $\ell = \sigma\sqrt{2\pi e}$ we obtain a constant entropy $S[\phi; m] = 0$.

In the former case we encounter the Lie parameter for the scaling subgroup in the same form as it occurs in Definition 3.7, Page 52, which is somewhat miraculous. The latter case reflects the self-similarity of the Gaussian point operator. See also Problem 4.6.

4.2 The Multiscale Local Jet

A convenient and robust way to capture local image structure in the context of the scale-space paradigm is by evaluating derivatives up to some order at any desired level of scale (recall Result 3.3, Page 64). However, partial derivatives depend upon the local coordinate frame relative to which they are computed. Typically one adopts a Cartesian coordinate convention in which the coordinate lines are aligned with the grid. A partial derivative taken in isolation therefore does not express any intrinsic image property.

As opposed to partial derivatives, the concept of a *local jet* [96, 165, 171, 172, 237] does encode an intrinsic image property. Basically the k-jet (or local jet of order k) of a function f at base point x can be identified with the low order part of its Taylor series, keeping only terms of orders $0, \ldots, k$. This is well-defined if f is k-fold differentiable at x. Thus we can evidently speak of the k-jet of a detector at any fiducial base point, and hence—by duality—also of the local k-jet of a source or an image. The k-jet of f at x is denoted $j^k f(x)$.

A local jet defines an equivalence class: all functions having the same derivatives up to order k (inclusive) at a given point are representative of the same k-jet. More generally, we can define a topology on the space of local jets, which enables

---

[4]The dimension may be absorbed into the parameter so as to make the rate of change of entropy a universal constant.

us to compare how "close" two functions are taking into account all of their k-th order local structure. This is the so-called $C^k(x, \mathbb{R}^n)$-topology defined by the following seminorm.

$$\|f\|_x^p \overset{\text{def}}{=} \sum_{i=0}^{k} \sum_{\mu_1 \ldots \mu_i = 0}^{d} |f_{\mu_1 \ldots \mu_i}(x)|^p \,, \tag{4.29}$$

with $p \geq 1$ (again, Greek indices denote derivatives). Different choices of $p \geq 1$ generate distinct topologies only in the non-realizable limit $k \to \infty$. Without loss of generality (as long as k remains finite) we could therefore take $p = 1$. However, there is a problem with this, viz. *lack of manifest covariance* (recall Trivialities 3.1 and 3.2). In order to obtain a covariant definition we must first of all consider *covariant derivatives*, and secondly, contract these into a covariant form. Assuming for the moment that we can indeed define such "covariant derivatives[5]", a more satisfactory one may be the following (cf. Problem 4.21).

Definition 4.1 ($C^k(x, \mathbb{R}^n)$-Topology)
Cf. Equation 4.29. The $C^k(x, \mathbb{R}^n)$-topology is defined on the basis of the seminorm

$$\|f\|_{x;k}^2 \overset{\text{def}}{=} \sum_{i=0}^{k} \lambda_i f_{\mu_1 \ldots \mu_i}(x) f^{\mu_1 \ldots \mu_i}(x) \,.$$

Indices represent covariant derivatives; λ_i are constant, positive weights.

(The inner summations are implicit by the Einstein convention.) The weights have no topological relevance, but have been added to make a nice connection with the functional of Equation 4.19. Of course there always remains a source of ambiguity, even in the limiting case $k \to \infty$: if $\|f - g\|_{x;k} = 0$, then f and g differ at most by a function having zero derivatives of orders $0, \ldots, k$ at x ($k \in \mathbb{Z}_0^+ \cup \{\infty\}$). If $g \in j^k f(x)$ but $g \notin j^{k+1} f(x)$, then k is called the *order of contact* of f and g at x.

The next step is to generalise the local jet concept to scale-space. For the sake of simplicity we disregard the temporal domain (it can easily be taken into account). If scale were an independent variable we could consider independent local jets for space and scale, requiring two independent orders of contact. However, scale-space structure is tightly coupled to high resolution spatial structure in a causal fashion (the diffusion equation); consequently the inclusion of scale is not trivial. We must therefore look for a *multiscale local jet* compatible with the scale-space paradigm and characterised by a *single* order of contact in scale-space [84].

Consider a formal expansion near the origin $(x; t) = (0; 0)$ of a coordinate system centred at some interior point in scale-space. The scale parameter t differs from the variance parameter s in Lemma 3.7, Page 65 (or Equation 3.10, Page 66) by an offset and a scaling factor: $2t = s - s_0 = \sigma^2 - \sigma_0^2$. That is, the scale of interest

---

[5] We return to covariant differentiation in Chapter 5; for the moment it suffices to ignore the attribute "covariant" while sticking to *rectilinear coordinates*.

is σ_0. We have

$$L_\infty(x;t) = \sum_{m=0}^{\infty} \sum_{j=0}^{\infty} \frac{1}{m!} \frac{1}{j!} L^{(j)}_{i_1\ldots i_m} x^{i_1} \ldots x^{i_m} t^j, \qquad (4.30)$$

in which a subscript of L refers to a covariant spatial derivative and a parenthesised superscript to a scale derivative, evaluated at the point of interest. More precisely, if ϕ is the Gaussian point operator corresponding to the point of interest, then

$$L^{(j)}_{i_1\ldots i_m} \stackrel{\text{def}}{=} F^{(j)}_{i_1\ldots i_m}[\phi] \stackrel{\text{def}}{=} (-1)^m F[\phi^{(j)}_{i_1\ldots i_m}]. \qquad (4.31)$$

(Remark: there is *no* minus sign for a scale-derivative.) The scale-space paradigm relates scale derivatives to spatial derivatives: $L^{(j)}_{i_1\ldots i_m} = \Delta^j L_{i_1\ldots i_m}$. Taking this into account, Equation 4.30 can be written as

$$L_\infty(x;t) = \exp\{t\Delta + x \cdot \nabla\} L, \qquad (4.32)$$

in which the exponential function "exp" is defined by its series expansion in the usual way. This reveals the role of the Laplacian as the infinitesimal generator of rescalings in addition to that of the gradient as the infinitesimal generator of translations. The expansion holds on a full scale-space neighbourhood.

Sofar things are fairly trivial. Our objective is to define "order of contact" by means of a scale-space consistent truncation.

Proposition 4.1 (Scale-Space Taylor Polynomials)
Define the k-th order polynomial L_k as

$$L_k(x;t) \stackrel{\text{def}}{=} \sum_{m=0}^{k} l_m(x;t),$$

with homogeneous polynomials l_m given by

$$l_m(x;t) \stackrel{\text{def}}{=} \sum_{j=0}^{[m/2]} \frac{1}{(m-2j)!} \frac{1}{j!} \Delta^j L_{i_1\ldots i_{m-2j}} x^{i_1} \ldots x^{i_{m-2j}} t^j,$$

in which $[p]$ denotes the entier of $p \in \mathbb{R}$, i.e. the largest integer smaller than or equal to p. Then for every $m = 0, \ldots, k$, l_m is a homogeneous polynomial solution to the diffusion equation. Consequently L_k is a k-th order polynomial solution as well.

Proof 4.1 (Proposition 4.1)
It is easily verified that the homogeneous polynomials l_m, and hence the L_k, indeed satisfy the diffusion equation. By taking the Laplacian it follows, after some dummy index manipulations, that

$$\Delta l_m(x;t) = \sum_{j=0}^{[m/2]-1} \frac{1}{(m-2j-2)!} \frac{1}{j!} \Delta^{j+1} L_{i_1\ldots i_{m-2j-2}} x^{i_1} \ldots x^{i_{m-2j-2}} t^j.$$

But it can readily be seen by computation that this equals $\frac{\partial}{\partial t} l_m(x;t)$.

Note that the l_m have a definite parity, viz. $(-1)^m$. Since the coefficients in Proposition 4.1 may have arbitrary values, they can be put in isolation so as to disclose a symmetric contratensor of rank m, which represents a fundamental, m-th order polynomial solution to the diffusion equation.

Corollary 4.1 (Scale-Space Taylor Polynomials)
The m-th order homogeneous polynomial solution to the diffusion equation can be written as

$$l_m(x;t) \stackrel{\text{def}}{=} L_{i_1 \ldots i_m} P^{i_1 \ldots i_m}(x;t),$$

in terms of the basis polynomials

$$P^{i_1 \ldots i_m}(x;t) = \sum_{j=0}^{[m/2]} \frac{1}{(m-2j)!} \frac{1}{j!} x^{(i_1} \ldots x^{i_m-2j} g^{i_m-2j+1 \, i_m-2j+2} \ldots g^{i_m-1 i_m)} t^j,$$

in which parentheses surrounding upper indices denote index symmetrisation.

An in-depth discussion of the so-called dual metric tensor g^{ij} showing up in the formula is deferred until Chapter 5; again one may be content for the moment with its Cartesian form, in which the components are given by the Kronecker symbol: $g^{ij} = 1$ if $i = j$, otherwise 0 (recall Page 63). As a result of symmetrisation, the number of *essential components* in d dimensions equals

$$\#P^{i_1 \ldots i_m} = \#L_{i_1 \ldots i_m} = \binom{m+d-1}{d-1}. \tag{4.33}$$

The polynomials $P^{i_1 \ldots i_m}$ reflect the scale-space model independent of actual measurement data. For $m \in \mathbb{Z}_0^+$ they form a complete, but non-orthogonal local basis, essentially the familiar Taylor basis for the spacetime domain, but with scale dependent "correction terms".

The coefficients $L_{i_1 \ldots i_m}$ potentially correspond to local measurements[6], but must be split into two groups. For $m = 0, \ldots, k$ (say) they correspond to *actual* measurements, or to *potential* measurements that could be reliably extracted from the raw image data. In vision they may correspond to neural activity induced by stimulation of receptive fields [47, 125] with profiles modeled by (linear combinations of) the low order members of the Gaussian family [167, 305]. Together these coefficients parametrise the image's local k-jet relative to their corresponding basis. Thus the k-jet is a physical object. The coefficients $L_{i_1 \ldots i_m}$ beyond order k are discarded, either because they cannot be reliably extracted, or because there exists no corresponding filter implementation for their computation. There is no *a priori* upper limit for k (one can always think of image data for which the k-jet makes sense), but there is always one in any specific case.

The discarded degrees of freedom give rise to metameric images [160, 161]. Since both L_∞ as well as L_k are scale-space images, so is their difference $R_k = L_\infty - L_k$. It is not a physical object, as it is not physically representable; one

---

[6]There exist nontrivial constraints given an upper and lower bound on the data: $0 \leq L \leq L_{\text{max}}$. Such a constraint on zeroth order induces boundary constraints on the local jet, *vice versa* [161].

could call it the *local ghost of order* k. The ghost image can be expressed in terms of the higher order polynomials $P^{i_1\cdots i_m}$ for $m \geq k+1$. The corresponding $L_{i_1\ldots i_m}$ constitute a local, hierarchical characterisation of an infinite-dimensional class of scalar field configurations that induce identical local measurements of orders $0,\ldots,k$.

Example 4.1 (Some Lowest Order Cases)
Up to fourth order, the homogeneous polynomial solutions to the diffusion equation are given by

$$
\begin{aligned}
l_0(x,t) &= L, \\
l_1(x,t) &= L_i x^i, \\
l_2(x,t) &= \frac{1}{2} L_{ij} x^i x^j + L_{ij} g^{ij} t, \\
l_3(x,t) &= \frac{1}{6} L_{ijk} x^i x^j x^k + L_{ijk} x^i g^{jk} t, \\
l_4(x,t) &= \frac{1}{24} L_{ijkl} x^i x^j x^k x^l + \frac{1}{2} L_{ijkl} x^i x^j g^{kl} t + \frac{1}{2} L_{ijkl} g^{ij} g^{kl} t^2.
\end{aligned}
$$

Addition of these yields a representative member of the image's multiscale local 4-jet. The symmetric tensors corresponding to the monomials above are given by

$$
\begin{aligned}
P(x,t) &= 1, \\
P^i(x,t) &= x^i, \\
P^{ij}(x,t) &= \frac{1}{2} x^i x^j + g^{ij} t, \\
P^{ijk}(x,t) &= \frac{1}{6} x^i x^j x^k + \frac{1}{3}\left(x^i g^{jk} + x^j g^{ik} + x^k g^{ij}\right) t, \\
P^{ijkl}(x,t) &= \frac{1}{24} x^i x^j x^k x^l \\
&\quad + \frac{1}{12}\left(x^i x^j g^{kl} + x^i x^k g^{jl} + x^i x^l g^{jk} + x^j x^k g^{il} + x^j x^l g^{ik} + x^k x^l g^{ij}\right) t \\
&\quad + \frac{1}{6}\left(g^{ij} g^{kl} + g^{ik} g^{jl} + g^{il} g^{jk}\right) t^2.
\end{aligned}
$$

One should note that adding a higher order term may bring in relatively little new information, as the corresponding detectors get increasingly correlated; neither the Gaussian family nor the local Taylor basis happens to be orthogonal.

The structure of the formula for the k-th order scale-space polynomial $P^{i_1\cdots i_k}$ in Corollary 4.1 can be captured by the following *diagrammatic rules*, cf. Figure 4.8:

- Represent each factor x^i by means of a single branch (symbolising the free tensor index i) emanating from a vertex point (the white disk).

- Represent each factor $2tg^{ij}$ by means of a (symmetric) double branch emanating from a vertex point (the shaded disk).

- Finally, connect such elementary pieces in all possible ways to obtain different vertices having exactly k branches, and add them together. The overall combinatorial factor for the diagram is $1/k!$.

x^i
$2\,t\,g^{ij}$

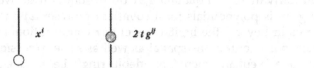

Figure 4.8: Elementary vertices used in the diagrammatic representation of scale-space polynomials.

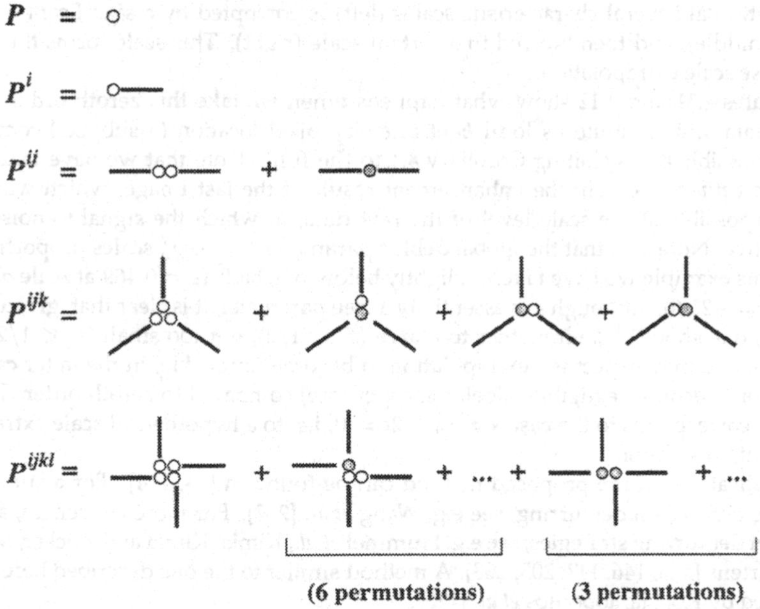

P =
P^i =
P^{ij} = +
P^{ijk} = + + +
P^{ijkl} = + + ... + + ...
(6 permutations)
(3 permutations)

Figure 4.9: Diagrammatic representation of the five lowest order polynomials of Example 4.1.

Figure 4.9 illustrates this "diagrammar" for the cases written out in Example 4.1.

We can use Corollary 4.1 for various purposes. If we can reliably extract all derivatives of Equation 4.31 up to some order, we can use the corresponding basis polynomials for a covariant representation of image structure which is valid beyond the limitations of our computational grid. In other words, we can make continuous spatial as well as scale excursions. A special case of this is "image enhancement", or "deblurring", i.e. a scale-space Taylor expansion in high-resolution direction, by taking $x = 0$ and $t < 0$. An illustration of this is given in Example 4.2 below.

Another role of Corollary 4.1 is in the study of *catastrophe theory* [96, 237, 250, 270] in the context of the scale-space paradigm, cf. [158, Chapter 9] and [41, 42, 84, 102, 101, 134, 135, 137, 138, 152, 156, 162, 170, 187, 188, 191]. This study is important in order to understand the deep structure of a scale-space image, and may thus be helpful in the design of multiscale algorithms, such as image segmentation, registration [63, 204], etc. (Recall Boxes 4.2 and CTSS.)

Example 4.2 (Deblurring Gaussian Blur)
An example of deblurring using Corollary 4.1 is presented in Figures 4.10–4.12. Figure 4.10 shows how the simulation data have been obtained: a noise-free, binary image carrying information at several characteristic scales (left) is corrupted by a significant amount of noise (middle), and then blurred to a certain scale (right). This scale forms the origin of pixelwise scale extrapolation.

Figures 4.11 and 4.12 show what happens when we take this zeroth order image as initial data and compute its local k-jet at every pixel location (a subpixel computation is also possible by exploiting Corollary 4.1 to the full). Note that we have taken *twelfth order derivatives* to obtain the enhancement result of the last image, which would have been impossible at the scale level of the raw data, at which the signal-to-noise ratio is prohibitive. Note also that the global deblur parameter $t = -\varepsilon\sigma_0^2$ scales proportionally to σ_0^2; in this example we have taken ε slightly below one half[7] ($\varepsilon = 0.469$ at scale $\sigma_0 = 7.39$, hence $t = -25.6$). Although ε is essentially a free parameter, it is clear that, given a certain order k, one should take it neither too large ($\varepsilon \gg 1/2$), nor too small ($\varepsilon \ll 1/2$). In the first case one may expect the extrapolation to become unreliable, in the latter case the ε-corrections become negligible (albeit very accurate) compared to zeroth order. The value $\varepsilon = 1/2$ corresponds to the case $s = s_0 + 2t = 0$, i.e. to a hypothetical scale extrapolation to "infinite resolution".

Applications of the proposed method can be found in [84, 104]. For a survey and a list of references on deblurring, see e.g. Wang *et al.* [287]. For more references, as well as different deblurring strategies, see e.g. Hummel *et al.*, Kimia, Kimia and Zucker, Mair *et al.*, and Martens [126, 146, 147, 205, 208]. A method similar to the one described here has been proposed by Konstantopoulos *et al.* [175].

---

[7]This was done by visual inspection after a quick trial-and-error by taking a starting value $t = -0.1$ and doubling this until a visually reasonable result was obtained. This explains the somewhat odd value $\varepsilon = 0.469$; no attempt has been made to rigorously optimise the choice of ε. Results are qualitatively similar for nearby values of $\varepsilon \approx 0.5$.

Figure 4.10: A synthetic, binary test image (intensity difference 255 units) of 512×512 pixels (left), corrupted by additive, pixel-uncorrelated Gaussian noise with a standard deviation of 255 intensity units (middle), and blurred to scale $\sigma = 7.39$ pixels (right).

Figure 4.11: Deblurring as calculated for the right image of Figure 4.10. The derivatives involved in the scale expansion have been evaluated at high resolution, scale $\sigma = 1.00$ pixels, on the low-resolution input image ($\sigma = 7.39$ pixels). **Top row:** First, second and third order deblurring, using $\varepsilon = 0.469$. **Bottom row:** fourth, fifth and sixth order expansion.

Figure 4.12: Upper right part of the last image in Figure 4.11 overlayed on top of the initial image for the sake of comparison. Note that further enhancement could have been obtained by exploiting also the spatial part of the jet bundle (spatial supersampling; experiment not carried out).

4.3 Temporal Causality

One might conjecture that the scale-space paradigm violates *temporal causality*, as part of the temporal support of a filter $\phi \in \mathcal{G}(\mathbb{R}^n)$ always extends into the remote future (the same applies to linear filters $\phi \in \mathcal{S}(\mathbb{R}^n)$ in general). This argument cannot be upheld though. Dividing time into "past" and "future" implies the existence of a preferred *epoch*, the "present moment". In retrospective image analysis this is precisely the kind of thing we set out to avoid. Attributing an *a priori* privileged status to any time frame in, say, a medical image sequence, would bias its diagnostic value. If such a bias is undesirable on such general grounds, then it should not be introduced in a syntactical description. All time instances are to be treated on equal foot, thus *no epoch* should be distinguished. But without an epoch, the problem of causality does not apply! One could also say that a movie is a recollection of equivalent epochs; the measurement procedure is identical at all acquisition times. For this reason we deliberately declined to incorporate a causal temporal structure in the design of our devices.

Infinite support of scale-space filters equally applies to space and time, and should not be confused with violation of causality. Recall that what matters in this respect is that the filters are "essentially compact", and that scales should be kept within physically sensible limits. Theory does not prescribe any specific bounds, but has been formulated for all scales—infinite scope, infinitesimal grain—so as not to muddle it up with technicalities. The concept of scale is purest in such a universal context. A scale-space implementation necessarily entails an approximation of the exact paradigm due to finite scope and grain.

We re-introduce an artificial epoch whenever we display a movie on a screen for visual inspection. The epoch is always *now* to the active observer. In general, any on-line system is confronted with such a "time horizon", beyond which it cannot collect input data (*vision*, *robotics*, and other systems operating in real-time). The example of a movie clearly demonstrates that causality is not necessarily intrinsic to the source; a video tape has been fully recorded at the time of inspection. The causality problem is introduced at the device level, i.c. *by the observer*. Again, this reminds us of Kant's conjecture.

In the context of real-time image processing causality clearly poses a fundamental problem, and we must seek to establish an image processing syntax that will make it *manifest*. However, a dilemma immediately presents itself. Koenderink states it quite clearly [155]:

> "It is clear that the scale-space method can only lead to acceptable results over the complete axis, but never over a mere semi-axis. On the other hand the diffusion equation is the unique solution that respects causality in the resolution domain. Thus there can be no hope of finding an alternative. The dilemma is complete."

In the approach adopted in this book we arrived at Gaussian scale-space from the requirement of duality and its consistency with linear image processing, plus the introduction of an unambiguous point concept. The result is the same, the assumptions are equally fundamental, and the conclusion drawn by Koenderink indeed appears inevitable:

> "The only way that is open to a possible solution appears to be to *map the past onto a complete axis*."

Note that one can do this *isomorphically*. Indeed, if we assume the "purpose" of time to be, as Wheeler passes on the graffiti on a toilet door [298], "to keep everything from happening all at once", then the only thing we need to insist on is that a remapping has a non-degenerate derivative on any *physically relevant* time window. This is the essence of Koenderink's *scale-time* theory [155], explained in more detail in the next section[8] (cf. also Lindeberg and Fagerstrom [196]).

4.3.1 Manifest Causality

The issue of causality arises when studying evolutionary systems that acquire and process data simultaneously. Causal systems are believed to be insensitive to events that have not happened *yet*. However, in the Newtonian picture of spacetime (recall Section 3.7, in particular Definition 3.7 on Page 52) there is no reference to the notion of causality. The classical spacetime model does not preclude its possibility, but neither supports it explicitly.

There is no reason to modify our base manifold in order to deal with real-time acquisition systems. In fact we can maintain all Newtonian spacetime symmetries as they are[9] (for these are quite plausible), and take causality into account in the construction of our device space. This means that causality becomes a syntactical element, i.e. causality is employed in order to structure measurement data. As usual, we operationally define causal state space as the topological dual of the causal device space thus obtained, once again avoiding the need for *ad hoc* assumptions about actual sources in the world.

---

[8]I will follow a slightly different approach that yields the same result, which I consider *a fortiori* equivalent. Please refer to the original paper to appreciate Koenderink's philosophy properly.

[9]Of course, a temporal shift implies a consistent transformation of the present moment.

For simplicity we will henceforth assume $n = 1$, unless stated otherwise, i.e. we consider 1-dimensional time sequences de-emphasising spatial information. The extension to spacetime movies is straightforward. The present moment partitions the time manifold $\mathbf{T} \sim \mathbb{R}$ into past, $\mathbf{T}_a^- \sim \{t \in \mathbb{R} \mid t < a\}$, and future, $\mathbf{T}_a^+ \sim \{t \in \mathbb{R} \mid t > a\}$, relative to the present moment $\mathbf{T}_a^0 \sim \{t \in \mathbb{R} \mid t = a\}$. The present moment is insofar nonphysical that it comprises only a single instant—an interval of measure zero—but is relevant as an evolution parameter in a real-time system. Let us now turn to the incorporation of manifest causality.

Let g_a denote a one-to-one smooth mapping from the past onto the real axis, $\mathbf{S} \sim \mathbb{R}$ say:

$$g_a : \mathbf{T}_a^- \to \mathbf{S} : t \mapsto s \stackrel{\text{def}}{=} g_a(t), \tag{4.34}$$

and let h_a be its inverse g_a^{-1}:

$$h_a : \mathbf{S} \to \mathbf{T}_a^- : s \mapsto t \stackrel{\text{def}}{=} h_a(s). \tag{4.35}$$

We have such mappings for every present moment, whence the label a. Of course, g_a (and therefore h_a) must have a nonnegative derivative, and without loss of generality we may assume that $g_a'(t) > 0 \ \forall t \in \mathbf{T}_a^-$ (and hence $h_a'(s) > 0 \ \forall s \in \mathbf{S}$). This implies the following limiting requirements:

$$\lim_{t \to -\infty} g_a(t) = -\infty \quad \text{and} \quad \lim_{t \uparrow a} g_a(t) = +\infty. \tag{4.36}$$

Evidently there must be a unique zero-crossing, which defines the origin for $\mathbf{S}$:

$$g_a(a - \tau) = 0. \tag{4.37}$$

(A natural choice of τ presents itself: the typical delay or integration time that characterises any real-time system, or the frame interval in a digitised sequence. This choice would map the most recent, still unresolved part of history onto the positive s-axis.) The question arises of how to single out a unique mapping g_a, satisfying the above limiting conditions.

Let us start by incorporating the desirable symmetries, notably scale invariance and shift invariance. We obtain a system with time invariant characteristics provided we measure all times relative to the present. Hence for every event t in the past, $a - t$ will be the relevant (time shift invariant) parameter. In other words, on the new time manifold $\mathbf{S}$ events are labelled by their *age*, or *elapsed time* relative to the present moment, rather than by the fixed read-out of some arbitrarily gauged clock. Scale invariance follows from a dimensional analysis (Pi theorem: Box 3.4, Page 77). On the basis of the three relevant time variables under consideration, viz. t (time coordinate of some historical event), a (present moment) and τ (time unit), we can form exactly one independent scale and shift invariant quantity[10]. We can take it to be

$$p = \frac{a - t}{\tau}. \tag{4.38}$$

---

[10]The reason for introducing a *fixed* time constant τ rather than a free time scale at this point will become clear shortly.

We conclude that the mapping $g_a(t)$ must be of the form $\gamma(p)$, where γ is some dimensionless, a-independent function. The analogues of Equations 4.36 and 4.37 are given by:

$$\lim_{p \to \infty} \gamma(p) = -\infty \quad \text{and} \quad \lim_{p \downarrow 0} \gamma(p) = +\infty \qquad (4.39)$$

and

$$\gamma(1) = 0. \qquad (4.40)$$

The limits of Equation 4.39 correspond to the mappings of infinite past and present moment, respectively, while Equation 4.40 corresponds to that of the "immediate past" (corresponding to the delay τ). Note that γ maps elapsed, not absolute times. A one unit delay is mapped to the origin of the s-axis. The negative s-axis corresponds to events that occurred more than one unit τ in the past. The positive s-axis corresponds to events that took place less than one τ-unit in the past. We have

$$s = \gamma(p) \quad \text{with inverse} \quad p = \eta(s). \qquad (4.41)$$

Unless stated otherwise we henceforth set $\tau \equiv 1$.

Koenderink argues [155] that temporal resolution for a real-time system is most sensibly defined as an independent parameter in the evolutionary S-domain rather than on the static time-axis **T**. The reason is that it allows us to follow the same arguments as previously applied to the off-line case. An independent resolution parameter for S is moreover more natural for an active (artificial or biological) vision system; on **T** one would need an *explicit* constraint in order to avoid interference with the unpredictable future $\mathbf{T}_a^+$. Besides this causality argument, it is clearly prudent to consider recent events more carefully than those of the remote past in order to generate efficacious action. It will be seen that a fixed level of resolution in the causal domain S implies a "fading memory" in the sense that the corresponding resolution in ordinary time becomes proportional to delay.

Let γ and η be admissible isomorphisms according to the above requirements. Then[11]

$$\Gamma \overset{\text{def}}{=} -\log(\eta \circ \gamma) \quad \text{and} \quad H \overset{\text{def}}{=} \exp(-\gamma \circ \eta) \qquad (4.42)$$

are admissible as well, i.e. Γ and H satisfy all limiting conditions of Equations 4.39 and 4.40. In fact, we have

Result 4.2 (Canonical Isomorphism)
See Equation 4.38. The pair (Γ, H), defined by

$$s = \Gamma(p) = -\log(p) \quad \text{and} \quad p = H(s) = \exp(-s),$$

defines a canonical isomorphism between the multiplicative group $\mathbb{R}^+$ and the additive group $\mathbb{R}$. That is, if $s_k = \Gamma(p_k)$ and $p_k = H(s_k)$ for $k = 1, 2$, then we have

$$\Gamma(p_1 p_2) = s_1 + s_2 \quad \text{and} \quad H(s_1 + s_2) = p_1 p_2.$$

---

[11] Take notice: log defaults to $\log_e$, or ln, in agreement with the usual convention in mathematics.

This induces a canonical isomorphism between $\mathbf{T}_a^-$ and S. "Canonical" refers to the fact that a scaling of T-time units $\tau \mapsto e^\varepsilon \tau$ corresponds to a shift in S-representation:

$$\Gamma(e^{-\varepsilon}p) = s + \varepsilon \quad \text{and} \quad H(s + \varepsilon) = e^{-\varepsilon}p. \tag{4.43}$$

The canonical representation appears to be a natural one in a physical sense. It is known from empirics that many natural phenomena manifest themselves through numerical realizations obeying *Benford's first-digit law*. This remarkable law states that the first digits of a large ensemble of natural numerical data are not uniformly distributed, but follow a logarithmic probability law; in decimal representation, a "1" is roughly six-and-a-half times more probable than a "9". Figure 4.13 illustrates Benford's law. A nice confirmation of this has been given by Buck *et al.* [29], who have verified Benford's law by considering the alpha decay half lives of 477 different atomic nuclei. Benford's law immediately follows from the assumption that random intervals are uniformly distributed in s-time[12]. To see this, consider the logarithm of p (Result 4.2) given in decimal notation for *arbitrary* time unit τ, say $\lambda \equiv -s \log_{10} e \ modulo \ 1$, with $\lambda \in [0,1)$ (note that the extraction of powers of 10 does not affect the first digit of a number, and recall that a change of unit $\tau \mapsto e^\varepsilon \tau$ amounts to a shift of λ). Benford's law then corresponds to a *uniform* distribution for λ on the interval $[0,1)$, indeed the only one that admits a change of units.

Since only s-shift invariant functions will be of interest (as these are insensitive to a substitution of t-clocks: $t \mapsto t' = \alpha t + \beta$ and thus comply with our spacetime model), we will henceforth use the canonical time representation to construct temporally causal scale-time filters. Let us introduce a free scale parameter δ for the causal manifold S. This defines a kind of dimensional unit[13], because treating all levels of resolution on equal foot amounts to scale invariance on S. By construction the filter families $\mathcal{S}(\mathbb{R})$ and $\mathcal{G}(\mathbb{R})$ make sense on S, and we can define temporally causal samples and images as usual, proceeding as follows.

Suppose we have a pre-recorded raw signal $f(t)$, say, in which $t \in \mathbf{T}$ labels consecutive time frames. Alternatively, in a real-time system, we may regard it as the source signal at acquisition time t. We introduce the epoch a that subdivides the time axis. Real-time signal processing always takes place on the submanifold $\mathbf{S} \sim \mathbf{T}_a^-$, hence it is natural to define $f_a(p) \equiv f(a - p)$. For fixed positive p and varying a this represents the *retarded signal p units ago*. Regarded as a function of p for fixed a we have the *signal history* relative to a given epoch.

Definition 4.2 (Absolute versus Relative Time)
Define absolute time as the mapping

$$t_a : \mathbb{R}^+ \to \mathbf{T}_a^- : p \mapsto t \stackrel{\text{def}}{=} t_a(p) \stackrel{\text{def}}{=} a - p,$$

and relative time as its inverse $p_a \equiv t_a^{\text{inv}}$, *i.e.*

$$p_a : \mathbf{T}_a^- \to \mathbb{R}^+ : t \mapsto p \stackrel{\text{def}}{=} p_a(t) \stackrel{\text{def}}{=} a - t.$$

---

[12]Note that this assumption does not by any means *explain* Benford's law. However, physical distributions other than Benford's type are excluded by scale invariance!
[13]But note that δ is actually dimensionless.

The map $t_a(p)$ is the read-out of an arbitrarily gauged clock p units rewound. It is fixed for a physical event (and a given clock); for instance, $t_a(p)$ returns your date of birth given your age p. Absolute time and our phenomenological notion of time are close-knit, because both provide invariant tags to events we may recollect from memory. Its inverse $p_a(t)$ is the amount of time past since the same clock indicated t; it thus represents the elapsed time since the event took place, or the age of a process initiated by the event. For instance, $p_a(t)$ returns your age given your date of birth. Clearly $t_a(p)$ and $p_a(t)$ convey the same information and presuppose that one agrees on the epoch a.

At every moment in time we can look at a source field in retrospection.

Definition 4.3 (Signal History)
See Definition 4.2. Let $f : \mathbf{T} \to \mathbb{R} : t \mapsto f(t)$ be a given signal. For every epoch a the signal history is defined as

$$f_a \stackrel{\text{def}}{=} t_a^* f .$$

(Recall that pull back of a source f is geometrical jargon for the reparametrisation effect induced by a substitution of arguments: $\theta^* f \equiv f \circ \theta$.) Because we can *independently process* a recorded signal history at different moments in time, we have—in an operational sense—a *two-dimensional time plane*, swept out by real-time a and historical event time t. For a *processed* signal history this is a crucial observation; indeed, if we define history as what we recall of it, then events (memories) will depend not only upon their actual event time t but also on the moment of recollection a! are doomed to fail.)

In daily practice we often index data records using absolute time tags. The canonical isomorphism H, however, is set up to map to elapsed time, so we will often need to convert the result to absolute time. The following definition expresses this concatenation of events, relating the canonical time domain to absolute time.

Definition 4.4 (The Isomorphism S $\sim$ T$_a^-$)
Recall Result 4.2 and Definition 4.2. The isomorphism $\mathfrak{t}_a : \mathbf{S} \to \mathbf{T}_a^- : s \mapsto t \stackrel{\text{def}}{=} t_a(s)$ is defined by

$$\mathfrak{t}_a \stackrel{\text{def}}{=} H^* t_a ,$$

and its inverse $\mathfrak{p}_a : \mathbf{T}_a^- \to \mathbf{S} : t \mapsto s \stackrel{\text{def}}{=} p_a(t)$ by

$$\mathfrak{p}_a \stackrel{\text{def}}{=} p_a^* \Gamma .$$

Figure 4.14 shows a graph of the isomorphism.

4.3.2 The "Specious Present": Real-Time Sampling

The mapping t_a can be used to translate sources and detectors between the isomorphic domains S and T_a^- in the same way as in Diagram 3.1 on Page 42:

$$
\begin{array}{ccc}
\phi \in \Delta & \xrightarrow{\;t_{a*}\;} & \varphi_a \in t_{a*}\Delta \\[4pt]
\pi \Big\downarrow & & \Big\downarrow \pi_a \qquad \pi_a \circ t_{a*} = t_a \circ \pi \\[8pt]
s & \xrightarrow{\;t_a\;} & t \\[4pt]
\pi_a \Big\uparrow & & \Big\uparrow \pi \qquad \pi_a \circ t_a^* = t_a^{\mathrm{inv}} \circ \pi \\[8pt]
f_a \in t_a^*\Sigma & \xleftarrow{\;t_a^*\;} & f \in \Sigma
\end{array}
\qquad (4.44)
$$

We have for instance:

Definition 4.5 (Causal Signal)
See Result 4.2 and Definition 4.2. The canonical time representation of a raw signal f at epoch a is given by

$$ f_a \stackrel{\text{def}}{=} t_a^* f . $$

Clearly f_a depends only upon signal history f_a (manifest causality: Problem 4.11). Figures 4.15–4.17 illustrate the mapping t_{a*}.

Definition 4.6 (Causal State Space and Causal Device Space)
See Diagram 4.44. We define causal state space as the space of causal sources:

$$ t_a^*\Sigma \stackrel{\text{def}}{=} \{t_a^* f \mid f \in \Sigma\} . $$

Similarly we define causal device space as the space of causal detectors:

$$ t_{a*}\Delta \stackrel{\text{def}}{=} \{t_{a*}\phi \mid \phi \in \Delta\} . $$

Note that causal sources $t_a^*\Sigma$ live in S, whereas causal detectors $t_{a*}\Delta$ live in T.

Result 4.3 (Causal Local Sample)
Recall Definition 3.1 on Page 40 and Definition 4.5. Denoting the tempered distribution corresponding to the function $f_a : S \to \mathbb{R}$ by $\mathfrak{F}_a \in t_a^*\Sigma$, i.e. $\mathfrak{F}_a = t_a^* F$, we can define a causal local sample as

$$ \lambda_a : t_a^*\Sigma \times \Delta \to \mathbb{R} : (\mathfrak{F}_a, \phi) \mapsto \mathfrak{F}_a[\phi] . $$

In integral form we have

$$ \mathfrak{F}_a[\phi] \stackrel{\text{def}}{=} \int ds\, f_a(s)\, \phi(s) , $$

for any $\phi \in \Delta$. We can take either $\Delta \equiv \mathcal{S}(\mathbb{R})$ or $\Delta \equiv \mathcal{G}(\mathbb{R})$.

By trivial extension, including the spatial domain, we obtain analogous definitions for causal local samples in the context of linear image processing, and "scale-spacetime" images, respectively. The latter is particularly easy by virtue of filter separability on the causal manifold $\mathbf{M}_c \equiv \mathbf{S} \times \mathbf{X}$.

A causal image is obtained from causal samples as the orbit under the action of the spacetime symmetry group, as usual (cf. Definition 3.5, Page 48), for all possible epochs $a \in \mathbb{R}$, i.e. for each a we have a "section" of spatiotemporal activities $\Lambda_a \equiv \{\mathfrak{F}_a, \phi, \Theta\} \equiv \{F, \varphi_a, \Theta\}$. As for Θ we can formally maintain the group parametrisation of Definition 3.7 on Page 52 if we replace $\mathbf{M} = \mathbf{T} \times \mathbf{X}$ by $\mathbf{M}_c = \mathbf{S} \times \mathbf{X}$. This is so because t-shift invariance is trivially accounted for in the definition of the s-coordinate, t-scale invariance boils down to s-shift invariance ("canonical coordinate"), and s-scale invariance has been explicitly postulated in order to stratify $\mathbf{S}$ into a scale-continuum; a simple exchange trick $t \leftrightarrow s$ should therefore work. S-scale invariance brings us to yet another conceptual picture: $\Lambda_a \equiv \{F, \phi, \Theta_a\}$ ("causal classical spacetime": Problem 4.10).

In practice a raw signal or video is typically recorded as a set of consecutively stored numbers or spatial "frames". Frame index then provides a practical and suitable **T**-clock. That is, we usually have resource to a record in absolute time format F, not $\mathfrak{F}_a$. We can work with the former yet simulate the latter by transposition into device space. This yields an equivalent, but more useful (because computationally realizable) interpretation of the "causal world" model of Result 4.3 in the spirit of Kant's philosophy.

Result 4.4 (Causal Local Sample)
See Definition 4.6 for notation. The following definition is equivalent to Result 4.3. If $\mathfrak{F}_a \equiv t_a^* F$ and $\varphi_a \equiv t_{a*}\phi$, then

$$\mathfrak{F}_a[\phi] = F[\varphi_a].$$

That is,

$$\lambda_a : \Sigma \times t_{a*}\Delta \to \mathbb{R} : (F, \varphi_a) \mapsto F[\varphi_a],$$

or

$$F[\varphi_a] = \int dt\, f(t)\, \varphi_a(t).$$

This means that, instead of a causal source field $\mathfrak{f}_a(s)$ and ordinary filter profiles $\phi(s)$, we now have an unconstrained source field $f(t)$—like a video tape not necessarily confined to the causal semi-axis—and causal filters $\varphi_a(t)$ defined on **T**. The result is fully determined by propagation of causal information only, because the support of the filters in $t_{a*}\Delta$ is contained in $\mathbf{T}_a^-$, as it should: the time horizon is, however, introduced at the processing level.

Results 4.3 and 4.4 can be summarised as follows.

Paradigm 4.2 (Manifest Causality)
Recall Paradigm 2.1. Causal state space is the topological dual of causal device space. On **T** *we have*

$$\Sigma \stackrel{\text{def}}{=} t_{a*}\Delta' \quad \text{"Kantian picture".}$$

On S *this is equivalent to*

$$t_a^* \Sigma \overset{\text{def}}{=} \Delta' \quad \text{"causal world picture"}.$$

In the light of duality either view provides a viable explanation for real-time interaction of sources and detectors. One could therefore adopt an unbiased view by saying that a real-time sample emerges by interaction involving *three* constituents: $\lambda_a = \{F, t_a, \phi\}$ (in quantum physics one would write $\langle F|t_a|\phi\rangle$: this so-called "bracket formalism" is briefly explained in Box 4.4, and illustrated in the context of Fourier transformation). If we denote the collection of isomorphisms for all possible epochs by $\mathfrak{T} = \{t_a \mid a \in \mathbb{R}\}$, and the Newtonian spacetime group by Θ as before, then we can again abstract by stating that a causal image ("activity") comprises a *quadruple* $\Lambda = \{F, \mathfrak{T}, \phi, \Theta\}$, in this way capturing both options of Paradigm 4.2 without bias. Recall that an off-line sample (image) arises in a similar way by replacing all nontrivial isomorphisms t_a by the identity map id (respectively $\mathfrak{T}$ by the trivial one-element group). It will turn out that it can also be regarded as a *limiting* case, viz. for $a \to \infty$ (Section 4.3.3). In this way transformations of time, or of "empty spacetime" in general, arise as abstractions in theories of duality.

It is instructive to consider the explicit form of the filters $\varphi_a(t)$; they are clearly of a different type than the ones we have encountered before. In particular, let us consider the temporal point operator $\phi \in \mathcal{G}^+(\mathbb{R})$, given by Result 3.2 for $n = 1$.

Result 4.5 (The Prototypical Causal Point Operator)
If $\phi \in \mathcal{G}^+(\mathbb{R})$ *is the prototypical point operator on* S *given by*

$$\phi(s) = \frac{1}{\sqrt{2\pi}} \exp\left(-\frac{1}{2}s^2\right),$$

then on T *we have* $\varphi_a \equiv t_{a*}\phi$ *given by*

$$\varphi_a(t) = \begin{cases} 0 & t \geq a \\ \frac{1}{\sqrt{2\pi}(a-t)} \exp\left(-\frac{1}{2}\log^2(a-t)\right) & t < a. \end{cases}$$

It is important to point out that $\varphi_a(t)$ is *not a linear correlation filter*; a time shift not only affects its argument t, but also its label a. For fixed epoch a the complete, symmetry-adapted filter looks as follows.

Result 4.6 (The Causal Point Operator)
Definitions as in Result 4.5. In Newtonian time, an arbitrary causal point operator is fully characterised by its base point and its temporal inner scale, and is given by

$$\phi(s'; s, \delta) = \frac{1}{\delta} \phi\left(\frac{s' - s}{\delta}\right),$$

when expressed as a function on S. *On* T *we have*

$$\varphi_a(t'; t, \delta) = \begin{cases} 0 & \max(t, t') \geq a \\ \frac{1}{\sqrt{2\pi}(a-t')\delta} \exp\left(-\frac{1}{2\delta^2}\log^2\left(\frac{a-t'}{a-t}\right)\right) & \max(t, t') < a. \end{cases}$$

(Primes are attached to the integration dummy.) This means that causal filtering of a signal requires *three* parameters:

- the time frame t of interest,

- the epoch a relative to which events are evaluated, and

- the level of (graded) resolution δ.

The parameter free prototypical filter φ_a in Results 4.4 and 4.5 corresponds to unit delay $a - t \equiv 1$ and unit scale $\delta \equiv 1$. The dimensionless scale parameter δ is the amount of Gaussian blur in the S-domain. It corresponds to a *relative* time scale on $\mathbf{T}$; fixing δ amounts to resolving events at time scales relative to age. Figure 4.15 compares the filters $\phi(s'; s, \delta)$ and $\varphi_a(t'; t, \delta)$. Some characteristics of the basic causal point operator are:

- its centre of gravity corresponds exactly to its base point t;

- its standard deviation approximately equals $(a - t)\delta/\sqrt{2}$ if $\delta \ll 1$;

- its "skewness" is negative, is independent of delay, and is approximately $-8/\sqrt{\pi} \approx -4.51$ if $\delta \ll 1$. The magnitude of skewness rapidly increases with increasing δ.

Skewness is a measure of asymmetry relative to the centre of gravity [2]. Negative skewness is the consequence of causality; it signifies an asymmetric tail of the filter towards the future. Figure 4.16 illustrates the characteristic features of the causal point operator. For precise definitions and exact results, see Problem 4.18.

By construction, the S-domain is the natural domain for the formulation of causal scale-time representations. Consequently we should carry over all known scale-space concepts to this domain, and *subsequently* transform—for the sake of presentation—back to ordinary time $\mathbf{T}$. In particular, Result 4.6 can be generalised to account for the complete Gaussian family (Result 3.3) of *causal differential operators*. One then obtains likewise asymmetric profiles, but with multimodal impulse responses; the number of zero crossings is preserved by the canonical isomorphism (Result 4.2), and therefore equals differential order.

Conclusion 4.1 (The Causal Gaussian Family)
Basically we can copy Result 3.3 on Page 64 and replace references to $\mathbf{M} = \mathbf{T} \times \mathbf{X}$ by $\mathbf{M}_c = \mathbf{S} \times \mathbf{X}$.

Smoothness of the filters is preserved to the extent that in the new time domain the filters are *still everywhere infinitely differentiable*, but they are *no longer analytical* on $\mathbf{T}$ due to the logarithmic singularity at $t = a$: *all* filters must vanish smoothly at the time horizon.

Observation 4.1 (Prediction of the Future)
Thus we cannot Taylor expand into the future in order to make predictions.

How unfortunate indeed!
Figures 4.18–4.20 and Problem 4.19 provide some more illustrations of scale-time.

4.3.3   Relation to "Classical" Scale-Space

To see how Result 4.6 relates to conventional Gaussian scale-space in the t-domain, consider the limiting case when "nullifying" the anticausal part of the time axis. In this limit there exists no causality problem and one would expect to retrieve conventional Gaussian scale-space. Indeed, if we consider a small time window $[t_-, t_+]$ in the remote past, i.e. such that $t_+ - t_-$ is small compared to, say, $a - t_+$, then we can approximate the causal point operator of Result 4.6 as follows. Let t', t be two time instances corresponding to s', s, respectively, relative to epoch a. Then $s' - s = \log(1 + \varepsilon) = \varepsilon + \mathcal{O}(\varepsilon^2)$, with $\varepsilon = (t' - t)/(a - t')$. Discarding higher order terms we then have

$$ds' \, \phi(s'; s, \delta) = \frac{ds'}{\delta} \phi\left(\frac{s' - s}{\delta}\right) \approx \frac{dt'}{\sigma} \phi\left(\frac{t' - t}{\sigma}\right) = dt' \, \phi(t'; t, \sigma), \qquad (4.45)$$

in which we have *renormalised scale* by setting $(a - t')\delta \approx \sigma$ constant[14]. Thus we have indeed re-obtained Result 3.2, Page 61 (for $n = 1$); see Problem 4.14.

4.4   Summary and Discussion

The essential degree of freedom that constitutes the pivot of Gaussian scale-space is *inner scale*. Whereas it is known from physics that inner scale is subject to the scaling group (as is any other "hidden scale parameter", or dimensional unit, associated with a physical quantity), a semigroup has been shown to lie at the core of an operational manifestation of scale, and consequently of an operational representation of image structure.

It has also been pointed out that there exists an alternative image processing framework, known as mathematical morphology, which differs only in the details of the probing method. Whereas linear image processing relies on linear, topological duality, mathematical morphology relies on nonlinear, "morphological duality". In the linear paradigm one essentially interprets an image as the result of a local integration given a measure in the form of a raw image. In the nonlinear paradigm one utilises a "tactile probe" to scan the raw image of interest (using tangential dilation this becomes a local probe as well; the result may be set-valued). The choice of paradigm is conventional and should be inspired by *generic* semantical considerations.

Both duality axioms could be captured within a single, abstract framework. This should be further explored for its unifying and generalising potential. "In-between" modes of probing are in principle feasible, expressing a kind of "density-tactile" probes of finite volumetric resolution within a surface layer of a "semi-transparent" object. Such an in-between duality principle may provide image processing frameworks different from the ones studied here.

It has been shown that the paradigms postulated in the previous chapter can be reconciled with the requirement of temporal causality if one incorporates the

---

[14]This is a slapdash cocktail-party argument that should be reasonably convincing if you are not a mathematician. A rigorous justification requires a scaling in combination with the right kind of limiting procedure.

current moment as a "logarithmic singularity", as proposed by Koenderink [155]. Causality becomes manifest after a canonical time parametrisation involving only the history part of the time domain relative to a fiducial epoch. In scale-spacetime theory one obtains a family of causal differential operators characterised by a delay in addition to an inner scale, which is just the Gaussian family when mapped to canonical coordinates. The resulting filters enable the application of differential calculus to signals and image sequences in those case where temporal causality is a prerequisite, and may provide a plausible taxonomy of temporal impulse responses found in biological visual systems.

Real-time processing depends upon a signal's history. This is obvious for a system with memory. It is perhaps less obvious, but *still* the case for one without memory: real-time sampling requires, like any measurement in spacetime, a finite sampling aperture, *ergo* a history of recorded events. In the absence of a memory one could say that the system only keeps record of a single moment in time, in the sense of employing temporal aperture functions with a fixed delay (the "specious present"). In particular, the "instantaneous" state of a signal at any fixed moment in time is always irrelevant, since "punctal" values of a source field are not measurable.

There is an increasing interest in so-called "nonlinear scale-spaces" [70, 87, 103, 225, 249, 291, 292, 299, 300, 301]. Unlike linear and morphological scale-space theories, most nonlinear models lack a transparent duality principle. Since image structure can only be appreciated if one understands the details of the underlying duality principle, nonlinear scale-spaces are less suited for the purpose of a generic structural representation of image data. However, by suitable choice of nonlinearity, they have turned out quite useful in relation to specific tasks. The intriguing question therefore remains of how to *connect* scale-space theories, both establishing a generic data format in the form of an explicit duality principle as well as incorporating an adaptation mechanism for focusing on specific tasks [87].

☞ Box 4.2 (Catastrophe Theory)

A critical point of a function is a point at which the gradient vanishes (Problem 4.7). Typically this occurs at isolated points, at which the Hessian (i.e. the matrix of second order derivatives) has nonzero eigenvalues. The *Morse lemma* states that the qualitative properties of a function at these so-called *Morse critical points* are essentially determined by the quadratic part of the Taylor series (the *Morse canonical form*). However, in many practical situations one encounters families of functions that depend upon *control parameters*. An example of a control parameter is scale in a scale-space image. Catastrophe theory is the study of how the critical points change as the control parameters change.

For each fixed set of control parameters we are back in the situation governed by the Morse lemma, but varying a control parameter may cause a qualitative change. This happens when a Morse critical point turns into a *non-Morse critical point*, i.e. when the Hessian degenerates. It is hard to record such an event (one has to pick values for the control parameters with infinite precision), but one can sandwich it in-between arbitrarily tight parameter bounds.

Having several control parameters to play with one can get into a situation in which l eigenvalues of the Hessian become zero, leaving $n - l$ nonzero. The *Thom splitting lemma* simplifies things: it states that, in order to study the degeneracies, one can simply discard the $n - l$ variables corresponding to the regular $(n - l) \times (n - l)$-submatrix of the Hessian that is obtained after bringing it to block-diagonal form (and relabelling variables). That is, one can split up the function into a non-Morse and a Morse part, and study the canonical forms of each in isolation, because the same splitting holds in a neighbourhood of a non-Morse function. Again, the Morse part can be canonically described in terms of the quadratic part of the Taylor series. The non-Morse part can also be put into canonical form, called the *catastrophe germ*, which is a polynomial of order 3 or higher. The word "germ" indicates "the part that determines the qualitative structure".

The Morse part does not change qualitatively after a small perturbation. Critical points may move, critical values may change, but nothing happens to their type: if i eigenvalues of the Hessian are negative prior to perturbation (a "Morse i-saddle"), then this will still be the case afterwards. Thus there is no need to scrutinise the perturbations.

The non-Morse part, on the other hand, does change qualitatively upon perturbation. In general, the non-Morse critical point of the catastrophe germ will split into a number of Morse critical points. This state of events is called *morsification*. The Morse saddle types of the isolated Morse critical points involved in this process are characteristic for the catastrophe. *Thom's theorem* provides an exhaustive list of "elementary catastrophes" $(1, \ldots, 5$ control parameters), with canonical formulas for the catastrophe germs as well as for the perturbations needed to describe their morsification. Box 4.3 describes catastrophe theory in the context of scale-space.

☞ **Box 4.3 (Catastrophe Theory and the Scale-Space Paradigm)**

Box 4.2 deals with catastrophe theory in general. No constraints are imposed on the family of functions or the admissible perturbations. However, one frequently encounters families that *are* subject to constraints. For example, an isotropic scale-space image—we disregard time—defines a versal family of $C^\infty(\mathbb{R}^d)$-functions that satisfy the diffusion equation.

Let L be an isotropic scale-space image; we consider the generic events we may encounter at the origin, corresponding to an arbitrary point and (positive) scale. Scale is our single control parameter. If the gradient is non-zero, then

$$L_{\text{non-critical}}(x; t) = L_i P^i(x; t)$$

is the stable germ; it fully determines the qualitative behaviour in a full neighbourhood of the origin, and satisfies the scale-space condition. Note that it is actually scale-independent, and that small perturbations do not matter.

If the gradient vanishes, which will occur at countably many points for any given scale, then qualitative image structure is governed by the Hessian. Generically this is a non-singular matrix, and the stable germs are constructed as follows. Split up the Hessian $L_{ij} = Z_{ij} + R_{ij}$ such that $Z_{ij} = L_{ij} - \frac{1}{d}\triangle L\, g_{ij}$ and $R_{ij} = \frac{1}{d}\triangle L\, g_{ij}$. Then we have germs in two varieties:

$$L_{\text{Morse-hypersaddle}}(x) = Z_{ij} P^{ij}(x; t),$$
$$L_{\text{Morse-extremum}}(x; t) = R_{ij} P^{ij}(x; t).$$

The first one is scale-independent (!) and describes a hypersaddle. The second one is scale-dependent and describes an extremum. Whether it is a maximum or a minimum depends upon the sign of the Laplacian, the only degree of freedom of the germ. Again, perturbations can be safely ignored.

Versality of the scale-space family implies that by varying scale while tracking a critical point we may pass scale values at which the Hessian degenerates. Again assuming this to be the case at the origin, we then encounter the so-called *fold catastrophe*. It describes the situation where critical points of opposite Hessian signature are either annihilated or created as scale increases $t \uparrow 0$. Both events are generic, and they are the only generic events in isotropic scale-space if no further constraints are imposed. In both cases the critical points collide from exactly opposite directions and tangential to the $t = 0$ plane. Corresponding canonical forms for the two events in a suitable coordinate system, assuming the Hessian degenerates in x_1-direction, are

$$L(x; t) = x_1^3 + 6x_1 t + Q(x_2, \ldots, x_d; t)$$

for the annihilation event, and

$$L(x; t) = x_1^3 - 6x_1 t - 6x_1 x_2^2 + Q(x_2, \ldots, x_d; t)$$

for the creation event (the latter requires $d \geq 2$), with, for some $a_i \in \mathbb{R}$,

$$Q(x_2, \ldots, x_d; t) = \sum_{i=2}^{d} a_i \left(x_i^2 + 2t \right).$$

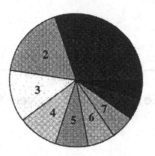

Figure 4.13: Benford's first-digit law.

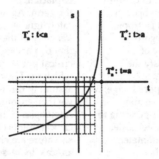

Figure 4.14: The isomorphism $T_a^- \sim S$ discards the unknown future. The vertical asymptote indicates the epoch or time horizon. The dashed box marks a typical time window for a real-time system. The system's active "life-time" and response time delimit the temporal field of view. Note that we obtain a *graded resolution history* if we assume uniform sampling in the S-domain.

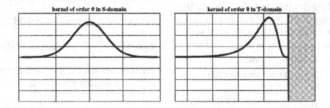

Figure 4.15: Comparison of causal point operator profiles in s and t domains (Result 4.6). The s-representation on the left corresponds to a fixed level of scale δ, and to a fixed base point s. The map t_{a*} brings us from there to the t-representation on the right, which corresponds to a fixed delay $a - t$. Time progresses from left to right; the causal filter vanishes rapidly but smoothly towards the time horizon, indicated by the vertical boundary line. (Of course there is no such moment in the left graph.) The shaded region indicates the unrevealed future.

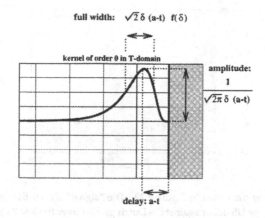

Figure 4.16: See Figure 4.15. Full width is defined as twice the standard deviation; $f(\delta)$ is a smooth function with $f(0) = 1$, cf. Problem 4.18.

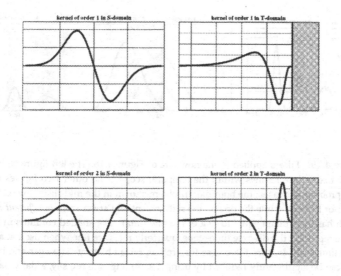

Figure 4.17: As Figure 4.15 but for the profiles of first and second order derivatives.

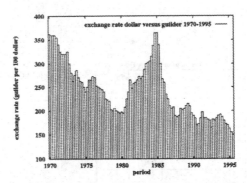

Figure 4.18: Raw data used in Figure 4.20. The "signal" shows the evolution of the exchange rate of the US dollar against the Dutch guilder over the last 25 years, sampled as quarterly averages (for convenience attributed to the start of a quarter). The epoch used in the scaletime analysis of this source is $a = 1995.25$, i.e. half-way into 1995.

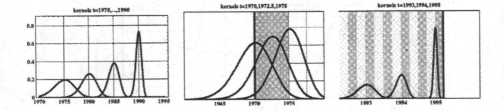

Figure 4.19: Filters applied to the raw data of Figure 4.18. The left figure shows a few relevant scaletime kernels within the temporal scope. The other two figures illustrate some problematic cases: the boundary problem shown in the middle is caused by *lack of data* prior to 1970, while the one on the right essentially arises from *insufficient resolution* (which has been taken equal to that of the raw data). The shaded columns in the right picture represent the discrete time-steps; quarterly rates are fine when used at an appreciable time scale, but are too coarse for probing recent history. The near-end boundary layer can be significantly reduced by using a sampling rate on, say, a daily basis. One has a *fundamental* boundary problem of the order $\mathcal{O}(\tau/\delta)$ if τ is the smallest delay; i.e. $\tau \approx 3$ months (potentially $\tau \approx 1$ day), yielding a layer of extent 2.5 years (or 10 days, respectively). Results are shown in Figure 4.20.

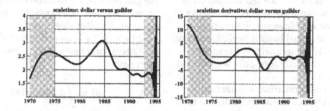

Figure 4.20: Scaletime analysis of the evolution shown in Figure 4.18. The left graph represents the basic, zeroth order scaletime signal, the right one shows the canonical time derivative. Parameter values: $\delta = 0.1$, $a = 1995.25$. The boundary layers, within which no accurate filter representation exists, are indicated by shading. In view of its extent and negligible truncation, the 1975-filter can be regarded as the remote interval boundary. The other boundary layer, caused by insufficient resolution, has been clarified by the shaded columns in Figure 4.19. Note that, by construction, one never has a boundary problem in the sense of filter overlap into the future.

☞ Box 4.4 (Bracket Formalism)

Fourier transformation boils down to a change of coordinates in a—possibly ∞-dimensional—linear space. But physically meaningful objects, i.c. sources (covectors) and detectors (vectors), must be covariant! Thus the details of Definitions 2.8 (Page 28) and 2.19 (Page 33) seem to obscure one subtle point: *nothing really happens...*

To appreciate this, consider Definition 2.8 with $n = 1$. By virtue of equivalence we may conceive of a functions and its Fourier transform as coordinate manifestations of one and the same physical object relative to different bases. However, in the usual notation the actual basis is *implicit*, and one needs to rely on an intelligent reader to interpret it correctly (typically with the aid of notation).

A similar situation arises in quantum physics, in which linear operators are used to represent physical quantities, such as energy, momentum, etc. Whether we choose to express these relative to a spatiotemporal or a frequency basis has no physical significance. This has led physicists to adopt the so-called *bracket formalism*. It is similar to our manifest covariant definition of local grey-value samples.

A covector is indicated by a "bra" $\langle w|$, a vector by a "ket" $|v\rangle$. Contraction yields a number, the "bracket": $\langle w|v\rangle$. In particular we can write spatiotemporal and frequency basis vectors (covectors) as $|x\rangle$ and $|\omega\rangle$ (respectively $\langle x|$ and $\langle \omega|$). Example: $\langle F|\phi\rangle = F[\phi]$, a special instance of which is $\langle x|\phi\rangle = \delta_x[\phi] = \phi(x)$, i.e. we can probe the *spatiotemporal structure* of a detector $|\phi\rangle$ by means of a source basis $\{\langle x| \, | \, x \in \mathbb{R}^n\}$, but we may just as well probe its *frequency structure* by using $\{\langle \omega| \, | \, \omega \in \mathbb{R}^n\}$ instead. Either way we obtain full knowledge about the detector; a detector is not a "spatiotemporal instrument"! The same can be said about a source field.

A dual basis comprises covectors $\langle v'|$ and corresponding vectors $|v\rangle$), such that for all labels v, v' we have

$$\langle v'|v\rangle \stackrel{\text{def}}{=} \delta_{v'v},$$

in addition to the so-called *closure property*, expressing completeness:

$$I \stackrel{\text{def}}{=} \int_{v'v} |v\rangle\langle v'|.$$

(A mechanism for converting vectors to covectors, *vice versa*, allows us to disregard dual covectors and talk about an orthonormal basis instead. Symbols are to be interpreted as in the condensed index notation: Box 3.1, Page 74.)

We can use closure to convert from one basis to another. For example, suppose $|\phi\rangle$ represents a filter object, and $\{\,|x\rangle \, | \, x \in \mathbb{R}^n\,\}$ is an orthonormal spatiotemporal basis. Using closure we then have

$$|\phi\rangle = \int dx \, |x\rangle \, \langle x|\phi\rangle.$$

Apparently the filter's spatiotemporal coordinates are $\phi(x) \equiv \langle x|\phi\rangle$. A similar expression holds relative to a frequency basis, which is defined as the complete set of normalised eigenvectors satisfying

$$i\frac{d}{dx}\langle x|\omega\rangle \stackrel{\text{def}}{=} \omega\,\langle x|\omega\rangle,$$

in other words, as the basis that diagonalises a spatiotemporal gradient operator. (The i makes the derivative operator Hermitian, admitting a complete basis of eigenvectors and a real eigenvalue spectrum; recall Problem 2.8, Page 37.) Normalisation yields

$$\langle x|\omega\rangle \stackrel{\text{def}}{=} \frac{1}{\sqrt{2\pi}}\,e^{-i\omega x},$$

whence $\phi(x) \equiv \langle x|\phi\rangle$ follows by closure:

$$\langle x|\phi\rangle = \int_\omega \langle x|\omega\rangle\,\langle\omega|\phi\rangle = \frac{1}{2\pi}\int d\omega\,\widehat{\phi}(\omega)\,e^{-i\omega x},$$

with inverse $\widehat{\phi}(\omega) \equiv \sqrt{2\pi}\,\langle\omega|\phi\rangle$:

$$\langle\omega|\phi\rangle = \int_x \langle\omega|x\rangle\,\langle x|\phi\rangle = \frac{1}{\sqrt{2\pi}}\int dx\,\phi(x)\,e^{i\omega x}.$$

Problems

4.1 (Wavelet Theory)

Consider Figure 4.21, symbolising the role of filters in wavelet theory, in comparison with Figures 2.2 on Page 22 and 3.3 on Page 78, and discuss conceptual differences. (For references on wavelet theory, see Page 91.)

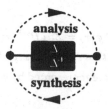

Figure 4.21: Filter paradigm in wavelet theory.

4.2

Page 92: why is Einstein's derivation of random walk "a bit shaky"? Can you give a more rigorous argument?

4.3

Page 98: "...*any* functional containing only quadratic terms of the form $u^{(i)}(x)u^{(j)}(x)$, can be written as in Equation 4.19, at least with suitable boundary conditions in effect." Prove.

4.4

Page 98: "In an isotropic n-dimensional space one does not need additional multipliers; the rule is one for each order." Prove.

4.5

Show that, with regard to the variational principle for ϕ, and *if constrained to the subspace* $f \in L^2(\mathbb{R})$, Equation 4.19 is equivalent to the one actually proposed by Nielsen *et al.* [221], viz.

$$E[u(x)\,;\,f(x)] = \int dx \left\{ \frac{1}{2}\,(u(x) - f(x))^2 + \sum_{i=1}^{\infty} \lambda_i\, u^{(i)}(x)u^{(i)}(x) \right\} .$$

In image processing one tends to think of the first term as a "penalty" measure, and of the remainder as a measure of "smoothness". Argue why Equation 4.19 is more general and also more natural.

4.6

Recall Definition 3.19, Page 66, and Equation 4.26. Compute the entropy of the Gaussian point operator relative to the measure $m = \mathcal{A}$.

4.7

Box 4.2: "A critical point of a function is a point at which the gradient vanishes." Sketch a method for detecting and localising Morse critical points in scale-space with sub-pixel precision in the 2D case.

4.8
Consider the following decomposition of third order derivatives:

$$Z_{ijk} = L_{ijk} - \frac{1}{d+2} \left(\Delta L_i g_{jk} + \Delta L_j g_{ik} + \Delta L_k g_{ij} \right),$$

$$R_{ijk} = \frac{1}{d+2} \left(\Delta L_i g_{jk} + \Delta L_j g_{ik} + \Delta L_k g_{ij} \right),$$

so that $L_{ijk} = Z_{ijk} + R_{ijk}$.

a. Show that the "Monkey saddle" $\phi(x) = Z_{ijk} P^{ijk}(x;t)$ does not depend upon scale.

b. Show that its complement $\psi(x;t) = R_{ijk} P^{ijk}(x;t)$ does depend upon scale.

c. How many degrees of freedom are represented by $\phi(x)$ and $\psi(x;t)$, respectively?

4.9
We have $\mathbf{T} = \cup_{a \in \mathbf{T}} \mathbf{T}_a^- = \cup_{a \in \mathbf{T}} \mathbf{T}_a^0 = \cup_{a \in \mathbf{T}} \mathbf{T}_a^+$. Which conceptual picture is most adequate for a real-time system? Why?

4.10
Page 115: "S-scale invariance brings us to yet another conceptual picture: $\Lambda_a \equiv \{F, \phi, \Theta_a\} \ldots$"

a. Express S-scaling in the T-domain.

b. Describe Θ_a in $(t;x)$-coordinates.

c. Describe Θ_a^* as well as Θ_{a*} likewise.

4.11
See Definition 4.5: show that $\mathfrak{f}_a = H^* \mathfrak{f}_a$.

4.12
Let $\theta : x \mapsto \theta(x) = (\theta^0(x), \ldots, \theta^d(x))$ be a spacetime reparametrisation, and $(\theta_* \phi, \theta^* f)$ the induced effect on a given filter-source pair (ϕ, f).

a. Prove:
$$\theta_* \nabla \phi = \partial_\mu \circ \theta_* \left(\nabla \theta^\mu \cdot \phi \right).$$

(○ denotes composition; summation convention applies.)

b. Ditto:
$$\nabla \circ \theta^* f = \nabla \theta^\mu \cdot \theta^* \left(\partial_\mu f \right).$$

4.13 (Bracket Formalism)
Rewrite the raw image formula of Definition 3.6 using the bracket formalism (Box 4.4).

4.14
Footnote 14, Page 118: justify Equation 4.45 by a rigorous motivation of the central equality. (Hint: consider an integrable source field with support interval $[t_-, t_+]$ as explained in Section 4.3.3.) Outline the basic idea of the proof in the form of graphical sketches of the filter $\varphi_a(t'; t, \delta)$ for various combinations of delay $a - t$ and scale δ (and fixed a).

4.15
See Result 4.6 for notation. We consider real-time samples obtained at fixed delay, $a - t \equiv \tau$ say, and for fixed inner scale δ. For simplicity we take $\delta \equiv \tau \equiv 1$, and define the

"specious present" $\chi(u)$ as the causal point operator of Result 4.6 under these constraints, with dummy argument $u \equiv a - t'$:

$$\chi(u) \stackrel{\text{def}}{=} \begin{cases} 0 & u \leq 0 \\ \frac{1}{u} \exp\left(-\frac{1}{2} \log^2 u\right) & u > 0. \end{cases}$$

Note that derivatives w.r.t. u are in fact derivatives w.r.t. a. We are interested in the rate of change of a real-time sample $\lambda(a) \equiv F[\chi] \equiv \int dt' \, f(t') \chi(a - t')$ with acquisition time a. Assume for simplicity that $f \in C^0(\mathbb{R})$ (an assumption that can be relaxed).

a. Show that in general (without using boundary properties of χ)

$$\dot{\lambda}(a) \stackrel{\text{def}}{=} \frac{d}{da} F[\chi] = F[\frac{\partial}{\partial a}\chi] + \delta_a F[\chi] \,,$$

in which $\delta_a F$ corresponds to a point source at acquisition time, viz. $\delta(a - t') f(t')$.

b. Show that $\delta_a F[\chi] = 0$.

c. Show that $\frac{\partial}{\partial a}\chi = -\dot{\chi}$, where the dot indicates a derivative w.r.t. the integration variable t'.

d. Generalise to arbitrary orders, and show that

$$\stackrel{(n)}{\lambda}(a) \stackrel{\text{def}}{=} \frac{d^n}{da^n} F[\chi] = \stackrel{(n)}{F}[\chi] \stackrel{\text{def}}{=} (-1)^n F[\stackrel{(n)}{\chi}] \,.$$

e. For a smooth source signal $f(t)$, show that $\stackrel{(n)}{F}$ is the distribution corresponding to the n-th order derivative of $f(t)$.

f. Apparently the rate of change of acquisition data with acquisition time depends only upon the *history* of the signal, and not at all on its present or future state. Explain this paradoxical result! (Hint: smoothness does not imply analyticity.)

g. Discuss the essential difference between the causal derivative operators $(-1)^n \stackrel{(n)}{\chi}$ in this problem and those of Conclusion 4.1. Which are most natural in a framework of scale-time theory?

4.16 (Continuation of Problem 4.15)
Problem 4.15 shows that the rate of change of acquisition data with acquisition time can be expressed in terms of a sampling with (transposed) derivatives of the causal point operator *in ordinary time*. In this problem we derive the explicit form of the corresponding filter profiles.

a. Show that

$$\chi'(u) = \chi(u) \, \psi(u) \,,$$

with

$$\psi(u) \stackrel{\text{def}}{=} -\frac{1}{u} (1 + \log u) \,.$$

b. Use **a** to derive the following recursive relation for the n-th order derivative of $\chi(u)$:

$$\chi^{(n)}(u) = \sum_{k=0}^{n-1} \binom{n-1}{k} \chi^{(n-k-1)}(u) \, \psi^{(k)}(u) \,.$$

c. Prove by induction that the derivatives of $\psi(u)$ are given by

$$\psi^{(k)}(u) = \frac{(-1)^k \, k!}{u^{k+1}} \left(\beta_k - \log u \right) ,$$

in which the constants β_k are defined as

$$\beta_k \stackrel{\text{def}}{=} \sum_{i=1}^{k} \frac{1}{i} - 1 .$$

d. Show, again by induction, that the derivatives of $\chi(u)$ are given in closed form by

$$\chi^{(n)}(u) = \chi(u) \sum_{m=1}^{n} \frac{1}{m!} \sum_{l_1=1}^{n} \cdots \sum_{l_m=1}^{n} (n; l_1, \ldots, l_m) \, \psi^{(l_1-1)}(u) \ldots \psi^{(l_m-1)}(u)$$

for all $n > 0$. Here, the *multinomial coefficient* $(n; l_1, \ldots, l_m)$ is defined as

$$(n; l_1, \ldots, l_m) = \left\{ \begin{array}{cc} \frac{n!}{l_1! \ldots l_m!} & \text{if } l_1 + \ldots + l_m = n \\ 0 & \text{otherwise} . \end{array} \right.$$

Result **d** is quite explicit, but somewhat complex. We study it in somewhat more detail.

e. Show that, if one attributes a degree of homogeneity to $\psi^{(l)}$ equal to $l + 1$ in the inner summation over $l_1, \ldots, l_m$ in **d**, then only homogeneous terms of degree n show up in the expression.

f. Write out the explicit expressions of result **d** of orders $n = 1, 2, 3, 4$.

4.17
Figures 4.15 and 4.17 compare canonical and ordinary time representations of scale-time filters. However, the effects of discretisation on the mappings is not taken into account. Suppose we want to implement the displayed operators in **T**-representation, such as to reflect a regular grid in **S**. How should we resample a digital signal defined on a regular grid?

4.18
See Result 4.6. Here we study some statistics of the causal point operator. Define

$$\text{erf } x \stackrel{\text{def}}{=} \frac{1}{\sqrt{2\pi}} \int_{-\infty}^{x} d\lambda \, e^{-\frac{1}{2}\lambda^2} ,$$

and, for convenience below,

$$I(x) \stackrel{\text{def}}{=} e^{\frac{1}{2}x^2} \text{erf } x .$$

Recall the definition of linear and central momenta: Definitions 3.12 and 3.14 (for $n = 1$).

a. Show that $m_0(t, \delta) = m_0[\varphi_a(\,.\,; t, \delta)] = 1$.

b. Define the centre of gravity of $\varphi_a(t'; t, \delta)$ as its first order linear momentum $c_a(t, \delta) = m_1[\varphi_a(\,.\,; t, \delta)]$. Show that $c_a(t, \delta) = t$. (Hint: no computation, symmetry!)

c. Define absolute inner scale or standard deviation, and variance, as usual: $\sigma_a(t, \delta) = \sqrt{s_a(t, \delta)}$ and $s_a(t, \delta) = \sigma_2[\varphi_a(\,.\,; t, \delta)]$, respectively. Argue that absolute inner scale cannot depend upon a or t separately, but must be a function of delay $a - t$.

d. Show that
$$s_a(t, \delta) = (a - t)^2 \left(I(0) - 2I(\delta) + I(2\delta) \right) .$$

e. Assume $0 < \delta \ll 1$. Show by expansion that
$$s_a(t, \delta) \approx (a - t)^2 \left(\frac{1}{2}\delta^2 + 2\sqrt{\frac{2}{\pi}}\delta^3 + \mathcal{O}(\delta^4) \right) .$$

f. Define skewness as the dimensionless quantity
$$\alpha(\delta) \stackrel{\text{def}}{=} \frac{\sigma_3[\varphi_a(\,.\,; t, \delta)]}{\sigma_2[\varphi_a(\,.\,; t, \delta)]^{3/2}} .$$

Argue why skewness cannot depend upon a or t, nor on delay $a - t$. (Hint: Box 3.4.)

g. Show that
$$\alpha(\delta) = \left(\frac{\sqrt{2}}{\delta} \right)^3 \left(I(0) - 3I(\delta) + 3I(2\delta) - I(3\delta) \right) .$$

h. Assume $0 < \delta \ll 1$. Show by expansion that
$$\alpha(\delta) \approx -\frac{8}{\sqrt{\pi}} - \frac{9}{\sqrt{2}}\delta + \mathcal{O}(\delta^2) .$$

What is the significance of this negative value?

4.19
In this problem we study the eigenfunctions and eigenvalues of the causal differential operators of Conclusion 4.1. An index μ denotes a derivative w.r.t. the μ-th Cartesian coordinate on $\mathbf{M_c} = \mathbf{S} \times \mathbf{X}$, i.e. $x = (x^0 \equiv s; \vec{x})$. We use corresponding Fourier coordinates $\omega = (\omega_0 \equiv \nu; \vec{\omega})$, and collectively denote temporal and spatial scales by the single parameter σ by employing a "cs-convention" analogous to the ct-convention of Definition 3.17, Page 62.

a. Recall Definitions 2.20 (Page 34) and 2.21 (Page 34). Show that the planar wave $\mathfrak{F}_\omega$, with function representation
$$\mathfrak{f}_\omega(x) = \exp\left(i\omega \cdot x \right) ,$$

is an eigenfunction for every correlation operator $\phi_{\mu_1 \ldots \mu_k} \in \mathcal{G}(\mathbf{M_c})$, with eigenvalue
$$\lambda_{\mu_1 \ldots \mu_k}(\omega; \sigma) = (-i\omega_{\mu_1}) \ldots (-i\omega_{\mu_k}) \exp\left(-\frac{1}{2}\sigma^2 \|\omega\|^2 \right) .$$

b. Show that the eigenfunctions are separable, and that (up to arbitrary amplitude) the temporal component is given in usual $\mathbf{T}$-parametrisation by
$$\mathfrak{f}_{\nu;a}(t'; t) = \exp\left(-i\nu \log \frac{a - t'}{a - t} \right) .$$

Here, $a - t > 0$ is the delay relative to the current moment $a \in \mathbf{T}$, and t' is the field's dummy argument; it is understood that $\mathfrak{f}_{\nu;a}(t'; t) = 0$ whenever $\max(t, t') \geq a$.

c. Interpret the "canonical temporal frequency" ν in terms of ordinary frequency. In what sense do we have a "fading of memory"? Sketch the graph of an eigenfunction in the $\mathbf{T}$-domain for two distinct values of a, but fixed scale δ and delay $a - t$.

d. A nitty-gritty detail: note that we have attached a label a to our source in T-time, and none in S-time; this is opposite to consistent practice throughout this chapter. Explain why it is nevertheless natural to do so.

4.20 (Benford's Law)
Page 112: "Benford's law immediately follows from the assumption that random events are uniformly distributed in s-time."

a. Show that, if this is the case, we have the following *a priori* distribution of first digits $k = 1, \ldots, 9$ in decimal notation:

$$P_k = \log_{10} \frac{k+1}{k} .$$

Footnote 12: "...physical distributions other than Benford's type are excluded by scale invariance!".

b. Prove that Benford's distribution is indeed invariant under rescaling of (ordinary) time units.

4.21 (Topology for Local Jets)
a. Show that the seminorms of Equation 4.29 are in general not covariant, and that Definition 4.1 does define a covariant seminorm for the space of local k-jets.

b. The seminorm of Definition 4.1 is closely related to the case $p = 2$ of Equation 4.29. Can you give a covariant analogue of the $p = 1$ case?

4.22
Show that the components of the k-th order local jet $j^k f(x)$ relative to a coordinate basis transform as a closed set under (i) arbitrary coordinate reparametrisations $x = \eta(y)$, and (ii) arbitrary (invertible) grey-scale transformations $f = H(g)$.

4.23 (Scale Derivatives)
Why is there no minus sign $(-1)^j$ in Equation 4.31 for a scale derivative?

4.24 (Essential Components of Local Jet)
Prove Equation 4.33.

The question about the applicability of the assumptions of geometry at infinitesimal distances is connected with the question about the inner reasons of the measure relationships of space. With this question, which one may perhaps still include in the study of space, the previous remark is used, that a discrete manifold contains the principle of the measure relationships already in the concept of this manifold, but it must come from somewhere else for a continuous manifold. Hence, the reality on which space is based must either be a discrete manifold, or the reasons for the measure relationships must be searched outside in attracting forces acting on them.

—BERNHARD RIEMANN

CHAPTER 5

Local Image Structure

Local image properties are those that can be associated with a single base point at a fixed inner scale, i.e. that can be defined in terms of derivatives taken at a fixed point in scale-space. Multilocal properties take into account multiple local neighbourhoods. (Note that certain multilocal properties may have a local interpretation at a coarser level of scale or at a higher order of differentiation.) Local properties are the building blocks on the basis of which multilocal expertise must be built. Indeed, the very scale-space construct is intended to encompass global structure as a manifestation of purely local entities that are causally connected in a continuous tree-like structure ("dynamic shape" [162]).

Established methods described in the mathematical literature can be applied to the analysis of local image structure in a technically straightforward way, at least within the context of the scale-space paradigm for any fixed value of inner scale. In this chapter we will address this "superficial structure" of an image, assuming $\sigma \equiv 1$, unless stated otherwise. This is clearly only a small step towards an understanding of deep structure, which is a new and largely unexplored field of mathematics that should immensely titillate contemporary geometers.

Local image operators must be well-posed and "meaningful". Well-posedness has been discussed *in extenso* in the previous chapters, where it was shown that one can compute any derivative of a scale-space image in a well-posed way with the help of the Gaussian family (Result 3.3, Page 64), as far as data quality permits, by means of a simple linear filtering. Without anticipation of final purpose or knowledge of image formation one must take "meaningful" in a pre-categorical sense, i.e. relative to the commitments laid down in our syntax conventions. In particular, local image descriptors must be *covariant* (Trivialities 3.1 and 3.2 on Pages 42 and 43, respectively). It has been argued that the principle of covariance is a trivial requisite for any theory, though not quite trivial *in form* if quantities are expressed in terms of their components relative to a coordinate frame (the usual case!). Covariance implies coordinate independence despite coordinatisation. In turn this implies that, if an object has multiple components, then these must transform *as a set* in a consistent way under coordinate transformations, viz. such that objective conclusions can be drawn independent of the chosen reference frame. Furthermore, local image descriptors must be compatible with the spacetime model, i.e. they must be *invariant under the spacetime symmetry group*.

Recall the essential difference between a mere change of coordinates and a change in geometry: Definitions 3.4 and 3.7 on Pages 45 and 52. The invariance group is a specific transformation group that defines the geometry of spacetime; the argument of covariance pertains to general automorphisms and applies equally to any other choice of geometry [92].

Section 5.1 is a preliminary on invariance as one encounters it in the image literature and on the extent of its use in this book. In the light of the above requirements, the objects of interest are *vectors*, *covectors* and, in general, *tensors*, i.e. multilinear mappings of vectors and covectors into the real numbers. A brief introduction on tensors has been given in Box 3.5 on Page 78. Section 5.2 provides a more systematic, more or less self-contained account of tensor calculus applied to differential image structure. The reader is encouraged to consult Appendix A for a summary and for pointers to relevant literature. Finally, in Section 5.3, tensor calculus is applied to the construction of differential invariants.

5.1 Groups and Invariants

If we say that "tensors are invariant", we tacitly refer to the classical spacetime geometry. It would be a pleonasm to say that "tensors are covariant" in the sense of Trivialities 3.1 and 3.2, Pages 42 and 43. However, "covariant tensors" is also classical parlance for a specific category of tensors, not to be confused with the principle of covariance! This, together with associated categories ("contravariant" and "mixed tensors"), will be explained in Section 5.2.

An essential feature of classical spacetime is that it does not possess a metric, i.e. one cannot sensibly define "distance in spacetime", although one can obviously define spatial distances and temporal durations. In other words, classical spacetime is a product space of Euclidean space and time[1], with some additional

---

[1]Note that pseudo-isotropic spacetime defined by the ct-convention of Definition 3.17 on Page 62 is merely a camouflage of this product space.

geometrical structure that expresses our common sense, but does not need to be scrutinised here [92]. For simplicity we shall henceforth consider only the spatial dimensions; time is a trivial instance obtained by setting dimension equal to unity, and combinations of spatial and temporal dimensions are trivial as well by virtue of the factorisation. This means that the invariance group of interest is essentially the *scale-Euclidean* or *self-similarity group*, the spatial subgroup of Definition 3.7 on Page 52 (recall also Problem 3.7 on Page 84).

A task oriented or knowledge based level of description generally calls for a restricted class of invariants by extension of the basic invariance group. Important in computer vision—but beyond the scope of this book—is the *projective group*, reflecting the geometry induced by a projective camera [64, 89, 216] (the *affine group* often arises as an approximation). The invariants of such an extended group can be realized as *specific* instances or combinations of invariants of the basic spacetime group. A nice illustration of how certain nontrivial group actions can be constructed in the context of projection data is described by Van Gool *et al.* and by Moons *et al.*, who propose the "stereo Lie groups" GS(4) and SS(4) [97, 213, 214]. A different framework for forming invariants in computer vision, based on the algebras of Grassmann [100] and Clifford [39], has been described by Lasenby *et al.* and by Bayro-Corrochano *et al.* [14, 179, 180], inspired by the work of Hestenes [115]. An extensive account of the use of Clifford algebra in projective geometry has been given by Hestenes and Ziegler [116].

Another example of a transformation group, which we shall encounter later, is the group of "general grey-scale transformations", which accounts for nonlinear, invertible transformations of grey-scales at fixed resolution [80, 82]. The requirement of invariance under this group means that one discards all grey-scale information except for the induced geometry of iso-grey-level contours and their nesting hierarchy.

From the above examples it is clear that "invariance" is no less polysemous than the notion of "transformation". Transformations may explicitly act on a grey-scale image, or on features extracted by pre-processing, such as shape induced curves or surfaces. The latter is a common approach in computer vision in an attempt to circumvent the notorious "shape from shading" problem [122]. It is clear that the latter type of action must be derivable from the former, but it is not always clear how that works. The problem in general is that, even if one knows the action of a group on "empty space", one does not automatically know the right *representation*, i.e. the way the group acts on various physical objects involved. This requires an awareness of all factors involved in the formation of an image, which is usually not feasible in the case of conventional imaging (light source details, object reflectance properties [220], camera geometry, etc.). This question of representation makes photography notoriously difficult compared to, say, tomography, in which measurement duality is not complicated by such factors.

Besides by reference to underlying group, invariants can be categorised as previously noted, viz. according to whether they describe local (or differential), multilocal, or global image properties. An example of a differential invariant is the well-known Laplacian $\triangle L$ (recall Notation 3.1, Page 65); it is of order 2, because it can be constructed on the basis of the image's local 2-jet [81]. The so-called "semi-differential invariants" fall in the multilocal category [28, 97, 98]. Global invariants have been proposed by, among many others, Brandt and Feng [27], and Segman and Zeevi [255]. Finally, differential invariants in Fourier space also correspond to global invariants in ordinary space, such as image momenta and the well-known Fourier power spectrum. *Et cetera*.

In brief, the reason for introducing an invariance group in the first place is to isolate image parameters one considers to be irrelevant *a priori* from potentially relevant ones. Requiring invariance guarantees that the irrelevant ones are discarded from the outset and only the residual information is taken into consideration.

In the absence of a task and without resort to a specific image formation model, the most general group of interest is the basic spacetime symmetry group, i.e. the scale-Euclidean group. Equivalently, all spatial relations of interest (lengths, angles) are scale invariant relations measured using a "flat", homogeneous and isotropic *Euclidean metric*. We henceforth consider only differential invariants in this primordial sense pertaining to the very structure of classical spacetime. These can be regarded as useful primitives if additional semantics is available.

In the next section we introduce tensors and explain the basics of tensor calculus, which provides the mathematical framework for the description of geometrically meaningful concepts. We outline both "classical" as well as contemporary formulations, but emphasize practical use over mathematical rigour. Appendix A contains a review and suggestions for further reading.

5.2 Tensor Calculus

The line of approach in this section is as follows. In Section 5.2.1 we start by introducing a tensor that plays a prominent role in spaces equipped with a concept of distance and angle, the so-called *metric tensor*. The existence of a metric implies a great deal of additional structure and is therefore rather restrictive. However, since space and time are metrical spaces, it is convenient to account for metrical relations from the outset. At the same time the metric tensor serves to illustrate tensors in general. General tensors are treated systematically in Sections 5.2.2 (without metrical relations) and 5.2.3 (with metrical relations). Differential calculus and tensor calculus are linked via the covariant derivative; this is explained in Section 5.2.4 (the need for covariant derivatives was mentioned *en passant* on Page 63 of Chapter 3, and several times in Chapter 4). The intrinsic curvature of a non-flat space (with or without metric) can be described in terms of a fundamental tensor, which is introduced in Section 5.2.5. Finally, in Section 5.2.6 one more fundamental tensor is introduced. It appears somewhat special, and can be understood in terms of a further generalisation of the tensor concept, presented in Section 5.2.7.

5.2.1 The Euclidean Metric

The classical formulation of a Euclidean metric is in terms of a "line element" $d\ell$, which measures the distance between two "infinitesimally close" points in space.

Definition 5.1 (Euclidean Metric)
The covariant Euclidean metric tensor is defined by a line element

$$d\ell^2 \overset{\text{def}}{=} g_{ij}(x)\, dx^i dx^j\,,$$

in which $g_{ij} = g_{ji}$ *and* $g \equiv \det g_{ij} > 0$, *for which there exists a coordinate frame such that, for all* $x \in \mathbb{R}^d$,

$$g_{ij}(x) = \begin{cases} 1 & \text{if } i = j \\ 0 & \text{otherwise.} \end{cases}$$

By symmetry and positivity there always exists a coordinate frame in which $g_{ij}(x)$ becomes the identity matrix *at any isolated point* x. In a Euclidean space coordinates exist for which this canonical form holds *globally*. Due to invariance such a coordinate system is not unique however; the canonical form is preserved under arbitrary *Cartesian coordinate transformations*. By default it will be assumed that the grid of a raw image is aligned with the grid points of a Cartesian frame, with the origin at the point of interest. The distinction between subscripts and superscripts, such as in the expression for $d\ell^2$ in Definition 5.1, is important for the considerations of classical tensor calculus, the reason for which will become clear soon. Recall for instance that the summation convention applies to pairs of upper and lower indices.

Definition 5.2 (Covariant and Contravariant Metric Tensors)
The contravariant, or dual metric tensor g^{ij} *is the matrix inverse of the covariant metric tensor* g_{ij} *of Definition 5.1:*

$$g^{ik}(x)\, g_{kj}(x) \overset{\text{def}}{=} \delta^i_j\,.$$

Note that also the dual metric simplifies to the identity matrix in a Cartesian coordinate system, thus there is no distinction between upper and lower index representations of the metric in that case.

Example 5.1 (Euclidean Metric: Various Representations)
In 2D we have, in a suitable Cartesian coordinate system $(x^1, x^2) = (x, y)$,

$$d\ell^2 = dx^2 + dy^2\,, \quad \text{so that} \quad g_{ij} \mathrel{\widehat{=}} \begin{pmatrix} 1 & 0 \\ 0 & 1 \end{pmatrix} \quad \text{as well as} \quad g^{ij} \mathrel{\widehat{=}} \begin{pmatrix} 1 & 0 \\ 0 & 1 \end{pmatrix}.$$

If we switch to polar coordinates $(x^1, x^2) = (r, \phi)$ defined by $(x, y) = (r \cos \phi, r \sin \phi)$, we obtain

$$d\ell^2 = dr^2 + r^2 d\phi^2\,, \quad \text{i.e.} \quad g_{ij} \mathrel{\widehat{=}} \begin{pmatrix} 1 & 0 \\ 0 & r^2 \end{pmatrix} \quad \text{and} \quad g^{ij} \mathrel{\widehat{=}} \begin{pmatrix} 1 & 0 \\ 0 & r^{-2} \end{pmatrix},$$

whereas log-polar coordinates $(x^1, x^2) = (\rho, \theta)$, defined by $(r, \phi) = (\ell_0 e^\rho, \theta)$ for some constant ℓ_0, yield the "conformal metric"

$$d\ell^2 = \ell_0^2 e^{2\rho} \left(d\rho^2 + d\theta^2 \right), \quad \text{or} \quad g_{ij} \cong \ell_0^2 e^{2\rho} \begin{pmatrix} 1 & 0 \\ 0 & 1 \end{pmatrix} \quad \text{and} \quad g^{ij} \cong \ell_0^{-2} e^{-2\rho} \begin{pmatrix} 1 & 0 \\ 0 & 1 \end{pmatrix}.$$

Example 5.2 (Scale-Invariant Euclidean Metric)
  - A scale invariant Euclidean metric is obtained by measuring spatial distances relative to inner scale:
$$d\widetilde{\ell} \stackrel{\text{def}}{=} \frac{d\ell}{\sigma}.$$
In this way we obtain a family of spatial metrics, one for every level of scale. Note that this does *not* define a scale-space metric (such a metric has been proposed by Eberly [60]).

  - Alternatively, one obtains a scale invariant metric by measuring distances relative to any fiducial yardstick that scales like a distance, ℓ_0 say:
$$d\widetilde{\ell} \stackrel{\text{def}}{=} \frac{d\ell}{\ell_0}.$$

Such an extrinsic parameter ℓ_0 was used in the log-polar transform of Example 5.1.

In order to explain what a tensor actually is, and why one distinguishes between upper and lower indices, it is instructive to take the metric as an example.

In practice it is most natural to conceive of a metric without reference to the nonsense of "infinitesimals", viz. as a bilinear mapping of two vectors into the real numbers, which is just the well-known scalar product of the two vectors (with defining properties as listed in Appendix A, Equation A.43). An important interpretation of the covariant metric tensor is then obtained if we fill only one of its two vector slots. The object that results is an operator that requires one more vector to be inserted. Thus it is a *covector* in the same sense as introduced on Page 19 of Chapter 2, but finite-dimensional. Recall that such covectors make up a linear space called the dual[2] of the vector space on which these covectors act. Thus the metric naturally provides us with a covector space for any given vector space. As a converter of vectors into covectors, the metric is also known as the *sharp operator* ♯. Its coordinate interpretation is that of an *index lowering operator*: if $\vec{v}$ is a vector with components v^i, then $♯\vec{v}$ is the covector with components $v_i \equiv g_{ij} v^j$. We can do similar things the other way around, using the inverse metric. When interpreted as a converter of covectors into vectors, the dual metric is also referred to as the *flat operator* ♭. Its classical interpretation is that of an *index raising operator*: if $\widetilde{v}$ is a covector with components v_i, then $♭\widetilde{v}$ is the vector with components $v^i \equiv g^{ij} v_j$. The importance of having a metric is thus that it allows us to put elements of dual spaces in one-to-one correspondence. Figure 5.1 illustrates the conceptual pictures of the metric in classical and modern form.

Since the Cartesian coordinate representations of the metric and its dual are identical, we find the following result.

---

[2]Unlike in the ∞-dimensional case, there is no distinction between algebraic and topological dual.

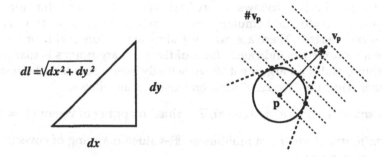

Figure 5.1: Classical and modern interpretations of the metric. **Left**: the classical "infinitesimal line element". **Right**: geometrical construction of the modern "sharp/flat" slot machine.

Result 5.1 (Special Case: Cartesian Tensors)
If one sticks to Cartesian coordinate frames, one can forget about the distinction between upper and lower indices.

This simplifies things a lot; in particular it helps to see how coordinate representations simplify in a Cartesian frame.

5.2.2 General Tensors

Covectors are vectors in the technical sense, i.e. elements of a linear space. The prefix reminds us of duality with a concomitant vector space.

Definition 5.3 (Dual Bases and Standard Dual Bases)
Dual bases of vectors and covectors are ordered sets, $\{\vec{e}_1, \ldots, \vec{e}_d\}$ and $\{\tilde{e}^1, \ldots, \tilde{e}^d\}$ say, satisfying

$$\tilde{e}^i[\vec{e}_j] \overset{\text{def}}{=} \delta^i_j \,.$$

In a standard coordinate frame,[3] the standard vector basis is given by $\vec{e}_i \equiv \partial_i$ and the standard covector basis by $\tilde{e}^i \equiv dx^i$.

(Observe that one does not need a metric in order to define dual bases.)
    We now define tensors in general.

Definition 5.4 (Tensors)
A tensor is a multilinear mapping of vectors and covectors. If $\{\vec{e}_1, \ldots, \vec{e}_d\}$ is a basis of vectors and $\{\tilde{e}^1, \ldots, \tilde{e}^d\}$ a basis of covectors, then we can decompose any mixed tensor of covariant rank k and contravariant rank l as

$$T = T^{i_1 \ldots i_l}{}_{j_1 \ldots j_k} \, \vec{e}_{i_1} \otimes \ldots \otimes \vec{e}_{i_l} \otimes \tilde{e}^{j_1} \otimes \ldots \otimes \tilde{e}^{j_k} \,.$$

---

[3]We consider only coordinate, or *holonomic bases*; noncoordinate, or *anholonomic bases*, can be taken into consideration as well, see e.g. Misner, Thorne and Wheeler [211].

The way to read this is as follows. Every basis vector defines a "slot" into which one can put a covector, and similarly, every basis covector expects a vector argument to be inserted. There are exactly k slots for vectors and l for covectors, and these are ordered as indicated. Each of the $k + l$ arguments is mapped in a linear fashion, and if all arguments have actually been provided, the result is a real number. Thus the "modern" view on tensors is as follows:

- a *covariant tensor* is a multilinear, $\mathbb{R}$-valued mapping of vectors ($l = 0$),

- a *contravariant tensor* is a multilinear, $\mathbb{R}$-valued mapping of covectors ($k = 0$), and, in general,

- a *mixed tensor* is a multilinear, $\mathbb{R}$-valued mapping of vectors as well as covectors ($k \neq 0 \wedge l \neq 0$).

The metric g_{ij} is an important example of a covariant tensor. For simplicity one often refers to a covariant tensor as a "cotensor", whereas a "tensor" is by default either contravariant ("contratensor") or of unspecified type. For the moment it suffices to think of a covariant tensor as a geometric object, the coordinates of which are denoted using only *lower indices*. Likewise, the dual metric g^{ij} is an instance of a contravariant tensor (or contratensor), which has only *upper indices*. Naturally, δ^i_j is an instance of a mixed tensor, characterised by both upper and lower indices (*mixed indices*). The *rank* of a tensor equals its number of indices, so all three metric representations are of rank 2. A general class of covariant tensors of arbitrary rank suggests itself, viz. the Gaussian filter family of Result 3.3 on Page 64. As a result, image derivatives of successive orders as defined in Notation 3.1, Page 65, constitute a family of covariant tensors, for which rank coincides with differential order. Apart from the dual metric g^{ij} we have also already encountered a general class of contravariant tensors, viz. the scale-space Taylor polynomials $P^{i_1 \cdots i_k}$ of Corollary 4.1 on Page 103, for which rank happens to coincide with monomial degree.

The convention of Definition 5.3 combined with Definition 5.4 makes the modern notation for the metric look quite similar to that of the classical line element:

$$G = g_{ij}(x)\, dx^i \otimes dx^j\,, \tag{5.1}$$

relative to a standard covector basis. Nevertheless, the interpretations of Definition 5.1 and Equation 5.1 differ in a fundamental way! A conceptual switch one has to be able to make off the top of one's head when translating classical into modern geometrical language is to forget about the dx^i as representing some kind of "infinitesimal vector". This should not be difficult, since it is nonsense anyway. Rather, the same expression dx^i in Equation 5.1 represents a covector, which, as we know, has nothing infinitesimal about it; it is simply an empty slot waiting for a vector. A C-programmer would probably write something like[4]

```
float Metric(vector_t FirstVector, vector_t SecondVector);
```

---

[4]In general one needs to account for the base point of the metric and its arguments.

instead of Equation 5.1, with implementation details consistent with the requirements for a metric.

If we want to insert a vector into the "machine" T of Definition 5.4, we have to tell in which of the k slots (basis covectors) it is supposed to be dropped. Likewise, inserting a covector requires us to specify which of the l basis vectors is supposed to handle it. If no confusion about ordering is likely to arise one usually dispenses with the white-space gap in the component form of a mixed tensor.

5.2.3 Tensors on a Riemannian Manifold

As it is, the tensor T only accepts a sequence of exactly k vectors and l covectors, in the order according to which it is "prototyped". However, being in the possession of a metric, we can view any covariant, contravariant or mixed tensor of a given *total* rank as a disguise of one and the same abstract tensor object, which "knows" how to handle any data type inserted into its slots. The abstract tensor with such built-in conversion rules could be symbolised by

$$\mathbf{T} \sim \{T; \sharp, \flat\}. \tag{5.2}$$

This is best appreciated by some examples.

Example 5.3 (Tensors on a Riemannian Manifold: Sharp and Flat)
- If $\vec{v}$ is a vector with components v^i, i.e. $\vec{v} = v^i \partial_i$, then $\tilde{v} \equiv \sharp\vec{v}$ is a covector with components $v_i = g_{ij}v^j$.
- If $\tilde{v}$ is a covector with components v_i, i.e. $\tilde{v} = v_i dx^i$, then $\vec{v} \equiv \flat\tilde{v}$ is a vector with components $v^i = g^{ij}v_j$.
- The *Cartesian* components of $\tilde{v}$ and $\vec{v}$ are the same; in general, however, $v_i \neq v^i$. Abstraction: $\mathbf{v} \sim \{\vec{v}, \tilde{v}\}$.
- A covector maps a vector to a number, i.c. $\tilde{v}[\vec{w}] = v_i w^i$, which is just the familiar scalar product $\vec{v} \cdot \vec{w}$ for $\vec{v} = \flat\tilde{v}$. In other words, a covector is a cotensor of rank 1.
- A vector can just as well be said to map a covector to a number: $\vec{v}[\tilde{w}] = v^i w_i$, which is again just the scalar product $\vec{v} \cdot \vec{w}$, but now with $\vec{w} = \flat\tilde{w}$. In other words, a vector is a contratensor of rank 1.
- In general: the last k input entries for T as defined in Definition 5.4 are declared as vectors, the first l as covectors. An error message results in case of a "type mismatch", unless we build implicit conversion rules with the help of the metric. If a vector is passed into a slot of $\mathbf{T}$ for which the prototype T expects a covector, the $\sharp$-operator is applied before it is released into the slot of T. Likewise, a covector is converted to its corresponding vector by applying the $\flat$-operator if type consistency demands it.

Thus any prototype similar to Definition 5.4 will do as long as covariant and contravariant ranks add up to the specified total rank: $k' + l' = k + l$. The recipe for $\mathbf{T}$ (Equation 5.2) allows us to talk unambiguously about tensors without the predicate "covariant", "contravariant", or "mixed". The scalar product is a familiar instance of this procedure, where we first convert one of the two vectors into a covector before we can actually perform the contraction: $\vec{v} \cdot \vec{w} = \sharp\vec{v}[\vec{w}] = \sharp\vec{w}[\vec{v}]$.

The covariant, contravariant, and mixed component representations of the metric, g_{ij}, g^{ij} and δ^i_j, respectively, are likewise manifestations of one and the same abstract tensor object **G**, into which we may choose to insert two vectors (the usual prototype of Equation 5.1), two covectors, or one of each, in any order and without ambiguity.

Despite the "democratic abstraction" of Equation 5.2, there exists a historical preference with regard to the component representations of several standard tensors. For example, a derivative, when given in component form, is usually represented as a covariant tensor, whereas a polynomial is most often written in terms of its contravariant components.

By definition, vectors and covectors can be contracted. The fact that all tensors can be decomposed relative to a basis of vectors and covectors (Definition 5.4) creates the possibility to contract one tensor onto another—or possibly onto itself—in as many different ways as one can pair covariant and contravariant entries. Each contraction is established by inserting a dual vector-covector pair $\{\vec{e}_i, \tilde{e}^j\}$ into two of its slots (this reduces the rank of the tensor by 2, and creates a matrix object with component labels i and j), and evaluating the trace over i and j. The generalised, metric-equipped tensor definition of Equation 5.2 widens the number of possibilities even further, as it allows *any* two indices to be contracted, even if they are of the same type.

Example 5.4 (Contractions)
- Using the same notation as in the previous example, we can form the mixed 2-tensor $\vec{v} \otimes \tilde{w} = v^i w_j \, \partial_i \otimes dx^j$. It can be contracted onto itself in a unique way, yielding $\vec{v} \otimes \tilde{w} \, [dx^k, \partial_k] = v^i w_j \, \partial_i [dx^k] \, dx^j [\partial_k] = v^i w_j \, \delta^k_i \delta^j_k = v^k w_k$. Classical recipe: equate upper and lower index, and apply the summation convention; thus contraction of $v^i w_j$ yields $v^k w_k$.

- The Hessian L_{ij} is purely covariant, hence cannot be contracted onto itself, unless one tacitly invokes the $\flat$-operator for one of its entries: $\flat dx^k = g^{km} \partial_m$. Contraction yields the Laplacian: $H = L_{ij} \, dx^i \otimes dx^j \to H \, [\flat dx^k, \partial_k] = g^{km} L_{ij} \, dx^i [\partial_m] \, dx^j [\partial_k] = g^{km} L_{ij} \, \delta^i_m \delta^j_k = g^{km} L_{km} = \triangle L$. Classical procedure: raise one index, then contract; $L_{ij} \to L^k_j \equiv g^{ik} L_{ij} \to L^k_k = g^{km} L_{km} = \triangle L$.

- We cannot compute the spatiotemporal Laplacian as a contraction of the spatiotemporal Hessian $L_{\mu\nu}$, because we lack a metric for spacetime. Recall Definition 3.17 (Page 62) and Example 3.1 (Page 64); a pseudo-isotropic Laplacian L^μ_μ in spacetime requires *two independent contractions* of spatial and temporal Hessian (the latter being trivial), plus a linear combination of the results, involving the parameter c: $L^\mu_\mu = \sigma^2 (\triangle L + c^{-2} L_{tt}) = L^\mu_\mu = \sigma^2 \triangle L + \tau^2 L_{tt}$.

The classical way of introducing tensors is best appreciated by studying how a change of coordinate basis affects their component representation. Consider vectors first. If we introduce a vector $\vec{v}$ by its components v^i relative to a vector basis $\{\vec{e}_1, \ldots, \vec{e}_d\}$, and apply an arbitrary (non-singular) linear transformation of basis vectors $\vec{e}_i = A^j_{\ i} \vec{e}_j{}'$, then we must guarantee the objectivity of our definition by transforming the components in such a way that $\vec{v}$ is left unaffected: $\vec{v} = v'^i \vec{e}_i{}' \equiv v^j \vec{e}_j = v^j A^i_{\ j} \vec{e}_i{}'$, whence

$$v'^i = A^i_{\ j} v^j \, . \tag{5.3}$$

In classical textbooks on tensor calculus one tends to *define* vectors by saying these are d-tuples transforming according to Equation 5.3. Knowing how vectors "transform", duality determines the transformation of covectors. Consider for instance the scalar product in the form of a vector-covector contraction, with $\vec{v} = v^i \vec{e}_i$ and $\tilde{\omega} = \omega_i \vec{e}^i$, and suppose we transform v^i as previously. We can invert Equation 5.3, expressing the old components as functions of the new ones, $v^i = B_j{}^i v'^j$ with matrix B defined as the inverse-transpose[5] of A,

$$B^\mathsf{T} \stackrel{\text{def}}{=} A^{\text{inv}} . \tag{5.4}$$

The principle of covariance then dictates that $\tilde{\omega}[\vec{v}] = \omega_i v^i = \omega_i B_j{}^i v'^j = \omega'_j v'^j$, with

$$\omega'_j = B_j{}^i \omega_i . \tag{5.5}$$

Comparing Equations 5.3 and 5.5, we encounter an inverse-transposed matrix in the transformation of a covector relative to that of a vector, *vice versa*. Again, classical geometers would say that this transformation property *defines* covectors. This is of course nothing but the passive view on covariance, (Triviality 3.2, Page 43), applied to the finite dimensional case, which could be given an active interpretation as well in analogy with Triviality 3.1 on Page 42 (physically transform vectors and covectors, and consider their components relative to a fixed basis).

It is now clear how an arbitrary tensor ought to "transform":

$$T'^{i_1 \ldots i_l}{}_{j_1 \ldots j_k} = A^{i_1}{}_{p_1} \ldots A^{i_l}{}_{p_l} B_{j_1}{}^{q_1} \ldots B_{j_k}{}^{q_k} T^{p_1 \ldots p_l}{}_{q_1 \ldots q_k} . \tag{5.6}$$

Again, this *tensoriality criterion* is the classical way of *defining* a tensor [1]. For every contravariant index i we encounter a matrix $A^i{}_p$, while for every covariant index j we have a $B_j{}^q$, in which p, q are summation dummies.

One can define tensors at every point in spacetime. The totality of all possible tensors at all points is called a *tensor bundle*. If one picks from this bundle exactly one tensor for each point, such that the transition between neighbours is smooth, one has a *tensor field*. By the same token a *vector field* and a *covector field* arise as smooth "cross-sections" of a *vector bundle* and a *covector bundle*, respectively. Of course, transformations may also vary from point to point on the base manifold, so that the matrices in Equations 5.3–5.6, as well as all tensor components, are functions of x. In that case the transformation matrices, expressing the change of the local frame, are fully determined by the Jacobian matrix of the coordinate transformation at the point of interest. That is, if $x \mapsto x'(x)$, then

$$A^i{}_j = \frac{\partial x'^i}{\partial x^j} \quad \text{whence} \quad B_j{}^i = \frac{\partial x^i}{\partial x'^j} . \tag{5.7}$$

It is understood that these quantities are evaluated at the point of interest.

---

[5]One interprets the leftmost index of $A^i{}_j$ and $B_j{}^i$ as the row index.

5.2.4 Covariant Derivatives

Recall that the components of the filters $\phi_{\mu_1\ldots\mu_k}$ (Result 3.3, Page 64), and therefore the image derivatives $L_{\mu_1\ldots\mu_k}$ (Notation 3.1, Page 65), have been defined as partial derivatives *in a Cartesian frame*. Once having established tensor components in one coordinate frame, the representation in any other frame can in principle be determined by straightforward transformation. As soon as we turn to curvilinear coordinates, however, the tensor components are no longer partial derivatives in the transformed coordinate frame. This follows from the chain rule, which relates derivatives in x and x' coordinates:

$$\frac{\partial}{\partial x'^{j_1}} \circ \ldots \circ \frac{\partial}{\partial x'^{j_k}} = B_{j_1}{}^{i_1} \frac{\partial}{\partial x^{i_1}} \circ \ldots \circ B_{j_k}{}^{i_k} \frac{\partial}{\partial x^{i_k}} , \tag{5.8}$$

with k instances of the Jacobian matrix as defined in Equation 5.7. Since the Jacobian matrix is a function of spacetime we cannot shuffle them to the left (unless $k = 0, 1$, or in the special case of an affine transformation, for which the Jacobian matrix is constant). In other words, a partial derivative is not manifest covariant (i.e. cannot be regarded as a covariant tensor). If we use partial derivatives, we have to specify the coordinate system, which runs counter to our objective to state everything in terms of coordinate independent tensor equations. What we therefore need is a generalised derivative operator, which reduces to a partial derivative in Cartesian coordinates, but complies with the principle of covariance. That is, we would like to think of the k-th rank covariant tensors $\phi_{\mu_1\ldots\mu_k}$ and $L_{\mu_1\ldots\mu_k}$ as k-th order *covariant derivatives* of ϕ and L, respectively.

The lack of covariance of a partial derivative is caused by the fact that a derivative arises from a *multilocal* computation. Since components of objects defined at neighbouring points are relative to their respective local frames, they cannot be compared in the naive way. For example, if $v^i(x) = v^i(y)$ for all vector field components, it is not at all guaranteed that the two vectors are equal or even parallel; disregarding instances of sheer luck this will only be the case for Cartesian coordinates, or, trivially, if $x = y$. No confusion arises when comparing *scalar fields* (tensor fields of rank 0), since they do not have components relative to a local frame. What one would intuitively want to do when defining a covariant derivative (and in general, any meaningful multilocal operator) is to evaluate components of quantities defined at distinct points after a *parallel transport* to a common base point, so that all components are given relative to the same basis (or, in the passive view, after a rectification of coordinate lines).

It is easiest to illustrate the problem of covariant differentiation by considering the gradient of a mixed 2-tensor $T = T^i{}_j\, \vec{e}_i \otimes \vec{e}^j$ (the general case is straightforward and will be stated below). Since the basis vectors and covectors may vary from base point to base point, there will be a contribution from these in addition to that from the tensor field's coefficients:

$$\nabla T = \nabla T^i{}_j\, \vec{e}_i \otimes \vec{e}^j + T^i{}_j\, \nabla \vec{e}_i \otimes \vec{e}^j + T^i{}_j\, \vec{e}_i \otimes \nabla \vec{e}^j . \tag{5.9}$$

One defines the so-called *connection coefficients* by decomposition of the gradient of a basis vector relative to the full basis:

$$\partial_k \vec{e}_i \stackrel{\text{def}}{=} \Gamma^j{}_{ik} \vec{e}_j . \tag{5.10}$$

Duality (Definition 5.3) then implies that

$$\partial_k \tilde{e}^j = -\Gamma^j{}_{ik} \tilde{e}^i . \tag{5.11}$$

With this definition we can rewrite the k-th component of ∇T (Equation 5.9) as

$$\partial_k T = \left\{ \partial_k T^i{}_j + \Gamma^i{}_{\mu k} T^\mu{}_j - \Gamma^\mu{}_{jk} T^i{}_\mu \right\} \vec{e}_i \otimes \tilde{e}^j . \tag{5.12}$$

For a general tensor (Definition 5.4) one simply gets one "correction term" per entry, with a + sign for each contravariant index (i.e. for each basis vector), and a − sign for each covariant index (each basis covector). In order to avoid confusion one usually reserves the symbol ∂_k for the old-fashioned partial derivative acting on a coefficient function, and writes D_k for the entire collection of terms making up the coefficients of the covariant derivative. The covariant derivative of a tensor of covariant rank q is itself a tensor, and has covariant rank $q+1$ (the contravariant rank is not affected).

Result 5.2 (Covariant Derivative of a Mixed Tensor)
The components of the covariant derivative of a mixed tensor are given by

$$D_k T^{i_1 \ldots i_p}_{j_1 \ldots j_q} = \partial_k T^{i_1 \ldots i_p}_{j_1 \ldots j_q} +$$
$$+ \Gamma^{i_1}_{l_1 k} T^{l_1 i_2 \ldots i_p}_{j_1 \ldots j_q} + \ldots + \Gamma^{i_p}_{l_p k} T^{i_1 \ldots i_{p-1} l_p}_{j_1 \ldots j_q} - \Gamma^{l_1}_{j_1 k} T^{i_1 \ldots i_p}_{l_1 j_2 \ldots j_q} + \ldots - \Gamma^{l_q}_{j_q k} T^{i_1 \ldots i_p}_{j_1 \ldots j_{q-1} l_q} .$$

The conclusion we have reached is that the components of the gradient of a tensor are in general not equal to the partial derivatives of the components of the tensor, but contain correction terms that compensate for the twisting, turning, expansion and contraction of the basic vectors and covectors. These additional terms are also known as "affinity terms", or "Γ-terms", and a set of connection coefficients $\Gamma^k{}_{ij}$ as an *affine connection*.

It is important to remember that an affine connection is *not* a tensor field; its coefficients depend upon the coordinate system. For this reason the affine connection is sometimes called a "gauge field". It is fairly easy to derive the following transformation rule for the affine connection.

Proposition 5.1 (Affinity Transformation Rule)
If $x' = x'(x)$ is a general coordinate transformation with Jacobian matrices as given by Equation 5.7, and if Γ^k_{ij} is an affinity given in the x representation, then in the x' coordinate system we have:

$$\Gamma'^k_{ij} = A^k{}_n B_i{}^l B_j{}^m \Gamma^n_{lm} + A^k{}_n X^n{}_{ij} \quad \text{with} \quad X^n{}_{ij} \stackrel{\text{def}}{=} \frac{\partial^2 x^n}{\partial x'^i \partial x'^j} .$$

The appearance of the inhomogeneous term on the r.h.s. shows the non-tensorial, coordinate dependent nature of the affinity.

Proof 5.1 (Proposition 5.1)
We suffice with a sketch of the proof. Start with a covector field $\tilde{\omega}$ (say). Its gradient $\nabla \tilde{\omega}$ is a cotensor of rank 2 and is given by Result 5.2 ($p = 0, q = 1$). The covariant derivative

of its coefficients, $D_k\omega_j$, contains exactly one affinity term. Transform and demand that the l.h.s., i.e. $D_k\omega_j$, satisfies the tensoriality criterion (Equation 5.6), apply the chain rule to the partial derivative term on the r.h.s., and mark the unknown, transformed connection coefficients by a prime. Solving for the latter then completes the proof.

Once we know the affinity in one coordinate system, we can calculate it in arbitrary coordinates by applying Proposition 5.1. Using Result 5.2 we can thus compute covariant derivatives of arbitrary tensors in any coordinate system. In particular one should now appreciate that $L_{i_1\ldots i_k}$ (Notation 3.1, Page 65) is nothing but the k-th order covariant image derivative: Problem 5.5.

An almost trivial, yet important result is that the metric is "covariantly constant".

Result 5.3 (Covariant Derivative of Metric Tensor)
Recall Equation 5.1. The metric is covariantly constant:

$$\nabla \mathbf{G} = 0 .$$

Proof 5.2
In a Cartesian frame, in which the components g_{ij} are constant, the components of $\nabla \mathbf{G}$ are $\partial_k g_{ij} = 0$. By its tensorial nature we must have $\nabla \mathbf{G} = 0$ in *any* coordinate frame.

If one looks at the covariant derivatives $D_k g_{ij}$ in an arbitrary coordinate system (three terms: the partial derivative, and two affinity terms), it is not immediately obvious that these should vanish identically. But they always do.

From Result 5.3 one can show that the affinity Γ_{ij}^k is symmetric w.r.t. i, j, and uniquely determined by the metric. One calls this unique form the *metric affinity*, or "connection compatible with the metric".

Definition 5.5 (Christoffel Symbols)
The so-called Christoffel symbol of the first kind, $[ij,k]$, is defined by

$$[ij,k] = \frac{1}{2} \left(\partial_i g_{jk} + \partial_j g_{ki} - \partial_k g_{ij} \right) .$$

The Christoffel symbol of the second kind, $\left\{ \begin{matrix} k \\ ij \end{matrix} \right\}$, is obtained by raising the third index of $[ij,k]$:

$$\left\{ \begin{matrix} k \\ ij \end{matrix} \right\} = g^{kl} \, [ij,l] .$$

Proposition 5.2 (Metric Affinity)
See Definition 5.5. There exists a unique connection compatible with the metric, or metric affinity Γ_{ij}^k:

$$\Gamma_{ij}^k = \left\{ \begin{matrix} k \\ ij \end{matrix} \right\} .$$

Proof 5.3
Although not difficult, we refer to the literature, e.g. Cartan [33], Lawden [181], Misner, Thorne and Wheeler [211], or Spivak [265, Volume II, Chapter 6].

Let us consider an example.

Example 5.5 (Covariant Derivatives and the Principle of Covariance)
Polar coordinates (Example 5.1) are frequently used curvilinear coordinates (with a relatively harmless singularity). In these coordinates, all Γ^k_{ij} vanish identically except $\Gamma^\phi_{r\phi} = \Gamma^\phi_{\phi r} = r^{-1}$, and $\Gamma^r_{\phi\phi} = -r$. If, for example, we want to write the Laplacian ΔL in polar coordinates, we may write it as $D_i D^i L$, and write out the covariant derivatives in terms of partial derivatives and affinity terms. This brings in a first order "correction term", viz. $\Gamma^\phi_{r\phi} L_r = r^{-1} L_r$:

$$D_i D^i L = L_{rr} + r^{-2} L_{\phi\phi} + r^{-1} L_r \, .$$

This is indeed the form one obtains by a straightforward application of the chain rule, starting from the familiar Cartesian expression $\Delta L = L_{xx} + L_{yy}$.

In non-Riemannian spaces, i.e. spaces without a metric, there is some leeway for different connections. We need not be concerned with this here; it suffices to know that in our case covariant derivatives commute with contractions, and that we can freely raise and lower indices in tensor expressions involving gradients without the risk of inconsistencies. For example, the divergence of a vector field, $\nabla \cdot \vec{v}$, having components $D_i v^i$, equals both $g^{ij} D_i v_j$ as well as $D_i(g^{ij} v_j)$. As a second example, consider the tensor $\vec{v} \otimes \vec{w}$, with components $v^i w^j$. Contraction before differentiation yields $D_k(g_{ij} v^i w^j)$, whereas contraction afterwards yields $g_{ij} D_k(v^i w^j)$. These would not be equal if it were not for Result 5.3.

5.2.5 ✳ Tensors on a Curved Manifold

Covariant derivatives can be defined in other than a Euclidean metric space, in which case they generally do not commute. A non-vanishing commutator indicates an intrinsic local curvature of the base manifold, captured by the so-called *Riemann curvature tensor*. It is defined as follows.

Definition 5.6 (Riemann Curvature Tensor)
The Riemann curvature tensor is defined by the following commutator:

$$R^i_{jkl}\omega_i = [D_l, D_k]\omega_j \, ,$$

for any covector ω_i.

Related to this are the *Ricci curvature tensor* $R_{ij} = R^k_{ijk}$ and the *Ricci curvature scalar* $R = R^i_i$.

Lemma 5.1 (Riemann Curvature Tensor)
See Definition 5.6. In terms of the metric affinity we have

$$R^i_{jkl} = \Gamma^i_{rk}\Gamma^r_{jl} - \Gamma^i_{rl}\Gamma^r_{jk} + \partial_k \Gamma^i_{jl} - \partial_l \Gamma^i_{jk} \, .$$

Proof 5.4
Left as an exercise to the reader: Problem 5.6.

It should be clear from Definition 5.6 and Lemma 5.1 that covariant derivatives do indeed commute in our case; the global vanishing of the Riemann tensor is the punctilio of Euclidean space. From this geometrical perspective, covariant derivatives are trivial, but we do need them for the sake of manifest covariance.

Example 5.6 (Flat Space)
By explicit computation one may verify that the Riemann curvature tensor expressed in the polar coordinates of Examples 5.1 and 5.5 indeed vanishes identically. The intelligent way to see this is to note that $R^l_{ijk} = 0$ is a tensor equation, hence valid in any coordinate system. Indeed, we can make the Γ^k_{ij} vanish globally simply by switching back to Cartesian coordinates, in which zero Riemann curvature boils down to commutativity of partial derivatives.

Problem 5.7 illustrates that a non-flat metric space is geometrically less trivial, yet technically equally straightforward as far as covariance is concerned. An instance of a curved manifold is the metrical scale-space proposed by Eberly [60], which is characterised by a "hyperbolic geometry" [16, 67].

5.2.6 The Levi-Civita Tensor

Before we turn to local image structure, we need one more tensor, the so-called *Levi-Civita tensor*. It is an instance of an *antisymmetric* tensor and is classically introduced in component form as follows.

Definition 5.7 (Levi-Civita Tensor)
Recall Definition 5.1. The Levi-Civita tensor $\varepsilon_{i_1 \ldots i_d}$ in d dimensions is defined as the unique, completely antisymmetric tensor

$$\varepsilon_{i_1 \ldots i_d} \overset{\text{def}}{=} \sqrt{g}\, [i_1 \ldots i_d]\,,$$

in which $[i_1 \ldots i_d]$ indicates the antisymmetric symbol

$$[i_1 \ldots i_d] = \begin{cases} +1 & \text{if } (i_1, \ldots, i_d) \text{ is an even permutation of } (1, \ldots, d) \\ -1 & \text{if } (i_1, \ldots, i_d) \text{ is an odd permutation of } (1, \ldots, d) \\ 0 & \text{otherwise.} \end{cases}$$

Lemma 5.2 (Dual of the Levi-Civita Tensor)
See Definition 5.7. We have

$$\varepsilon^{i_1 \ldots i_d} = \frac{1}{\sqrt{g}}\, [i_1 \ldots i_d]\,.$$

Proof 5.5
By definition, $\varepsilon^{i_1 \ldots i_d} = g^{i_1 j_1} \ldots g^{i_d j_d} \varepsilon_{j_1 \ldots j_d} = \sqrt{g}\, [j_1 \ldots j_d]\, g^{i_1 j_1} \ldots g^{i_d j_d}$. Note that if $(i_1, \ldots, i_d) = (1, \ldots, d)$, then the antisymmetric symbol together with the last d factors is just the determinant of the contravariant metric matrix, i.e. $1/g$. For any other permutation of indices (interchange of rows in the matrix) one gets a sign factor corresponding to the permutation. This completes the proof.

The Levi-Civita tensor has d^d components. By antisymmetry one can deduce all unspecified components from a single one ($\varepsilon_{1...d}$, say) by repeatedly exchanging index pairs and toggling signs; in particular any component labelled by two or more equal index values must vanish identically, leaving only $d!$ nontrivial entries, of which the only independent one is normalised to unity in a Cartesian frame.

Again, the classical way of introducing the Levi-Civita tensor, Definition 5.7, is a bit awkward. The motivation is that it is a tensor, and therefore represents something geometrically meaningful. Note that its components (!) are not affected by volume-preserving coordinate transformations. If this were not the case, Definition 5.7 would be inconsistent, since it does not tell us relative to which coordinate frame the components are supposed to be evaluated. A similar argument, by the way, holds for the *Kronecker tensor*, the components of which are given by the Kronecker symbol δ^i_j (Page 63) in *any* reference frame.

However, there is one subtle property of the ε-tensor not accounted for in the tensor transformation rule of Equation 5.6. In order for its definition to be truly coordinate independent, one must account for a minus sign in the case of a mirror transformation: in the notation of Equation 5.6, we have

$$\varepsilon'_{i_1...i_d} = \text{sgn}\,(\det B)\, B_{i_1}{}^{j_1} \ldots B_{i_d}{}^{j_d} \varepsilon_{j_1...j_d}\,. \tag{5.13}$$

(See Problem 5.8.) The sign factor expresses an intrinsic orientation property[6]. For this reason the ε-tensor is sometimes referred to as a *pseudo-tensor*.

In modern mathematical accounts the Levi-Civita tensor arises more naturally as the so-called "unit d-form of positive orientation", or, in physical terms, as the natural unit of reference for spatial densities. One can picture it as an oriented, closed surface enclosing a region of unit volume and arbitrary shape [158]. To avoid unnecessary overhead in this chapter we will not go into the details of this; the interested reader may want to consult Section A.2.6 of Appendix A, where the Levi-Civita tensor is introduced as the "Hodge star of unity[7]" [1, 18, 36, 158]:

$$\varepsilon \overset{\text{def}}{=} *1\,. \tag{5.14}$$

This concise statement is in fact the key to understand the relation between scalars and densities, and justifies the terminology "scalar image" even if a "density image" is understood: the two are isomorphically related by "Hodge duality", so that the distinction is merely a matter of interpretation. Equation 5.14 will become relevant in Chapter 6.

5.2.7 Relative Tensors and Pseudo Tensors

To appreciate tensors in all generality one has to extend their defining tensoriality criterion to account for so-called *even* and *odd relative tensors* We start by an example. One easily verifies (Problem 5.8) that

$$\sqrt{g'} = |\det B|\,\sqrt{g}\,, \tag{5.15}$$

---

[6]Of course $\text{sgn}\,(\det B) = \text{sgn}\,(\det A)$.
[7]The "Hodge star operator" $*$ is also simply referred to as the "star operator".

from which it is obvious that $\sqrt{g}$ is not a scalar field (0-tensor). It is in fact an instance of an *odd relative scalar of weight* 1, or an *odd scalar density*. The antisymmetric symbol[8] $[i_1 \ldots i_d]$ is an instance of an *even relative tensor of weight* -1: same formula as Equation 5.6, but now with an extra factor $(\det B)^{-1}$ on the r.h.s. (Problem 5.8). If one wishes to stress that no such determinant factors are involved one sometimes speaks of "absolute tensors" and "absolute scalars".

In general, **P** is called an *odd relative tensor of weight w* if

$$P'^{i_1\ldots i_l}{}_{j_1\ldots j_k} = |\det B|^w A^{i_1}{}_{p_1} \ldots A^{i_l}{}_{p_l} B_{j_1}{}^{q_1} \ldots B_{j_k}{}^{q_k} P^{p_1\ldots p_l}{}_{q_1\ldots q_k} , \qquad (5.16)$$

and an *even relative tensor of weight w* if instead of $|\det B|^w$ one finds a factor $(\det B)^w$ on the r.h.s.:

$$P'^{i_1\ldots i_l}{}_{j_1\ldots j_k} = (\det B)^w A^{i_1}{}_{p_1} \ldots A^{i_l}{}_{p_l} B_{j_1}{}^{q_1} \ldots B_{j_k}{}^{q_k} P^{p_1\ldots p_l}{}_{q_1\ldots q_k} . \qquad (5.17)$$

If one multiplies relative tensors, weights add up. If weights as well as even/odd signatures on left and right hand sides of a tensor equation match, it is equivalent to an ordinary (absolute) tensor equation, thus manifest covariance is guaranteed (cf. Problem 5.10).

In general one obtains an *absolute pseudo-tensor* by multiplying an even and an odd relative tensor with opposite weights (in the case of the ε-tensor this factorisation is apparent from Definition 5.7). It should not be difficult now to figure out what a *relative pseudo-tensor* means: Problem 5.9. *Absolute* and *relative pseudo-scalars* are defined accordingly.

In classical tensor calculus one encounters an important absolute pseudo-scalar that is often decomposed into relative scalars of opposite weights, the so-called "invariant volume element" (Problem 5.11),

$$dV \stackrel{\text{def}}{=} \sqrt{g}\, dx^1 \ldots dx^d , \qquad (5.18)$$

which is the natural integration measure in a volumetric integral (wherever the metric determinant has been left out, it is understood that one uses Cartesian coordinates).

Taking a covariant derivative of a relative tensor requires an additional affinity term not present in Result 5.2.

$$D_k P^{i_1\ldots i_p}_{j_1\ldots j_q} = [\text{r.h.s. of Equation 5.2 with } T \to P] - w\,\Gamma^i_{ik} P^{i_1\ldots i_p}_{j_1\ldots j_q} . \qquad (5.19)$$

One way to see this is to write $P = \sqrt{g}^w T$, taking the covariant derivative (Leibniz's product rule!) according to Result 5.2, and noting that the metric is covariantly constant. This brings in all the usual affinity terms of Result 5.2, with T simply replaced by P. The partial derivative term, however, is now $\sqrt{g}^w \partial_k T^{i_1\ldots i_p}_{j_1\ldots j_q}$, which equals $\partial_k P^{i_1\ldots i_p}_{j_1\ldots j_q} - \partial_k \sqrt{g}^w T^{i_1\ldots i_p}_{j_1\ldots j_q}$. It is not so difficult to show (Problem 5.13) that $\partial_k \sqrt{g}^w = w\Gamma^i_{ik}\sqrt{g}^w$, so that we conclude that there is a

---

[8]In Spivak's account, the Levi-Civita tensor is identified with this symbol, i.e. without the $\sqrt{g}$, thus making it an even relative tensor of weight -1; this is natural in a geometric context where no metric is implied.

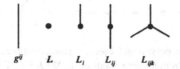

$$g^{ij} \qquad L \qquad L_i \qquad L_{ij} \qquad L_{ijk}$$

Figure 5.2: Metric tensor, filter and image derivatives, and scale-space polynomials all share a special property; they are *symmetric*. Thus it makes no difference if we interchange two indices. This allows us to make pictorial representations similar to the ones introduced for $P^{i_1 \cdots i_k}$ (Page 104 and Figures 4.8 and 4.9). This figure shows the diagrammatic representation of the dual metric tensor g^{ij}, and of the image derivatives $L_{i_1 \ldots i_k}$ for the lower orders $k = 0, 1, 2, 3$.

single extra term depending upon the weight w, indeed given by the last term in Equation 5.19.

A comprehensive account of tensor calculus including the generalisations in this section has been given by Spivak [265, Volume I, Chapter 4].

5.3 Differential Invariants

Let us now apply the established tensor formalism to the construction of differential invariants. We proceed as follows. In Section 5.3.1 we consider the construction of differential invariants from a technical viewpoint. It will be shown that there exists a straightforward procedure for constructing infinitely many differential invariants. In Section 5.3.2 we address the question of "structural sufficiency" and "necessity" of certain finite subsystems. We approach the same problem from an alternative perspective in Section 5.3.3. Finally we consider an extended invariance group in Section 5.3.4, in which we account for invariance under monotonic transformations of grey-values at fixed resolution.

5.3.1 Construction of Differential Invariants

The theory of tensors explained in Section 5.2, together with the diagrammar of Figure 5.2, greatly facilitates the construction of differential invariants. The basic principle is contraction.

Diagrammatically, a contraction is realized by connecting 2 external branches in a product of image vertices with the help of the (dual) metric element. The result of a complete contraction, which obviously requires an even number of external branches, is a scalar valued differential invariant. It is also called a *polynomial* (differential) *invariant*, since it is a polynomial in terms of image derivatives. Examples of this diagrammatic procedure are shown in Figures 5.3 and 5.4.

The examples in Figure 5.3 are relatively simple, and are all familiar from the image literature. The trivial diagram on the left, for instance, is just the zeroth order grey-value L, the simplest scalar one can think of (recall that scale and base point are implicit). The second one is the squared gradient magnitude $\|\nabla L\|^2$, the

Figure 5.3:   Some basic polynomial 2-jet invariants, explained in the text: L, $L_i L^i$, $L_i L_j^i L^j$, L_i^i and $L_{ij} L^{ij}$. In index-free notation: L, $\nabla^T L \nabla L$, $\nabla^T L \nabla \nabla^T L \nabla L$, $\triangle L$ and $\mathrm{tr}\,(\nabla \nabla^T L)^2$.

Figure 5.4:   Some higher order invariants: $L_{ijk} L^{ijk}$, $L_{ij}^i L_k^{jk}$ and $L_{ijk}^i L_l^{jl} L_m^k L^m$. For such complex invariants it becomes a cumbersome job to keep track of terms and co-variance if the summation convention were not used, even in 2D. But even the many indices used in this convention may be hard to keep track of, while there exists no viable alternative in terms of index-free matrix notation. The diagrammatic representation is particularly convenient here: manifest covariance is a plain consequence of the fact that we have a closed diagram.

simplest first order polynomial differential invariant. The one in the middle contains first and second order derivatives and denotes the second order directional derivative along the gradient direction, multiplied by the squared gradient magnitude; in matrix notation[9]: $\nabla^T L \nabla \nabla^T L \nabla L$, or, equivalently, $\mathrm{tr}\,(\nabla L \nabla^T L \nabla \nabla^T L)$. The fourth diagram equals the Laplacian $\triangle L$, the prototypical example of a pure second order invariant. Finally, on the right, we have the trace of the squared Hessian matrix $\mathrm{tr}\,(\nabla \nabla^T L)^2$. Unless differential order, in relation to inner scale, noise, grain and scope, becomes prohibitive [20], there is no reason to stop at second order or to consider trivial cases such as these. Figure 5.4 illustrates that the construction of polynomial differential invariants is technically as easy as putting together Lego building blocks. For example, the diagram on the right is a condensed description of what looks like an impossible expression when written out in Cartesian coordinates, even in the simple 2D case:

$$L_{ijk}^i L_l^{jl} L_m^k L^m = \tag{5.20}$$
$$(((((L_{xxxx} + L_{xxyy})(L_{xxx} + L_{xyy})) + ((L_{xxxy} + L_{xyyy})(L_{xxy} + L_{yyy})))L_{xx})$$
$$+((((L_{xxxy} + L_{xyyy})(L_{xxx} + L_{xyy})) + ((L_{xxyy} + L_{yyyy})(L_{xxy} + L_{yyy})))L_{xy}))L_x$$
$$+(((((L_{xxxx} + L_{xxyy})(L_{xxx} + L_{xyy})) + ((L_{xxxy} + L_{xyyy})(L_{xxy} + L_{yyy})))L_{xy})$$
$$+((((L_{xxxy} + L_{xyyy})(L_{xxx} + L_{xyy})) + ((L_{xxyy} + L_{yyyy})(L_{xxy} + L_{yyy})))L_{yy}))L_y.$$

(This expression, by the way, has been generated fully automatically from the l.h.s.; no reason to exhaust oneself!)

Clearly, any (not necessarily polynomial) function of polynomial invariants

---

[9]One interprets ∇L as a column vector, $\nabla^T L$ as a row vector, etc.

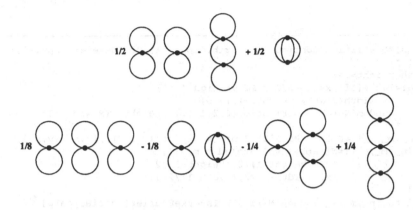

Figure 5.5: Diagrammatic representation of Equations 5.21 (top) and 5.22 (bottom).

is again an invariant. In fact, *any* differential invariant can be written as a function of polynomial invariants. Thus the set of polynomial invariants suffices to produce all other differential invariants. Figure 5.5 shows some more differential invariants, corresponding to the fourth order expressions

$$I = \frac{1}{2}L_{ij}^{ij}L_{kl}^{kl} - L_{ijk}^{i}L_{l}^{jkl} + \frac{1}{2}L_{ijkl}L^{ijkl} \tag{5.21}$$

and

$$J = \frac{1}{8}L_{ij}^{ij}L_{kl}^{kl}L_{mn}^{mn} - \frac{1}{8}L_{ij}^{ij}L_{klmn}L^{klmn} - \frac{1}{4}L_{ij}^{ij}L_{klm}^{k}L_{n}^{lmn} + \frac{1}{4}L_{ijk}^{i}L_{lm}^{jk}L_{n}^{lmn} , \tag{5.22}$$

i.e., in Cartesian coordinates,

$$I = L_{xxxx}L_{yyyy} - 4L_{xxxy}L_{xyyy} + 3L_{xxyy}L_{xxyy} \tag{5.23}$$

and

$$\begin{aligned}
J = \ & L_{xxxx}(L_{xxyy}L_{yyyy} - L_{xyyy}L_{xyyy}) - L_{xxxy}(L_{xxxy}L_{yyyy} - L_{xxyy}L_{xyyy}) + \\
& + L_{xxyy}(L_{xxxy}L_{xyyy} - L_{xxyy}L_{xxyy}) .
\end{aligned} \tag{5.24}$$

Figure 5.6 contains a self-contained Mathematica™ program[10] that implements the homogeneous, fourth order differential invariant $D = -I^3 + 27J^2$ in 2D, known as the *discriminant* of the fourth order binary form $L_{ijkl}\, x^i x^j x^k x^l$ [289].

Let us look into polynomial differential invariants in somewhat more detail. To begin with, we have to form a tensor product of covariant image derivatives, which must then be contracted so as to yield a closed diagram (no dangling branches). By construction there is a constraint on such diagrams.

---

[10]Mathematica™ is a registered trademark of Wolfram Research, Inc. See Wolfram [304] for details.

```
Needs["Statistics`ContinuousDistributions`"]; Off[General::spell1];
(* 1 *)
width=height=64;
data:=Table[If[(x<width/2+1 && y<height/2+1) ||
     (x>width/2 && y>height/2),10,0]+
     Random[NormalDistribution[0,2.1]],{y,height},{x,width}];
(* 2 *)
gd[n_,m_,s_]:=D[1/(2 Pi s^2) Exp[-(x^2+y^2)/(2 s^2)],{x,n},{y,m}];
gdk[n_,m_,s_]:=N[Table[Evaluate[gd[n,m,s]],
                {y,-(height-1)/2,(height-1)/2},
                {x,-(width-1)/2,(width-1)/2}]];
(* 3 *)
GD[data_,n_,m_,s_]:=Chop[N[Pi s^2 InverseFourier[Fourier[data]
                Fourier[RotateLeft[gdk[n,m,s],
                {height/2,width/2}]]]]];
(* 4 *)
scale=3;
Lyyyy=GD[data,0,4,scale]; Lxyyy=GD[data,1,3,scale];
Lxxyy=GD[data,2,2,scale]; Lxxxx=GD[data,4,0,scale];
Lxxxy=GD[data,3,1,scale];
(* 5 *)
discriminant = -(Lxxxx Lyyyy - 4 Lxxxy Lxyyy + 3 Lxxyy^2)^3 +
                27 (Lxxxx (Lxxyy Lyyyy - Lxyyy^2) -
Lxxxy (Lxxxy Lyyyy -
                Lxxyy Lxyyy) + Lxxyy (Lxxxy Lxyyy - Lxxyy^2))^2
(* 6 *)
ListDensityPlot[data,PlotRange->All]
ListDensityPlot[discriminant,PlotRange->All]
```

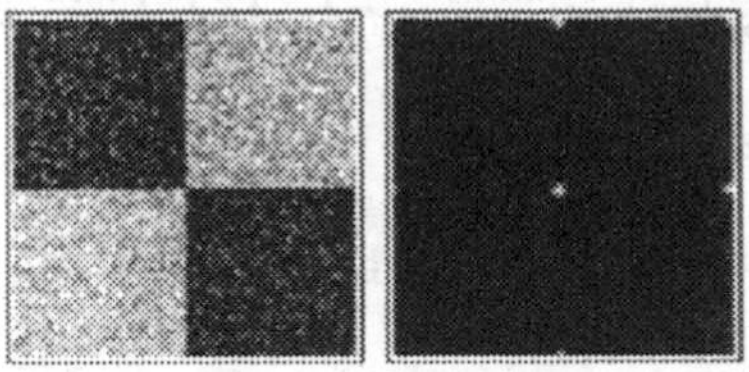

Figure 5.6: This program, written in Mathematica™, generates a 2D test image (bottom left) and computes the fourth order differential invariant $D = -I^3 + 27J^2$ (bottom right). Generalisation to arbitrary invariants or other programming languages is straightforward. The following steps can be distinguished: (1) generation of a 64×64 checkerboard pattern perturbed by pixel-uncorrelated Gaussian noise with standard deviation 2.1 (data); (2) definition of Gaussian derivatives (gd) and computation of discrete kernels (gdk); (3) expression for convolution product of initial data and filters by standard Fourier technique (GD); (4) evaluation of the fourth order derivatives at inner scale 3 that are actually needed (Lyyyy, Lxyyy, Lxxyy, Lxxxy, Lxxxx); (5) computation of the invariant discriminant; (6) visualisation of results. (Code fragment replicated by courtesy of Ter Haar Romeny [103].)

Proposition 5.3 (Constraints on Tensorial Diagrams)
A homogeneous polynomial (not necessarily connected) diagram, composed of $H \equiv \sum_k V_k$ vertices, where V_k is the number of k-vertices, having E external and I internal lines and C closed loops, satisfies the following constraints:

$$\begin{cases} C + H & = & I + 1 \\ B & = & E + 2I, \end{cases}$$

in which $B \equiv \sum_k kV_k$ is the total number of vertex branches. In particular we have $E = 0$ for a scalar diagram.

Proof 5.6 (Proposition 5.3)
Proposition 5.3 follows from a counting argument based on diagrammatic consistency: Problem 5.15.

Example 5.7 (Diagrammar)
A trivial example is an elementary k-vertex (Figure 5.2), for which we have $C = 0$, $V_k = 1$, all other $V_{l \neq k} = 0$, whence $H = 1$ and $B = k$. It is a tensor of rank $E = k$ without internal contractions: $I = 0$. Less trivial is the diagram on the right in Figure 5.4, corresponding to Formula 5.20. For this case we have, respectively, $C = 2$, $V_0 = 0$, $V_1 = 1$, $V_2 = 1$, $V_3 = 1$, $V_4 = 1$, all other $V_{k \geq 5} = 0$, whence $H = 4$ and $B = 10$, and $I = 5$. Furthermore it is a scalar, i.e. a tensor of rank $E = 0$. All these figures are consistent with Proposition 5.3.

The Levi-Civita tensor can be incorporated in the construction of invariants. In principle we need not include more than one ε-factor, because there is a statement that says that pairs of such factors can always be rewritten without reference to the Levi-Civita tensor.

Definition 5.8 (Generalised Kronecker Tensor)
The generalised Kronecker tensor of rank $2k$ ($k = 0, \ldots d$) is defined by:

$$\delta^{\nu_1 \ldots \nu_k}_{\mu_1 \ldots \mu_k} = \det A^{\nu_1 \ldots \nu_k}_{\mu_1 \ldots \mu_k},$$

in which the $k \times k$ matrix $A^{\nu_1 \ldots \nu_k}_{\mu_1 \ldots \mu_k}$ is given by:

$$A^{\nu_1 \ldots \nu_k}_{\mu_1 \ldots \mu_k} = \begin{pmatrix} \delta^{\nu_1}_{\mu_1} & \cdots & \delta^{\nu_k}_{\mu_1} \\ \vdots & & \vdots \\ \delta^{\nu_1}_{\mu_k} & \cdots & \delta^{\nu_k}_{\mu_k} \end{pmatrix}.$$

The determinant $\delta^{\nu_1 \ldots \nu_k}_{\mu_1 \ldots \mu_k}$ is an $k!$-sum of k-products of the fundamental Kronecker tensor, and so indeed defines a tensor of rank $2k$ (Problem 5.12). By construction, this tensor is antisymmetric with respect to its covariant and contravariant indices. For $k = 0$ we define $\delta = A = 1$ and for $k = 1$ we regain the familiar Kronecker tensor. In modern jargon one would define the generalised Kronecker tensor as the multilinear mapping

$$\delta_k : \underbrace{V \otimes \ldots \otimes V}_{k} \otimes \underbrace{V^* \otimes \ldots \otimes V^*}_{k} \to \mathbb{R} : \delta(\vec{v}_1, \ldots, \vec{v}_k; \tilde{\omega}^1, \ldots, \tilde{\omega}^k) \stackrel{\text{def}}{=} \det \omega[v],$$

(5.25)

in which $\omega[v]$ is the matrix with elements $\tilde{\omega}^i[\vec{v}_j]$; see e.g. Spivak [265, Volume I, Chapter 4].

Proposition 5.4 (Product of Levi-Civita Pseudo Tensors)
Recall Definition 5.8. A double product of ε-tensors can be written as a polynomial of δ-tensors:

$$\varepsilon_{\mu_1...\mu_k \lambda_1...\lambda_{d-k}} \varepsilon^{\nu_1...\nu_k \lambda_1...\lambda_{d-k}} = (d-k)!\, \delta^{\nu_1...\nu_k}_{\mu_1...\mu_k}\,.$$

Proof 5.7 (Proposition 5.4)
The proof is somewhat tedious and can be found in Appendix C.

A full contraction of a d-tensor onto the Levi-Civita tensor is also called an *alternation*. Tensors consisting of an odd number of ε-factors express properties characterised by an intrinsic orientation (a reversal of sign will occur after a mirror-transformation).

Example 5.8 (Differential Invariants Involving the Levi-Civita Tensor)
A 2D example of a polynomial invariant involving the Levi-Civita tensor is (in Cartesian coordinates)

$$X \stackrel{\text{def}}{=} \varepsilon^{ij} L_i L_{jk} L^k = (L_x^2 - L_y^2) L_{xy} - L_x L_y (L_{xx} - L_{yy})\,.$$

Since only one ε-tensor is involved it must express some second order differential image property with an intrinsic orientation. Indeed, one may verify that it corresponds to the curvature μ of the gradient integral curve at the point of interest, up to a factor $(L_i L^i)^{-3/2} = (L_x^2 + L_y^2)^{-3/2}$ (and possibly a conventional minus sign). Orientation is given by the bending of steepest descend lines (which is either clockwise or counterclockwise). A second example, involving two ε-factors, is

$$Y \stackrel{\text{def}}{=} \varepsilon^{ij} \varepsilon^{kl} L_i L_{jk} L_l = 2 L_x L_y L_{xy} - (L_x^2 L_{yy} + L_y^2 L_{xx})\,.$$

The interpretation is again facilitated if we divide it by the third power of the image gradient magnitude. We then obtain isophote curvature κ. However, since the number of ε-factors is now even, no intrinsic orientation is involved this time; with a myopic view limited to the isophote one cannot distinguish any preferred direction, hence no definite "spin". The pair of ε's can be removed by substitution according to Proposition 5.4:

$$\varepsilon_{ij} \varepsilon^{kl} = \delta_i^k \delta_j^l - \delta_j^k \delta_i^l\,,$$

which yields the apparently equivalent expression

$$Y = L_i L_j^i L^j - L_i L^i L_j^j\,.$$

The invariant Y has been thoroughly studied in the context of "corner detection" [19]: Figures 5.10–5.11. See also Llacer [199] for an application of isophote and flow-line curvature to medical imaging.

5.3.2 Complete Irreducible Invariants

A simple counting argument shows that even polynomial invariants for any given differential order are redundant; after all, one can always construct infinitely many polynomials of a finite number of variables. A nontrivial result obtained by Hilbert is the existence of a finite set of so-called *complete irreducible invariants*,

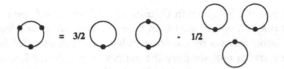

Figure 5.7: An example of reducibility in 2D: $L_{ij}L_k^jL^{ki} = 3/2\,L_i^iL_{jk}L^{jk} - 1/2\,L_i^iL_j^jL_k^k$. This can be most easily verified in a Cartesian frame in which (say) $L_{xx} = a, L_{xy} = 0, L_{yy} = b$, in which case it boils down to the algebraic identity $a^3 + b^3 = \frac{3}{2}(a+b)(a^2+b^2) - \frac{1}{2}(a+b)^3$. Note that this particular reducibility equation does *not* hold in higher dimensions.

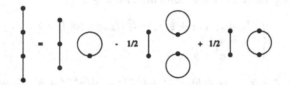

Figure 5.8: Another example of reducibility: $L_iL_j^iL_k^jL^k = L_i^iL_jL_k^jL^k - 1/2\,L_i^iL_j^jL_kL^k + 1/2\,L_{ij}L^{ij}L_kL^k$, again only valid in 2D. In the Cartesian frame described in the caption of Figure 5.7, in which the gradient is given by $L_x = p, L_y = q$, say, this boils down to the algebraic identity $a^2p^2 + b^2q^2 = (a+b)(ap^2+bq^2) - \frac{1}{2}(a+b)^2(p^2+q^2) + \frac{1}{2}(a^2+b^2)(p^2+q^2)$.

a subset of polynomial invariants in terms of which any other polynomial invariant can be expressed by polynomial combination (completeness, "structural sufficiency"), and which cannot be further reduced (irreducibility, "structural necessity") [117, 139, 148]. Figures 5.7 and 5.8 illustrate the possibility of reducing certain polynomial differential invariants in terms of simpler ones, a nontrivial game that has no obvious counterpart in Lego (notice the particular rational coefficients). In general, an irreducible set will contain *syzygies*, i.e. dependencies among its basic polynomial invariants that cannot be removed unless one forms non-polynomial combinations.

It is not easy to construct irreducible sets and their syzygies (if any) for the general case of arbitrary order and dimension. A brute-force method can however be used for the lower orders. For instance, if we restrict ourselves to the second order local jet, we can state the following proposition.

Proposition 5.5 (Complete Set of Irreducible 2-Jet Invariants)
A complete and irreducible system of polynomial 2-jet invariants in d dimensions is given by all connected diagrams built out of 0-,1- and 2-vertices with at most d internal and no external lines. There are no syzygies.

Note that Figure 5.3 is just this irreducible set for the 2D case.

Proof 5.8 (Proposition 5.5)
The brute-force method makes use of algebraic identities for the factorisation of polyno-

mials, similar to the examples given in Figures 5.7 and 5.8, which reveal the reducibility properties of most polynomial invariants in terms of a small set of elementary ones. Irreducibility of the latter follows by counting independent degrees of freedom. A smart choice of coordinate frame will simplify the expressions without loss of generality. An alternative proof, based on the so-called *Cayley-Hamilton theorem*, is given in Appendix D.

Example 5.9 (Reducibility)
Let I_k be the pure second order loop diagram with k 2-vertices and S_k the mixed chain diagram with k 2-vertices sandwiched between two 1-vertices. According to Proposition 5.5 I_k is reducible if $k > d$ and S_k is reducible if $k > d - 1$. Consider the 2D case for the sake of illustration ($d = 2$).

By diagonalising the Hessian it follows that, for $k > 2$,

$$\begin{cases} I_k = I_1 I_{k-1} + \frac{1}{2}(I_2 - I_1^2)I_{k-2} & \text{for } k > 2 \\ I_1 = a + b \\ I_2 = a^2 + b^2 \end{cases}$$

This follows from $a^k + b^k = (a+b)(a^{k-1} + b^{k-1}) - ab(a^{k-2} + b^{k-2})$, with $ab = \frac{1}{2}((a+b)^2 - (a^2 + b^2))$.
Some examples:

- $I_3 = \frac{3}{2}I_1 I_2 - \frac{1}{2}I_1^3$ (Figure 5.7);
- $I_4 = \frac{1}{2}I_2^2 + I_1^2 I_2 - \frac{1}{2}I_1^4$;
- $I_5 = \frac{5}{4}I_1 I_2^2 - \frac{1}{4}I_1^5$;
- in general: $I_k = \sum_{i+2j=k} \alpha_{ij} I_1^i I_2^j$, i.e. a polynomial in I_1 and I_2 in which the individual terms are given, apart from some numerical factor α_{ij}, by all possible disconnected diagrams with exactly k vertices that can be formed by multiplying the irreducible ones. The coefficients α_{ij} add up to unity.

The restriction on the indices in the above sum is a homogeneity constraint on the intensity function; only terms homogeneous of degree k contribute. Finally, consider the reducibility of the mixed invariants S_k for $k > d - 1 = 1$:

$$\begin{cases} S_k = I_1 S_{k-1} + \frac{1}{2}(I_2 - I_1^2)S_{k-2} & \text{for } k > 1 \\ S_0 = p^2 + q^2 \\ S_1 = ap^2 + bq^2 \end{cases}$$

Again this follows by decomposition: $a^k p^2 + b^k q^2 = (a^{k-1}p^2 + b^{k-1}q^2)(a+b) - (a^{k-2}p^2 + b^{k-2}q^2)ab$, with $ab = \frac{1}{2}((a+b)^2 - (a^2 + b^2))$, for $k > 1$. Examples:

- $S_2 = I_1 S_1 + \frac{1}{2}(I_2 - I_1)^2 S_0$ (Figure 5.8);
- $S_3 = \frac{1}{2}(I_1^2 + I_2)S_1 + \frac{1}{2}(I_1 I_2 - I_1^3)S_0$;
- $S_4 = I_1 I_2 S_1 + \frac{1}{4}(I_2^2 - I_1^4)S_0$;
- in general: $S_k = \sum_{i+2j+2p+3q=k+2} \alpha_{ijpq} I_1^i I_2^j S_0^p S_1^q$ with numerical coefficients α_{ijpq} that add up to unity.

Again the sum above extends over only those combinations that satisfy the intensity homogeneity constraint (of degree $k + 2$).

The conclusion is that the second order differential structure of an image is completely captured by the invariants given in Proposition 5.5. If we leave out any member of the irreducible set we are discarding information. This observation is particularly noteworthy for *feature detection*, such as "edge detection", "corner detection", and the extraction of "zero-crossings". The vastness of the literature on the subject bears witness to the fact that it is considered of great interest in image analysis and computer vision (and precludes the possibility of an exhaustive list; the reader is advised to search for typical keywords).

It is common practice to associate a local feature with some kind of local neighbourhood operator or differential invariant (some familiar instances of which can be recognised directly in the representation of the second order irreducible system, Proposition 5.5). Since such an operator or invariant can be designed at liberty, this essentially boils down to the vicious circularity that "a feature is the output of a feature detector", whereas the latter must be designed such as to detect the former. This implies that features cannot be mutually compared, except in a semantical context, viz. by their degree of support to the performance of a practical application or by reference to a postulated set of desirable criteria.

That this is intuitively not quite satisfactory is manifest in our use of a very limited vocabulary of distinguished features ("edges", "corners", "bars", "blobs"; this is not far from an exhaustive summary of what one encounters in practice). For this reason Koenderink proposes to identify the notion of a feature with an *equivalence class* on a local jet. The number of distinct equivalence classes naturally depends upon differential order, i.e. the level of scrutiny with which we wish to resolve local image structure. Ambiguity is resolved by virtue of completeness of the filter ensemble needed to represent the local jet up to some specified order. In other words, there is a complete set of equivalence classes associated with a complete set of filters. This is an intuitive and truly syntactical definition of a "feature"; the only semantics involved is the characterisation of the equivalence classes of source field configurations that induce identical representations when sampled by means of a complete set of detectors, for instance in the form of the irreducible system of Proposition 5.5 in the second order case, or in the form of the output of the underlying Gaussian filter derivatives *modulo* rotation. Irrespective the representation of the local jet, the metameric classes are invariably the same, which shows that a consistent (local) feature definition is indeed possible. For details the reader is referred to the literature [161].

5.3.3 Gauge Coordinates

Although a constructive method to form complete systems of irreducible invariants appears to exist [117], it is much easier to construct complete systems of non-polynomial differential invariants following an alternative approach, based on so-called *gauge coordinates*. The idea is to exploit knowledge of local image structure in order to select a particular Cartesian coordinate frame[11] ("gauge fixing"). The directional derivatives relative to such an intrinsic frame constitute a natural basis of non-polynomial differential invariants.

---

[11] I.e. gauge coordinates are not curvilinear, but refer to a pointwise selected Cartesian reference frame.

To get the gist of it, let us consider the 2D case. An admissible choice of gauge is to align one coordinate axis, the y-axis say, with the image gradient; the x-axis is then fixed (up to an optional orientation of basis; of course the resulting frame varies from point to point in the image plane). This implies that the directional derivative along the y-direction produces the magnitude of the image gradient, while the orthogonal x-derivative vanishes identically.

To avoid confusion with arbitrary Cartesian coordinates we shall henceforth denote these "gradient gauge" coordinates by (v, w) instead of (x, y). That is, at the origin of the local frame we have

$$L_v \stackrel{\text{def}}{=} 0 . \tag{5.26}$$

One easily verifies that, in manifest covariant notation,

$$\partial_v = \frac{\varepsilon^{ij} L_j}{\sqrt{L_k L^k}} D_i \quad \text{and} \quad \partial_w = \frac{g^{ij} L_j}{\sqrt{L_k L^k}} D_i , \tag{5.27}$$

in which all quantities are evaluated at the origin, and in which D_i denotes, as usual, the covariant derivative operator. In particular we have, in an arbitrary Cartesian frame,

$$\begin{pmatrix} \partial_v \\ \partial_w \end{pmatrix} = \begin{pmatrix} \cos \alpha & -\sin \alpha \\ \sin \alpha & \cos \alpha \end{pmatrix} \begin{pmatrix} \partial_x \\ \partial_y \end{pmatrix} , \tag{5.28}$$

expressing a rotation of the frame over an angle α with $\cos \alpha = L_y / \|\nabla L\|$. Since a rotation matrix equals its inverse-transpose, a similar relation holds for the covector basis (dv, dw) relative to (dx, dy), and also for the gauge coordinates (v, w) relative to (x, y), etc. First order gauge coordinates (v, w) are almost everywhere well-defined. The only problematic points are critical points, where the gradient is zero.

If, by way of example, we impose this gauge on the differential invariant Y of Example 5.8, we find $Y = -L_{vv} L_w^2$. This immediatley shows that it represents a trade-off between "edge strength" L_w and isophote curvature $-L_{vv}/L_w$ (the derivation of which will be given in Example 5.12 on Page 166). Thus it is indeed something one could interpret as "corner strength". One of Blom's motivations [19] for this particular combination was—as one readily verifies—that it is invariant under area preserving affine transformations.

Other gauges are possible. The "Hessian gauge" is a second order gauge obtained by aligning a local Cartesian frame with second order image structure, in such a way that the mixed second order derivatives vanish. Denoting the corresponding gauge coordinates in 2D by (p, q) we thus require

$$L_{pq} \stackrel{\text{def}}{=} 0 . \tag{5.29}$$

This choice of gauge is always admissible, since the Hessian is symmetric, but the corresponding frame is ambiguous if the eigenvalues of the Hessian happen to coincide ("umbilical points"). If $L_{xy} \neq 0$ in our Cartesian frame of choice, then we need to perform a rotation

$$\begin{pmatrix} \partial_p \\ \partial_q \end{pmatrix} = \begin{pmatrix} \cos\beta & -\sin\beta \\ \sin\beta & \cos\beta \end{pmatrix} \begin{pmatrix} \partial_x \\ \partial_y \end{pmatrix}, \qquad (5.30)$$

with $\tan 2\beta = L_{xy}/(L_{yy} - L_{xx})$, in order to realize the Hessian gauge.

Gauge coordinates facilitate the construction of complete systems of (possibly pseudo) differential invariants.

Proposition 5.6 (Complete Sets of (Pseudo) Invariants in 2D)
A complete set of (pseudo) invariants for the local k-jet in 2D is given in (v, w)-gauge by the set

$$G = \left\{ \frac{\partial^{m+n} L}{\partial v^m \partial w^n} \right\}_{m+n \leq k}.$$

Alternatively, in (p, q)-gauge,

$$H = \left\{ \frac{\partial^{m+n} L}{\partial p^m \partial q^n} \right\}_{m+n \leq k}.$$

The cases $(m, n) = (1, 0)$ and $(m, n) = (1, 1)$ are trivial for the respective choices of gauge.

Similar gauges can be used in higher dimensions. E.g. in 3D we may introduce gauge coordinates (u, v, w) by the gauge conditions $L_u = L_v = L_{uv} = 0$ ($L_w = \|\nabla L\|$, (u, v, w) positively oriented), or gauge coordinates (p, q, r) associated with a pure second order gauge $L_{pq} = L_{pr} = L_{qr} = 0$, etc. More generally, in a d-dimensional image we may introduce gauge coordinates $(u^1, \ldots, u^{d-1}, w)$ by the $d-1$ gauge conditions $L_{u^a} = 0$ ($a = 1, \ldots, d-1$) and the $\frac{1}{2}(d-1)(d-2)$ gauge conditions $L_{u^a u^b} = 0$ ($a \neq b = 1, \ldots, d-1$), making up a total of $\frac{1}{2}d(d-1)$ equations, i.e. the same as for a system of gauge coordinates $(p_1, \ldots, p_d)$ associated with a pure second order gauge $L_{p_i p_j} = 0$ ($i = 1, \ldots, d$).

In the next section gauge coordinates will turn out to be quite convenient.

5.3.4 Geometric or Grey-Scale Invariants

An interesting subclass of differential invariants arises if one imposes one more invariance constraint in addition to that induced by the classical group, viz. defined by the *group of general grey-scale transformations*. This group acts on the image's grey-scales through arbitrary, strictly monotonic transformations (note that without monotonicity it would not be a group):

$$\mathcal{T} : \mathbb{R} \to \mathbb{R} : L \mapsto \overline{L} \stackrel{\text{def}}{=} \mathcal{T}(L) \quad \text{with } \mathcal{T}'(L) > 0 \text{ (say) for all } L. \qquad (5.31)$$

One can think of several reasons why this group is of interest in perception, machine vision, and image analysis. For instance, it is well-known that nonlinear transformations of the retinal irradiance distribution (if not pushed to the extreme) have little or no effect on recognition competences. "Gamma corrections" and "histogram equalisations" are frequently employed to obtain a satisfactory rendering of an image for the purpose of display; the ultimate justification for doing so must lie in grey-scale invariance (at least, of the *relevant* structures). Shading variations that occur under everyday circumstances are to a significant extent accounted for by nonlinear grey-scale transformations. In the context of medical imaging, finally, after having registered two images from different modalities ("multimodality matching"), one may often account for a large part of their difference by means of such a transformation (put differently, the component *not* accounted for in this way expresses an *essential* difference in information content).

It is important to note that nonlinear grey-scale transformations do not commute with Gaussian blurring; when carried out at some inner scale such transformations have nontrivial consequences for the entire deep structure. For this reason we will, as before, concentrate on a fixed slice of the image's scale-space, without making the scale explicit in the notation.

The aim is to find a complete set of differential grey-scale invariants in any dimension and to any order. Clearly, the relevant geometry in this context is that of the image's iso-grey-level contours (isophotes) and their nesting hierarchy. Consequently the invariants of interest must provide a complete, local characterisation of the geometry of these contours[12].

That the concept of a local jet remains a meaningful one in the context of Equation 5.31 can be seen as follows.

Proposition 5.7 (Transformation of Spatial Derivatives)
Let $\bar{L}$ denote the image obtained by applying a grey-scale transformation to L (Equation 5.31). Then we have, for all $m \geq 1$:

$$\frac{1}{m!}\bar{L}_{i_1\ldots i_m} = \sum_{q=1}^{m} \frac{1}{q!} \mathcal{T}^{(q)}(L) \sum_{l_1\geq 1}^{\star} \cdots \sum_{l_q\geq 1}^{\star} S \circ \Lambda \left[\prod_{k=1}^{q} \frac{1}{l_k!} L_{i_1^k\ldots i_{l_k}^k} \right],$$

in which the q inner summations are restricted to indices that add up to m, i.e.

$$\star : \sum_{p=1}^{q} l_p = m,$$

S denotes the index symmetrisation operator, and Λ is the lexicographical index ordering operator that "unravels" or "flattens" a doubly-indexed list $\{a_l^k\}_{l=1,\ldots,l_k}^{k=1,\ldots,q}$ into a singly-indexed list $\{a_i\}_{i=1,\ldots,m}$ of length $m = l_1 + \ldots + l_q$. In this procedure, the upper index k takes lexicographical precedence over the lower index l. In other words, Λ concatenates the q rows labelled by k, the k-th row of which has l_k entries:

$$\Lambda \left[\{a_l^k\}_{l=1,\ldots,l_k}^{k=1,\ldots,q} \right] = \{a_i\}_{i=1,\ldots,m} \quad \text{with} \quad i = l + \sum_{p=1}^{k-1} l_p.$$

---

[12]One could alternatively study their orthogonal trajectories, the gradient integral curves.

Proof 5.9 (Proposition 5.7)
The proof of Proposition 5.7 relies entirely on the chain rule and Leibniz's product rule, and follows by induction with respect to m.

This is a bit of a mess, but there is no reason to panic (yet). It helps to write out some lowest order results, either by inspection of Proposition 5.7, or by calculation from scratch.

Example 5.10 (Some Lowest Order Results)
Up to order four we have the following relationship between the derivatives before and after transformation:

$$
\begin{aligned}
\overline{L}_i &= L_i \mathcal{T}'(L), \\
\overline{L}_{ij} &= L_{ij} \mathcal{T}'(L) + L_i L_j \mathcal{T}''(L), \\
\overline{L}_{ijk} &= L_{ijk} \mathcal{T}'(L) + [L_{ij} L_k + L_{jk} L_i + L_{ki} L_j] \mathcal{T}''(L) + L_i L_j L_k \mathcal{T}'''(L), \\
\overline{L}_{ijkl} &= L_{ijkl} \mathcal{T}'(L) + [L_{ijk} L_l + L_{ijl} L_k + L_{jkl} L_i + L_{ij} L_{kl} + L_{ik} L_{jl} + L_{il} L_{jk}] \mathcal{T}''(L) \\
&\quad + [L_{ij} L_k L_l + L_{ik} L_j L_l + L_{il} L_j L_k + L_{jk} L_i L_l + L_{jl} L_i L_k + L_{kl} L_i L_j] \mathcal{T}'''(L) \\
&\quad + L_i L_j L_k L_l \mathcal{T}''''(L).
\end{aligned}
$$

Some patterns can be discerned: to transform a k-th order derivative it apparently suffices to know the derivatives of orders less than or equal to k as they were prior to transformation (as well as, of course, information concerning the transformation itself). That all lower orders are involved while higher ones are irrelevant suggests that the structural entities of interest are indeed local jets, not derivatives as such.

The transformation of a local jet is nevertheless complicated. This is not surprising, since it pertains to the structure of the image's grey-scale landscape; its relation to the induced isophote topography is at best indirect. A special case arises if we consider the affine subgroup of Equation 5.31, $\overline{L} = \alpha L + \beta$, in which case things become much simpler; in particular it is then straightforward to construct differential invariants and invariant relations.

Whereas the local image jet introduced in Section 4.2 is a convenient descriptor by virtue of its operational nature (as it can be obtained through plain linear filtering), the observations made above argue for a more natural one that directly captures the order-by-order local structure of an isophote. If one imagines walking around in its tangent plane at a given point x, one can describe the isophote by a height map $w(u)$, in which the $d - 1$ coordinates $u = (u^1, \ldots, u^{d-1})$ parametrise the ground plane, and w indicates elevation relative to this: Figure 5.9. The coordinates $(u; w)$ thus define a local frame at x subject to a *partial gradient gauge*: the w-axis is aligned with the image gradient (the isophote's normal), but the remaining $d - 1$ coordinates u are left unconstrained. Consequently there is a residual gauge degree of freedom in the tangent plane, which is a (scale invariant) Euclidean subspace of dimension $d - 1$. It is therefore natural to employ tensors confined to this space in order to describe its geometrical objects, notably the *local isophote jet*.

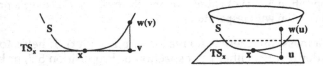

Figure 5.9: Local Monge patch parametrisation $w : \mathbb{R}^{d-1} \to \mathbb{R} : u \mapsto w(u)$ of an isophote S relative to its tangent plane TS_x at a regular point x, in this case $d = 2$ (left) and $d = 3$ (right). The function $w(u)$ can be interpreted as a "height map" for S at all points $(u; 0) \in \mathbb{R}^d$ in the tangent plane through x.

Definition 5.9 (Local Isophote Jet)

Let $S : \mathbb{R}^{d-1} \to \mathbb{R}^d : u \mapsto \phi(u)$ be a Monge patch parametrisation of the isophote surface S through a regular point as depicted in Figure 5.9:

$$\phi : \mathbb{R}^{d-1} \to \mathbb{R}^d : \begin{cases} \phi^a(u) &= u^a \quad (a = 1, \ldots, d-1) \\ \phi^d(u) &= w(u) \end{cases} .$$

Furthermore, let $w_{a_1 \ldots a_k}$ be the tensor of rank k formed by the k-th order partial derivatives of w w.r.t. partial gradient gauge coordinates $u^{a_1}, \ldots, u^{a_k}$ ($a_j = 1, \ldots, d-1$ for each $j = 1, \ldots, k$). Then the local jet of order m for the image's isophote S at base point x, having Monge patch parameters $u = 0$ in the local frame, can be represented by the set of all k-tensors $w_{a_1 \ldots a_k}$ up to rank m (inclusive), evaluated at that point:

$$j^m S(x) = \{ w_{a_1 \ldots a_k}(0) \mid k = 1, \ldots, m \} .$$

Completeness of the representation of Definition 5.9 follows from the fact that, in a full neighbourhood of the point of interest, the isophote is completely characterised by its Taylor series:

$$w(u) = \sum_{k=1}^{m} \frac{1}{k!} w_{a_1 \ldots a_k}(0) \, u^{a_1} \ldots u^{a_k} + \mathcal{O}(\|u\|^{m+1}) . \tag{5.32}$$

(It is understood that the Einstein summation applies to the a_j-indices relating to TS_x.)

We have arrived at a definition of geometric or grey-scale invariant local image structure (Definition 5.9), but its relation to the local image jet (Section 4.2) remains to be established. More specifically, we would like to have the tensors $w_{a_1 \ldots a_k}$ expressed in terms of $L_{i_1 \ldots i_m}$. A natural way to proceed is by taking derivatives along the isophote surface, i.e. total derivatives of L given the constraint $L = const$. These are most easily evaluated using the Monge patch parametrisation in the partial gradient gauge.

Corollary 5.1 (Implicit Differentiation along Isophotes)

Notation as in Definition 5.9. Define the k-tensor $L_{a_1...a_k}^{(j)}$ as the $(j+k)$th order derivative w.r.t. the local, partial gradient gauge coordinates $u^{a_1},\ldots,u^{a_k}$ and w, respectively Furthermore, let $\delta_{a_1...a_m} L$ denote the m-th order total derivative of L along the isophote surface induced by parameter variations $\delta u^{a_1},\ldots,\delta u^{a_m}$, then we have

$$\frac{1}{m!}\delta_{a_1...a_m} L = S \circ \Lambda \left[\sum_{q=1}^{m} \sum_{k=0}^{q} \frac{1}{k!} \frac{1}{(q-k)!} L_{a_1^0...a_{q-k}^0}^{(k)} \overset{\star}{\sum_{l_1 \geq 1}} \cdots \overset{\star}{\sum_{l_k \geq 1}} \prod_{p=1}^{k} \frac{1}{l_p!} w_{a_1^p...a_{l_p}^p} \right],$$

in which the k l_p-summations $(p = 1,\ldots,k)$ are subject to the constraint

$$\star : \quad q - k + \sum_{p=1}^{k} l_p = m,$$

Λ is the lexicographical index ordering operator as introduced in Proposition 5.7, and S denotes spatial index symmetrisation. By convention, if $k = 0$ the p-sum in $\star$ vanishes identically, and the k inner summations over l_p evaluate to unity.

Example 5.11 (Some Lowest Order Cases)

Up to order three we have (the indices a, b, c refer to the components of $u \in \mathbb{R}^{d-1}$)

$$\begin{aligned}
\delta_a L &= L_a + L_w w_a\,, \\
\delta_{ab} L &= L_{ab} + L_{aw} w_b + L_{bw} w_a + L_{ww} w_a w_b + L_w w_{ab}\,, \\
\delta_{abc} L &= L_{abc} + L_{abw} w_c + L_{acw} w_b + L_{bcw} w_a + L_{aww} w_b w_c + L_{bww} w_a w_c + L_{cww} w_a w_b \\
&\quad + L_{aw} w_{bc} + L_{bw} w_{ac} + L_{cw} w_{ab} + L_{ww} w_{ac} w_b + L_{ww} w_{bc} w_a + L_{ww} w_{ab} w_c \\
&\quad + L_{www} w_a w_b w_c + L_w w_{abc}\,.
\end{aligned}$$

We have the following *generating equation* for the grey-scale invariant tensors $w_{a_1...a_k}$:

$$\delta_{a_1...a_m} L = 0 \quad \text{for all } m \geq 1. \tag{5.33}$$

Combined with Corollary 5.1 this gives us an inductive scheme for the local isophote jet in terms of the local image jet.

Result 5.4 (Inductive Scheme for Local Isophote Jet)

The k-tensors $w_{a_1...a_k}$ $(k = 1,\ldots,m)$ are determined inductively by

$$L_w w_{a_1...a_m} =$$

$$-L_{a_1...a_m} - S \circ \Lambda \left[\underset{(q,k)\neq(1,1)}{\sum_{q=1}^{m} \sum_{k=1}^{q}} \frac{1}{k!} \frac{1}{(q-k)!} L_{u_{a_1}^0...u_{a_{q-k}}^0}^{(k)} \overset{\star}{\sum_{l_1 \geq 1}} \cdots \overset{\star}{\sum_{l_k \geq 1}} \prod_{p=1}^{k} \frac{1}{l_p!} w_{a_1^p...a_{l_p}^p} \right],$$

in which the k l_p-summations $(p = 1,\ldots,k)$ are subject to the same constraint $\star$ as in Corollary 5.1, and in which S and Λ are defined as before.

Proof 5.10 (Result 5.4)
Set the l.h.s. of the equation in Corollary 5.1 equal to zero, and solve for the $(q = 1, k = 1)$-term, i.e. the only term containing the m-th order derivative $w_{a_1...a_m}$, which is found to be multiplied by a factor L_w. The $(q = m, k = 0)$-term, the only effective one in the $(q, k = 0)$-sum, has been made explicit here.

Example 5.12 (Isophote Curvature in 2D)
Consider a point on an isophote in a 2D image, defined (generically a.e.) as a contour of constant grey-value, $L = const$. Impose the (v, w)-gauge: $(L_v, L_w) = (0, \|\nabla L\|)$ at the point of interest, $(v, w) = (0, 0)$ say. The isophote is described by a function defining the w-coordinate in terms of v. Taking the (total) first and second order derivatives of the defining equation yields

$$L_w w' + L_v = 0 \quad \text{and}$$
$$L_{ww} w'^2 + L_w w'' + 2L_{vw} w' + L_{vv} = 0,$$

in which a prime denotes d/dv. At the origin, where the (v, w)-gauge holds, this immediately yields $w' = 0$ and $w'' = -L_{vv}/L_w$. But in the case of zero slope ($w' = 0$), w'' is just the curvature κ of the isophote.

Although we have exploited properties that only hold in a special frame, there is no loss of generality, for we can use Equation 5.27 to find the expression for isophote curvature in any other frame (cf. Example 5.8):

$$\kappa = \frac{\varepsilon^{ij} \varepsilon^{kl} L_i L_{jk} L_l}{(L_p L^p)^{3/2}}.$$

Written out in an arbitrary Cartesian frame, this reads

$$\kappa = \frac{2 L_x L_y L_{xy} - (L_x^2 L_{yy} + L_y^2 L_{xx})}{(L_x^2 + L_y^2)^{\frac{3}{2}}}.$$

The derivation would have been much more involved if we had tried to arrive at this result *directly*!

One will perhaps appreciate the power of gauge coordinates better from the following example in 3D.

Example 5.13 (Isophote Curvature in 3D)
In 3D, an isophote is (generically a.e.) a surface. Its lowest order deviation from its tangent plane can be expressed by two independent invariants, the *principal curvatures* or, alternatively, the *mean* and the *Gaussian curvature*. We obtain a curve on this isophote surface by taking a normal section, i.e. the intersection with a plane containing the surface normal (the image gradient). While rotating around this normal we can measure the curvature of the resulting curve as a function of the rotation angle ("Dupin's indicatrix"). One then encounters two extrema in mutually perpendicular directions: these are the principal curvatures. Degeneracies may occur at isolated points ("umbilical points" or "flat points"). The (u, v, w)-gauge defined on Page 161 is well-defined, except at these degeneracies. (Details: any introductory book on differential geometry of curves and surfaces, e.g. Lipschutz [198].)

To derive mean and Gaussian curvature, we make the following two observations. The first is that in (u, v, w)-gauge the u and v axes are tangent to the principal directions (Problem 5.18). Secondly, if this is indeed true, then it is clear from the previous 2D example that the principal curvatures are given by $\kappa_1 = -L_{uu}/L_w$ and $\kappa_2 = -L_{vv}/L_w$. Mean curvature h and Gaussian curvature k are defined as their average and product, respectively: $h = \frac{1}{2}(\kappa_1 + \kappa_2), k = \kappa_1\kappa_2$.

Here is a "hack" to find the corresponding expressions in general coordinates; write down a rational invariant with the appropriate degrees of homogeneity for first and second order derivatives (e.g. for the principal curvatures: -1 and $+1$, respectively). One can take a guess—with a modest amount of foresight—among the few simplest second order invariants one can imagine that meet the appropriate homogeneity constraints:

$$h = -\frac{1}{2}\frac{\varepsilon^{ijk}\varepsilon^{lmn}L_iL_lL_{jm}g_{kn}}{(L_pL^p)^{3/2}} \quad \text{and} \quad k = \frac{1}{2}\frac{\varepsilon^{ijk}\varepsilon^{lmn}L_iL_lL_{jm}L_{kn}}{(L_pL^p)^2}.$$

Indeed, evaluating the expressions in gauge coordinates significantly reduces the number of effective terms and precisely yields the expressions that hold in the (u, v, w)-gauge: Problem 5.21.

In Examples 5.12 and 5.13 (notably Problem 5.21) we have used the full gradient gauge, and related the resulting expressions to ones that hold in an arbitrary frame [82, 212].

It may be noted that Result 5.4 actually specifies the *relative grey-scale invariants* $L_w\,w_{a_1...a_m}$. The attribute "relative" should now be interpreted having in mind the group of grey-scale transformations, i.e. the definition is similar to that in the context of the scale-Euclidean group (Equations 5.16–5.17), but now a Jacobian factor shows up after a grey-scale transformation (Equation 5.31), which is some power of $\mathcal{T}'(L)$.

Proposition 5.8 (Relative Grey-Scale Invariants)
Let $w_{a_1...a_m}$ be as in Definition 5.9. Then the tensors $w^{[p]}_{a_1...a_m}$, defined by

$$w^{[p]}_{a_1...a_m} = L^p_w\,w_{a_1...a_m},$$

are relative grey-scale invariants of weight p. This means that, if $\overline{L} = \mathcal{T}(L)$, then

$$\overline{w}^{[p]}{}_{a_1...a_m} = (\mathcal{T}'(L))^p\,w^{[p]}_{a_1...a_m}.$$

Proof 5.11 (Proposition 5.8)
This follows from the fact that L_w is a relative $\mathcal{T}$-invariant of unit weight: $\overline{L}_w = \mathcal{T}'(L)L_w$ (cf. Example 5.10).

Depending upon their weight p, relative invariants may be computationally more robust than absolute invariants, while still all relations of the type $I = J$ are invariant in absolute sense if weights on both sides correspond (a necessary consistency demand). A distinction between even and odd relative invariants is relevant if we allow for inversions ($\mathcal{T}'(L) < 0$).

Example 5.14 (Absolute and Relative Grey-Scale Invariants)
- Examples of absolute grey-scale invariants in 2D and 3D are the isophote curvature measures discussed in Examples 5.12 and 5.13.

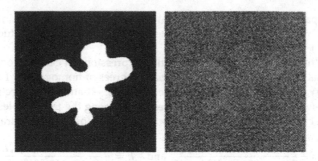

Figure 5.10: "Amoeba" image used as a test image in Figures 5.11 and 5.12, created as a binary, 256×256 image (left), and perturbed by additive, pixel-uncorrelated, Gaussian noise, with a standard deviation equal to initial contrast (right).

- A second order relative grey-scale invariant is the "corner strength" $Y = L_w^3 \kappa = -L_{vv} L_w^2$ of Example 5.8; see Figure 5.11. Its weight of 3 is motivated by the fact that this particular choice turns it into a polynomial invariant (robustness) with an additional invariance as argued on Page 160.

- Another second order relative grey-scale invariant is $X = -L_w^3 \mu = -L_{vw} L_w^2$, also defined in Example 5.8; see Problem 5.19.

- Figure 5.12 shows the associated relative invariant of weight 6, $Z \equiv -L_w^6 w'''$, "inflection strength". We have, in arbitrary, Cartesian, and gradient gauge coordinates, respectively:

$$
\begin{aligned}
Z &= \varepsilon^{il} \varepsilon^{jm} \varepsilon^{kn} L_l L_m L_n (L_{ijk} L_p L^p + 3 L_{ij} L_{kp} L^p) \\
&= 3(L_x^2 L_{yy} - 2 L_x L_y L_{xy} + L_y^2 L_{xx})(L_x L_y L_{yy} - L_y^2 L_{xy} + L_x^2 L_{xy} - L_x L_y L_{xx}) \\
&\quad - (L_x^2 + L_y^2)(L_x^3 L_{yyy} - 3 L_x^2 L_y L_{xyy} + 3 L_x L_y^2 L_{xxy} - L_y^3 L_{xxx}) \\
&= L_{vvv} L_w^5 - 3 L_{vv} L_{vw} L_w^4 .
\end{aligned}
$$

- Figures 5.13, 5.14 and 5.15 illustrate the relative invariants $h^{[3]} = L_w^3 h$, and $k^{[4]} = L_w^4 k$ associated with the mean and Gaussian curvatures h and k of Example 5.13 for a 3D torus. The invariants are presented slice-by-slice in 2D, slicing the torus along a plane parallel to its symmetry axis.

- With Result 5.4 we can systematically proceed to higher orders, e.g.

$$
w''' = -\frac{L_{vvv}}{L_w} + 3 \frac{L_{vv}}{L_w} \frac{L_{vw}}{L_w}
$$

is an absolute invariant.

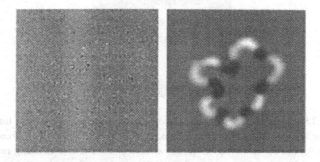

Figure 5.11: The invariant $-Y$ (Examples 5.8 and 5.14) applied to the perturbed amoeba image of Figure 5.10 at two different scales, $\log \sigma = \frac{1}{2}$ (left) and $\log \sigma = \frac{9}{4}$ (right), respectively. This invariant expresses a shear-invariant trade-off between gradient magnitude and isophote curvature and could therefore be called "corner strength". The curved arcs ("corners" in the extreme case) exist as measurable entities over a certain range of scales, gradually emerging from the noise as scale exceeds some lower threshold ("inner scale"), to finally agglutinate above some upper bound ("outer scale").

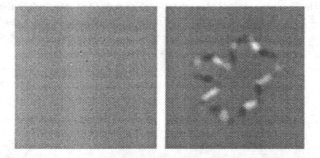

Figure 5.12: The invariant Z of Example 5.14 applied to the perturbed amoeba image of Figure 5.10 at two different scales, $\log \sigma = \frac{1}{2}$ (left) and $\log \sigma = \frac{9}{4}$ (right), respectively. This third order invariant measures a trade-off between gradient magnitude and rate of change of isophote curvature along the isophote, and could thus be called "inflection strength". Note that its extrema correspond to points of inflection and its zero-crossings to curvature extrema on the boundary.

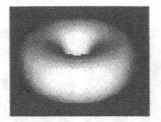

Figure 5.13: Visualisation of a 3D, 128 × 128 × 64-voxel image of a torus. The torus is defined by the parametric equation $(x, y, z) = ((R + r \sin \phi) \cos \theta, (R + r \sin \phi) \sin \theta, r \cos \phi)$ with $(\theta, \phi) \in [0, 2\pi)$ and $R > r$. The two defining radii r and R are 18 and 32 units relative to voxel size. The interior and exterior of the torus each have a uniform grey-value L_i and L_e, with $L_i - L_e > 0$.

Figure 5.14: Four typical slices showing the relative grey-value invariant $h^{[3]}$ for the torus image. Background grey corresponds to $h^{[3]} = 0$. The major part of the torus in this example has positive mean curvature. In this example, there is a slightly negative mean curvature at the inside (because $r < R < 2r$), reaching an absolute minimum on the circle $(x, y, z) = ((R - r) \cos \theta, (R - r) \sin \theta, 0)$ $(0 \leq \theta < 2\pi)$; in this example $h_{\min} = \frac{R-2r}{2r(R-r)} = -\frac{1}{126}$ in inverse voxel units. The maximum value is reached at the outside, i.e. on the circle $(x, y, z) = ((R + r) \cos \theta, (R + r) \sin \theta, 0)$ $(0 \leq \theta < 2\pi)$, viz. $h_{\max} = \frac{R+2r}{2r(R+r)} = \frac{17}{450}$ inverse voxel units. Scale: $\sigma = 2.25$ voxel units.

Figure 5.15: Four typical slices showing the relative grey-value invariant $k^{[4]}$ for the torus image. Background grey corresponds to $k^{[4]} = 0$. The image clearly shows the elliptic points on the outside of the torus $(0 < \phi < \pi)$, the hyperbolic points on the inside $(\pi < \phi < 2\pi)$, and the parabolic points separating the elliptic and hyperbolic parts of the surface $(\phi = 0, \pi)$. Scale: $\sigma = 2.25$ voxel units.

Problems

5.1

a. Verify the statements in Example 5.1 by computation.

b. What is the form of the covariant and contravariant metric tensor in a coordinate system (ξ, η) given by $(x, y) = (f(\xi, \eta), g(\xi, \eta))$? (Assume the inverse is $(\xi, \eta) = (\phi(x, y), \gamma(x, y))$.)

5.2

Derive explicit coordinate forms for the 2D Laplacian ΔL in the three coordinatisations of Example 5.1, as well as for the general coordinate reparametrisation of Problem 5.1.

5.3

Footnote 2, Page 138: "...Unlike in the ∞-dimensional case, there is no distinction between algebraic and topological dual." Why not?

5.4

Show that if S and T are tensors with components $S^{i_1 \ldots i_l}_{j_1 \ldots j_k}$ and $T^{i_1 \ldots i_l}_{j_1 \ldots j_k}$ relative to a given coordinate system x, respectively, for which

$$S^{i_1 \ldots i_l}_{j_1 \ldots j_k} = T^{i_1 \ldots i_l}_{j_1 \ldots j_k},$$

then

$$S'^{i_1 \ldots i_l}_{j_1 \ldots j_k} = T'^{i_1 \ldots i_l}_{j_1 \ldots j_k}$$

remains valid in *any* other coordinate system $x' = x'(x)$.

5.5

Page 146: "...In particular one should now appreciate that $L_{i_1 \ldots i_k}$ (Notation 3.1) is nothing but the k-th order covariant image derivative". Prove this statement.

5.6

Prove Lemma 5.1.

5.7 (An Intrinsically Curved Space)

One of the simplest examples in which we have a non-Euclidean metric is provided by the surface of a sphere: we can use the polar angles θ and ϕ in which we have $d\ell^2 = d\theta^2 + \sin^2 \theta d\phi^2$.

a. Show that $g_{\theta\theta} = g^{\theta\theta} = 1$, $g_{\phi\phi} = \sin^2 \theta$, $g^{\phi\phi} = \sin^{-2} \theta$ and $g_{\theta\phi} = g_{\phi\theta} = g^{\theta\phi} = g^{\phi\theta} = 0$.

b. Compute the metric affinity Γ^k_{ij}.

c. Compute Riemann tensor R^i_{jkl}, Ricci tensor R_{ij}, and Ricci scalar R, and show that $\det R^i_j = 1$ (what does this signify?).

d. Can you find Cartesian coordinates for a sphere?

5.8

a. Prove Equation 5.15:

$$\sqrt{g'} = |\det B| \sqrt{g},$$

i.e. $\sqrt{g}$ is an odd relative scalar of weight $w = 1$.

b. Show that, in d dimensions, the components of the antisymmetric symbol $[i_1 \ldots i_d]$ transform as an even relative tensor in accordance with Equation 5.16, with $w = -1$ and $(k, l) = (d, 0)$:

$$[i_1 \ldots i_d]' = (\det B)^{-1} B_{i_1}{}^{q_1} \ldots B_{i_d}{}^{q_d} [q_1 \ldots q_d].$$

c. Conclude that the Levi-Civita tensor transforms as an absolute pseudo-tensor, i.e. according to Equation 5.13.

5.9
Page 150: "It should not be difficult now to figure out what a *relative pseudo-tensor* means..."

5.10
a. Show that $\varepsilon_{i_1 \ldots i_d} = g \, \varepsilon^{i_1 \ldots i_d}$ holds in any coordinate system.
Despite its validity in any coordinate system, this is *not* a tensor equation, since covariant and contravariant orders do not match on both sides. We consider the equation in more detail below.

b. Show that raising and lowering indices in a relative (possibly pseudo) tensor does not affect its weight.

c. According to Problem 5.8, the l.h.s. is an absolute tensor, whereas the factor g on the r.h.s. is a relative scalar of weight $w = 2$. According to the homogeneity argument that "weights add up", the contravariant Levi-Civita tensor must therefore be a relative tensor of weight $w = -2$, in violation of the previous result. What is wrong with this argument?

d. What is the correct tensor equation relating covariant and contravariant Levi-Civita tensors?

5.11
Show that the "invariant volume element" $\sqrt{g} \, dx^1 \ldots dx^d$ is a pseudo-scalar.

5.12
a. Show that the determinant of a mixed 2-tensor A^i_j is an absolute scalar.

b. What about the determinants of A^{ij} and A_{ij}?

5.13
Prove: $\partial_k \sqrt{g} = \Gamma^i_{ik} \sqrt{g}$.

5.14
a. Consider the usual spherical coordinate basis in 3D: $\{\partial_r, \partial_\theta, \partial_\phi\}$, where $(r, \theta, \phi) = (\sqrt{x^2 + y^2 + z^2}, \arccos(z/\sqrt{x^2 + y^2 + z^2}), \arctan(y/x))$ in terms of Cartesian coordinates. What is the corresponding dual basis?

b. Compute the metric components g_{ij} in 3D spherical coordinates (r, θ, ϕ), and invert the resulting matrix to obtain g^{ij}.

c. Let f be a scalar field in 3D space. Write ∇f in terms of the spherical coordinate basis.

d. As c, but for the Laplacian Δf.

e. Specify the Levi-Civita tensor in the spherical coordinate basis.

5.15
a. Verify Proposition 5.3 for the diagrams of Figures 5.3 and 5.4.

b. Can you think of other diagrams having the same values of C, V_k, H, B, I, and E as the one on the left in Figure 5.3? How many are there?

c. Prove Proposition 5.3.

5.16

a. Prove the 2D reducibility relation shown in Figure 5.7.

b. Give a similar diagrammatic representation for the reducibility of $L_{ij} L_k^j L_l^k L^{li}$ in 2D.

c. As b, but now in 3D.

d. What about 4D?

5.17

Give the expressions corresponding to the three diagrams below using summation convention. Give also their explicit form in 2D and in 3D Cartesian coordinates.

5.18

Example 5.13, Page 166: "...the u and v axes are tangent to the principal directions."

5.19

An elaboration of Example 5.14: in the formula for w''' one recognises the grey-scale invariant isophote curvature κ of Example 5.12, as well as the factors μ and ν given by

$$\mu = \frac{\varepsilon^{ij} L_j L_k L_i^k}{(L_m L^m)^{3/2}} \quad \text{and} \quad \nu = -\frac{\varepsilon^{il} \varepsilon^{jm} \varepsilon^{kn} L_l L_m L_n L_{ijk}}{(L_p L^p)^2}.$$

Prove that μ and (consequently) ν are absolute grey-scale invariants as well. Show in addition that μ is the curvature of the gradient integral curve, as claimed in Example 5.8.

5.20

An extension/generalisation of Examples 5.13 and 5.14. Prove:

$$w_{uuu} = -\frac{L_{uuu}}{L_w} + 3\frac{L_{uu}}{L_w}\frac{L_{uw}}{L_w},$$

$$w_{uuv} = -\frac{L_{uuv}}{L_w} + \frac{L_{uu}}{L_w}\frac{L_{vw}}{L_w},$$

$$w_{uvv} = -\frac{L_{uvv}}{L_w} + \frac{L_{vv}}{L_w}\frac{L_{uw}}{L_w},$$

$$w_{vvv} = -\frac{L_{vvv}}{L_w} + 3\frac{L_{vv}}{L_w}\frac{L_{vw}}{L_w}.$$

5.21

Recall Example 5.13 for the notation.

a. Prove the following identity:

$$\varepsilon_{ijk}\,\varepsilon^{lmk} = \det\begin{pmatrix} \delta_i^l & \delta_i^m \\ \delta_j^l & \delta_j^m \end{pmatrix}.$$

b. Show that the mean curvature h can be written as

$$h = \frac{1}{2} \frac{L_i L_j^i L_j - L_i L^i L_j^j}{(L_k L^k)^{3/2}}.$$

c. Express mean and Gaussian curvature h and k in gradient gauge coordinates (u, v, w).

d. Ditto in terms of an arbitrary Cartesian coordinate system.

CHAPTER **6**

Multiscale Optic Flow

In the previous chapters we have established two definitions of spatiotemporal images with fundamentally different temporal aspects. The first one does not account for temporal causality. All time instances are treated on equal foot. As such it is useful for off-line image processing on pre-recorded data. The second definition does incorporate causality, and is therefore appropriate for on-line image processing of real-time acquisition data. Both definitions reflect the symmetries that pertain to the classical picture of space and time. However, both models could be called "pseudo-static" in the sense that none of them explicitly accounts for a *kinematic relation* between local image samples. Such relations naturally arise as a consequence of apparent *conservation laws*. For this reason we will define a kinematic concept known as *optic flow*, again—of course—in terms of its actual computation.

Introduced by Gibson in the context of optical pilot navigation [95, 164], optic flow has become a widely appreciated, but also somewhat confusing concept in computer vision and image analysis. The problem is that a meaningful definition always depends upon subjective factors, in other words, on semantics. It is therefore important to study objective data evidence of optic flow *in isolation*. Thus we are led to search for a *syntactical optic flow paradigm* that is fully compatible with our previous paradigms.

Since the scale-space representation is complete, a syntactical definition of optic flow should in principle boil down to a "reformatting" of image data according to some convention that captures the idea of a flow in a plausible way. The convention will be inspired by the assertion of a conservation law, in relation to which it becomes a "meaningful" convention (proto-semantics), but as a data reformatting principle its definition does not actually require anything to be conserved. Recall that the argument leading to our image definition was quite similar; we *produce* a Newtonian picture of spacetime, but we do not claim spacetime to *be* Newtonian.

The ideal context for the proposed definition will be that of grey-value images in which grey-scale represents a relevant physical parameter. We must distinguish two cases: either grey-scale reflects the density of a conserved scalar quantity (such as proton density cine-MR), or it represents the scalar quantity itself (such as the retinal irradiance under the assumption of Lambertian reflection). With some extra care one may include line-integral projection data [71].

In brief, the purpose of this chapter is threefold:

- to manifestly separate the optic flow d.o.f.'s[1] as far as these are supported by data evidence from external insight (syntax versus semantics): Section 6.1;

- to establish a conceptual definition of optic flow consistent with the scale-space paradigm: Section 6.2; and

- to set up an equivalent operational definition enabling actual computation: Section 6.3.

Thus we de-emphasize the semantics of the flow; as such, the theory is of general interest. We will point out how things can be made to work in practice, i.e. when given a particular imaging situation, but the details of this are beyond the scope of this book. However, one semantical case is particularly simple and illustrates the general computational strategy, viz. that of so-called "normal flow". It can also be used to define a "default" representation of syntactical flow, and is itself of interest in certain applications. We will use this case to verify the theoretical model by means of an analytically tractable flow simulation and extraction (Section 6.4), and conclude with a summary and a discussion (Section 6.5).

6.1 Towards an Operational Definition of Optic Flow

It is interesting to see how the concept of "flow" developed over the centuries, and how an operational definition of optic flow emerges through syncretism of seemingly opposing tenets: Box 6.1.

The present-day view that flow can be described by a vector field enjoys public consensus, and therefore provides a good point of departure. It reflects the desire to link "corresponding points"—whatever these may be—separated by arbitrarily small temporal intervals. The motivation for this is of course that in the physical world such pointwise connections are actually meaningful; ideally they correspond to an underlying motion of material points, or perhaps to wave phenomena.

The mathematical definition of a vector field as a "cross-section of the tangent bundle" [264, 265] states that one is to pick one vector at each base point, such that

---

[1]d.o.f.('s) = "degree(s) of freedom".

transitions to neighbouring points show no abrupt changes. Such a "speedometer" type of model (one velocity per base point), however, runs counter to physical intuition: surely one cannot obtain a vector by means of a measurement confined to a point of measure zero, following the same argument that defies existence of "punctal" grey-scale samples. Some (*a priori* arbitrary) spatiotemporal aperture must be involved. In particular, within the context of scale-space theory, the deep structure of an image must carry over to optic flow. Thus we are led to consider a *multiscale optic flow* field, characterised by a pair of scale parameters in addition to a spacetime base point. Of course, optic flow should somehow be induced by the source field. In this sense we can speak of a speedometer type of flow: the fiducial velocity at "infinite resolution" tightly coupled to that source.

Before we turn to a precise definition in Section 6.2, there are some conceptual problems to be addressed.

6.1.1 The "Aperture Problem"

A notorious problem is the so-called *"aperture problem"* [118, 207, 284]. The problem is one of *ambiguity*; simply counting d.o.f.'s suffices to see that one cannot extract an unambiguous vector field from a scalar conservation law. Nevertheless, confusion about this phenomenon still pervades the image literature.

For example, one frequently encounters the argument that "the aperture problem is a false problem", one that can be overcome as soon as "enough structure" is present in the image brightness. The argument is deceptively convincing: it is possible to extract a unique optic flow field [217, 218, 219, 234, 272, 273, 282, 290]. It is nevertheless a misconception; in one way or another, *one has to bring in semantics*. We could rephrase the ambiguity problem as follows: if we combine different assumptions with the same image data, we obtain different vector fields. The viability of such assumptions is independent of actual data, so that none of the solutions can be falsified on the basis of these. One has to appreciate that this kind of optic flow ambiguity—the co-existence of multiple consistent definitions—is nothing but the aperture problem in disguise!

Instead of concealing the problem in this way, let us try to get to the core of it. The basic observation is that any hypothetical motion confined to an isophote is *a priori* feasible (Figure 6.1). Put differently, if only for the data, one may seek to solve for the *homotopy* that links spatial isophotes over time *as a whole*, but one cannot hope to establish any *pointwise* connections between them. A "default" representation of the homotopy between spatial isophotes is *normal flow*, sometimes called *the* optic flow field (to be distinguished from the physically induced image velocity or *motion field*) [119, 121].

Generic isophotes are said to have "codimension one", which is just the dimension of the normal flow vector space (for fixed base point and resolution); in particular, their *shape* is immaterial. The intrinsic ambiguity of optic flow measurement has consequently nothing to do with the existence of "straight edges", another confusing *vox populi*. The association of aperture problem and straight edges again suggests that the ambiguity can be resolved on the basis of image intrinsic factors, e.g. by taking into account "corner points", or the image's "higher order differential structure" [217, 218, 234, 272]. In reality, however, one always adds semantics. "*Hineininterpretieren*" is basically the only way out of ambiguity, yet it is essential to understand the semantical content of any optic flow disambiguation scheme. Optic flow literature sometimes fails to be explicit on this matter.

Another way of expressing that isophote automorphisms have no observable effect on the flow is to say that optic flow is a *gauge field*: Figure 6.2. A gauge field is a descriptor of some physical field with non-observable local aspects, and is defined in terms of a a model (or *gauge theory*) that has a characteristic *local invariance*. Gauge theories are popular in physics because one can often simplify the description of a physical system by adding virtual d.o.f.'s as well as a symmetry that effectively cancels their physical effect (the traditional example is Maxwell's theory; the electromagnetic vector potential is the gauge field, with four components but only two physical photon "polarisations"). In the case at hand, the syntactical optic flow field can be regarded as a gauge field; it contains local d.o.f.'s that do not manifest themselves in any observable way, and hence can be fixed arbitrarily. All mysteries that have come to surround the aperture problem ever since it was introduced can be avoided if one appreciates the purely syntactical view of optic flow and its role in the framework of a gauge theory.

Source fields of interest are almost never chaotic distributions of unknown entities in a state of random motion, and even if one has no *a priori* knowledge whatsoever, one can always raise hypotheses to *fix the gauge* by imposing an (admissible) *gauge condition*. If this is done in an *ad hoc* way, the result will be a unique optic flow field that, although compatible with the data, lacks any physical meaning. Therefore external considerations beyond the evidence contained in the data, such as the details of the image formation process and the physics of the scene, should be taken into account in order to arrive at a sensible choice of gauge [119]. Thus the gauge condition expresses the semantics, and the gauge-fixed solution could be called the *semantical optic flow field*, a "meaningful" member of the metameric class of syntactically equivalent fields. It is important to appreciate that its significance is only relative to the interpretation implied by the gauge condition. For this reason it is equally important to keep syntax and semantics nicely apart until we know exactly what we are after, and even at that stage we may want to re-interpret, change our minds, in other words, alter our gauge in a feedback kind of fashion until we get things to work properly. After all, any *a priori* assertion we make may be wrong!

Examples of gauge conditions are rigidity or non-elasticity constraints for solid objects [55, 65], (in)compressibility and continuity conditions for fluids [4, 53], etc. A mathematical smoothness constraint is often used to fix the gauge (sometimes implicitly). Typically neither one of these provides a globally valid constraint; the gauge is confounded with the local semantics. For example, in the context of machine vision the assertion may be that motion is induced by projection of a shaded, sufficiently smooth, rigid surface patch, etc. [119, 123, 277, 283]; indeed, quite a number of assumptions (Problem 6.3). In medical imagery it may be asserted that blood flow satisfies incompressibility and continuity constraints, whereas bone tissue induces rigid motion, soft tissue deforms nonrigidly, etc.

Note that one would need a segmentation and classification of tissues in order to be able to tell in advance which motion constraint is a reasonable one to apply at any given location. That optic flow may in turn provide strong cues for segmentation indeed votes for the idea of successively refining the gauge by a kind of hermeneutic circle.

The term "optic flow" is henceforth taken in purely syntactical sense. In other words, it will be assumed to entail essentially all possible flows as far as these are compatible with the data, so optic flow equals normal flow modulo arbitrary hypothetical tangential flows [124]. In this terminology, normal flow can be regarded as a default representation of optic flow, subject to a "default gauge condition" that nullifies tangential flow. Thus in a loose sense we may even identify syntactical optic flow and normal flow.

6.1.2 Computational Problems

Although subject to immense effort, a satisfying operational definition of optic flow is by no means firmly established. It is not only a matter of semantics, but also one of plain computation. There are many possible approaches to optic flow measurement, not all of which have been investigated in-depth [13, 151, 262].

A classical, local approach is based on the "Optic Flow Constraint Equation" [7, 8, 9, 10, 13, 68, 123, 217, 218, 234, 252, 272, 282, 294], which defines optic flow locally on the basis of a conservation principle; invariant grey-values are attributed to points which are dragged along the flow. Its traditional formulation is, however, problematic, since it bypasses the indispensable role of device space in the realization of a grey-scale image. A related computational problem is of course ill-posedness, since it relies crucially on the differential structure of ordinary functions. Nevertheless, the method is basically useful as long as we account for the role of filters.

The OFCE can be formulated for scalar as well as for density (n-form) images; see Figure 6.3. The density case is well-known from fluid dynamics; in the context of optic flow it is also known as the "generalised motion constraint equation" [252].

6.2 The Optic Flow Constraint Equation

Without loss of generality we take $(x; \sigma) = (0; 1)$, and adopt the ct-convention of Definition 3.17, Page 62. It is straightforward to re-introduce the free scale parameters if desired (dimensional analysis!), and in reality one can (indeed *must*) exploit the scale d.o.f. in the computation [222, 224, 227]. Please do not confuse resolution of computation with that of the flow field itself; in fact, we shall aim for an "approximately infinite" resolution of the latter, meaning a resolution comparable to that of the raw image data (it will be seen that lower levels of resolution are derivable from this).

An important tool is the so-called *Lie derivative*. Lie derivatives capture variations of spacetime quantities along the integral flow of some vector field. To take a Lie derivative, one therefore needs to know this vector field. Actually, we shall only consider the 1-st order Lie derivative of an image, which will give us a *linear* model of optic flow; this is *not* a restriction and should not be confused with the spatiotemporal differential order of the flow field we might be interested in. We have no reason to impose any *a priori* restrictions on this. In the line of Horn and Schunck's approach [252] the vector field will, of course, be the optic flow field, which will be denoted by $v^\mu \equiv (v^0; v^i)$. Notice that, in principle, we allow for a nontrivial temporal component. However, unless stated otherwise, the following assumption will be made.

Assumption 6.1 (Temporal Gauge)
For all $x \in \mathbb{R}^n$ we assume throughout that

$$v^0(x) = 1 .$$

This is a usual—locally weak, but globally not necessarily realistic—assumption stating that the flow is everywhere nonvanishing and transversal to constant-time frames, in other words, that structural details are enduring despite possible deformations. It therefore expresses *conservation of topological detail*. It is in fact an instance of a gauge condition enforced on the basis of an *a priori* physical principle. See Figure 6.4.

The Lie derivative of a scalar function f w.r.t. a vector field v^μ is given by the directional derivative $\mathcal{L}_v f = D_\mu f v^\mu$. For a density we have $\mathcal{L}_v f = D_\mu (f v^\mu)$. However, both are ill-posed just like any other classical derivative (and, besides, only work if f is differentiable), thus one should resist the temptation to use them in implementations. They can, however, be turned into well-posed operators in the same way as before, viz. by defining their actions in a carefully constructed dual space, notably by interpreting the raw image f as a distribution in $\mathcal{G}'(\mathbb{R}^n)$.

Definition 6.1 (Lie Derivative of a Local Sample)
We have

$$\mathcal{L}_v F[\phi] \stackrel{\text{def}}{=} F[\mathcal{L}_v^T \phi] ,$$

in which the transpose $\mathcal{L}_v^T \phi$ is defined by $\mathcal{L}_v^T \phi = -D_\mu (v^\mu \phi)$ if the source represents a scalar, and by $\mathcal{L}_v^T \phi = -v^\mu D_\mu \phi$ if it represents a density.

In other words, if we transform our filters as densities we take it for granted that our source fields are scalars w.r.t. Lie derivation, *vice versa*. More generally, the nature of state space determines how sources transform under the action of a flow field, and thus, by the requirement of duality, how detectors in device space ought to transform [265]. For the sake of simplicity we will not consider any other cases than those of scalar and density sources.

Proof 6.1 (Consistency of Definition 6.1)
The only thing to prove is consistency with Definition 2.18, Page 31 (recall also Notation 3.1 on Page 65). We recall the motivation for this, at the same time applying it to Lie derivation. The easiest way is to consider the integral expression for a local sample, $F[\phi] = \int dx f(x) \phi(x)$. We must first of all decide on the geometric flavour of the source

field of interest f (scalar, density, etc.). This will determine its behaviour under the action of the flow, i.e., which coordinate form its Lie derivative has (assuming, for the moment, $f \in C^1(\mathbb{R}^n)$). Then it is simply a matter of partial integration to find the dual representation, explaining, in particular, how the minus signs in Definition 6.1 arise. Having done that, one carefully covers one's tracks, and simply takes the result as a definition.

Apart from this argument of partial integration, it may be intuitively clear that the transposed Lie derivative has to be the Lie derivative w.r.t. the opposite flow field.

Gauge invariant optic flow can now be defined as *any* flow field preserving $F[\phi]$:

Assumption 6.2 (Syntactical Optic Flow)
Recall Assumption 6.1. The syntactical optic flow field is defined as the equivalence class of vector fields v^μ that satisfy the OFCE:

$$\mathcal{L}_v F[\phi] = 0 .$$

In integral form, using Definition 6.1, this reads

$$\int dx \, f(x) \, D_\mu \left(v^\mu(x) \, \phi(x) \right) = 0$$

if the source is a scalar, and

$$\int dx \, f(x) \, v^\mu(x) \, D_\mu \phi(x) = 0$$

if it is a density.

This assumption couples the vector field v^μ to the raw image f in a natural way.

Postponing computational aspects to the next section, let us consider the connection between optic flow and scale-space theory. To this end it is convenient to introduce the notion of a *filter current* and that of a *source current*.

Definition 6.2 (Filter Current and Source Current)
The filter current ω^μ corresponding to a filter $\phi \in \Delta$ and a (smooth) vector field v^μ is defined as

$$\omega^\mu(x) = v^\mu(x) \, \phi(x) .$$

The source current J^μ corresponding to a source field $F \in \Sigma$ and a (smooth) vector field v^μ is defined as the vector-valued distribution with Riesz representation $j^\mu = v^\mu f$, i.e.

$$J^\mu[\phi] = \int dx \, j^\mu(x) \, \phi(x) .$$

Source and filter currents inherit their density or scalar properties from the underlying source f and filter ϕ, respectively. Note that, by virtue of temporal gauge (Assumption 6.1), the $\mu = 0$ component of the filter current equals the basic filter, whereas that of the source current is just the raw image. Notice also the duality $J^\mu[\phi] = F[\omega^\mu]$.

Like the raw data function f, the "naked" vector field v^μ, and consequently the source current j^μ are abstractions that exist only "under the integral", i.e. by virtue of a measurement. The following definition of a vector-valued distribution V^μ corresponding to v^μ presents itself.

Definition 6.3 (Optic Flow as a Distribution)
See Definition 6.2. The vector-valued distribution V^μ is defined as follows:

$$V^\mu[\phi] \stackrel{\text{def}}{=} \frac{J^\mu[\phi]}{F[\phi]} . \quad \text{Alternatively,} \quad V^\mu[\phi] \stackrel{\text{def}}{=} \frac{F[\omega^\mu]}{F[\phi]} .$$

The distribution V^μ is well-defined as long as $F[\phi] \neq 0$. In integral form we have

$$V^\mu[\phi] = \frac{\int dx\, j^\mu(x)\, \phi(x)}{\int dx\, f(x)\, \phi(x)} = \frac{\int dx\, f(x)\, \omega^\mu(x)}{\int dx\, f(x)\, \phi(x)} .$$

We adhere to a similar notation as before: capital V^μ denotes the vector valued distribution, small v^μ the corresponding vector field. Note that the temporal gauge of Assumption 6.1 is respected: for all $\phi \in \Delta$ we have

$$V^0[\phi] = 1 . \tag{6.1}$$

If $\Delta = \mathcal{G}^+(\mathbb{R}^n)$ then $V^\mu[\phi]$ represents the *multiscale optic flow* field induced by f. By construction, source and filter currents (but *not* the flow field itself!) satisfy the diffusion equation; for instance for the source current we have

$$\begin{cases} \partial_s J^\mu &= \tfrac{1}{2}\Delta J^\mu \\ \lim_{s\downarrow 0} J^\mu &= j^\mu . \end{cases} \tag{6.2}$$

This shows that optic flow inherits its deep structure from that of its underlying source, as one might have expected. Of course, the linear functional $J^\mu[\phi]$ is always well-defined ("linear flow"), as opposed to the nonlinear functional $V^\mu[\phi]$ ("geometrical flow"). In terms of currents, the Lie derivatives and corresponding conservation principles (Assumption 6.2) can be restated as

$$F[D_\mu \omega^\mu] = 0 \quad \text{and} \quad J^\mu[D_\mu \phi] = 0 , \tag{6.3}$$

in the scalar, respectively density case.

In addition to the appropriate form of conservation, Definition 6.3 requires us to decide on the gauge. In the absence of *a priori* knowledge (from which the gauge must be determined in practice) we may take the "default gauge" that nullifies tangential flow and in addition asserts that topological detail is conserved. This is in some sense the "least committed" solution with an interpretation in terms of a conservation law.

Definition 6.4 (Default Gauge: Normal Flow)
Cf. Assumption 6.2. Normal flow is defined as optic flow subject to the default gauge

$$\mathcal{L}_{*v} F[\phi] = 0 ,$$

in addition to the temporal gauge of Assumption 6.1. Here, $*v^\mu$ represents any spatial isophote tangent vector: $*v \cdot v \equiv 0$ and $*v^0 \equiv 0$. Note the two distinct interpretations of this equation corresponding to those in Assumption 6.2.

Note that this adds $d - 1$ independent constraints per base point and per level of resolution to the usual one for the temporal gauge.

Although formulated for one sample point only, Assumption 6.2 and Definition 6.4 should be understood as *globally valid* constraints (as is, trivially, Assumption 6.1); they retain their validity if we extend the model from local samples to images (i.e. if we push forward $\phi \mapsto \theta_*\phi$ under a spacetime symmetry transformation θ). This can be expressed in terms of a countable set of *local* constraints (valid at $\theta = \text{id}$) by stating that all spatiotemporal derivatives at the point of interest vanish as well[2].

Definition 6.5 (Syntactical Optic Flow)
See Definition 6.1 and Assumptions 6.1 and 6.2.

$$D_{\mu_1 \ldots \mu_k} \mathcal{L}_v F[\phi] = 0 \quad \text{for all} \quad k \in \mathbb{Z}_0^+,$$

in which the spatiotemporal derivatives of $\mathcal{L}_v F[\phi]$ are defined as usual:

$$D_{\mu_1 \ldots \mu_k} \mathcal{L}_v F[\phi] = (-1)^k \int dx\, f(x)\, \mathcal{L}_v^T \phi_{\mu_1 \ldots \mu_k}(x),$$

with $\mathcal{L}_v^T \phi_{\mu_1 \ldots \mu_k} = -D_\mu(v^\mu \phi_{\mu_1 \ldots \mu_k})$ or $\mathcal{L}_v^T \phi_{\mu_1 \ldots \mu_k} = -v^\mu \phi_{\mu\mu_1 \ldots \mu_k}$ if the source is a scalar or a density, respectively.

Consistency is proven in the same way as for Definition 6.1. When transposing Lie and covariant derivatives one has to take their ordering into account, applying the composition rule $(A \circ B)^T = B^T \circ A^T$.

6.3   Computational Model for Solving the OFCE

The distribution $V^\mu[\phi]$ of Definition 6.3 depends causally and continuously upon the product of vector field v^μ and raw data function f "under the integral". In order to construct it, it therefore suffices to find a representation for the "infinite resolution" field v^μ. In the stationary case, i.e. if the source is time-independent, a generic solution is readily given by $v^\mu = (1; \vec{0})$. The strategy in the general case will be to get the field v^μ, defined implicitly by Assumption 6.2, out of the integral. In fact we shall concentrate on the locally equivalent form of Definition 6.5 at a fixed sample point (the origin at unit scale), and try to construct an approximate local jet representation of v^μ on the same grid and at the same implicit scale as that of the raw image f.

---

[2]Caution: Lie derivatives and ordinary derivatives do not commute. In the literature one sometimes encounters a reversed order of Lie and ordinary derivative(s), in which case one has postulated a (usually not rigorously motivated) *independent* conservation law.

In principle, the optic flow field v^μ contains an infinite number of d.o.f.'s. This is inconvenient; moreover, most of them are irrelevant or computationally inaccessible anyway. But, depending upon one's task, a 0-th order approximation is usually too restrictive. For example, in the case of real-world movies, 1-st order properties of the vector field may reveal relevant information such as qualitative shape properties, surface slant [169], and time-to-collision [182, 183]. Unlike 1-st order, 2-nd order is quantitatively related to intrinsic surface properties of an object [173]. Also in tomographic images it may be useful to consider the flow's higher order differential structure. Moreover, there is no *a priori* limit to the highest order that is still accessible and significant; this depends very much upon matters such as image quality (noise and sampling characteristics), resolution of interest, etc.

Consider a *formal expansion* of v^μ near the origin, truncated at some arbitrary order M, v_M^μ. This is a polynomial intended to capture a finite number of local d.o.f.'s of the vector field.

Definition 6.6 (M-th Order Formal Expansion)
The formal expansion of order M of the vector field v^μ at the origin, denoted v_M^μ, is an M-th order polynomial

$$v_M^\mu(x) = \sum_{l=0}^{M} \frac{1}{l!} v_{M;\rho_1\ldots\rho_l}^\mu \, x^{\rho_1} \ldots x^{\rho_l} \,,$$

the coefficients of which may depend upon M.

The finite set of coefficients corresponds exactly to the d.o.f.'s we shall be looking for. It is essential to appreciate that v_M^μ is *not* required to be the M-th order Taylor polynomial of v^μ; the intention is that we obtain an approximation in the sense that

$$v_M^\mu(x) = v^\mu(x) + \mathcal{O}(\|x\|^{M+1})\,. \tag{6.4}$$

This implies that the coefficients are approximations of optic flow derivatives in the following sense:

$$v_{M;\rho_1\ldots\rho_l}^\mu = D_{\rho_1\ldots\rho_l} v^\mu(0) + \mathcal{O}(\|x\|^{M-l+1})\,. \tag{6.5}$$

Replacing v^μ by v_M^μ according to Definition 6.6 yields the following.

Result 6.1 (M-th Order Optic Flow Approximation)
See Definitions 6.5 and 6.6. Using $\mathcal{L}_{v_M}$ instead of $\mathcal{L}_v$ we have, for all $k = 0, \ldots, M$,

$$D_{\mu_1\ldots\mu_k} \mathcal{L}_{v_M} F[\phi] = -\sum_{l=0}^{M} v_{M;\rho_1\ldots\rho_l}^\mu \int dx\, f(x)\, D_\mu \Phi_{\mu_1\ldots\mu_k}^{\rho_1\ldots\rho_l}(x) = 0\,,$$

if the raw image is a scalar, and

$$D_{\mu_1\ldots\mu_k} \mathcal{L}_{v_M} F[\phi] = +\sum_{l=0}^{M} v_{M;\rho_1\ldots\rho_l}^\mu \int dx\, f(x)\, \Phi_{\mu_1\ldots\mu_k\mu}^{\rho_1\ldots\rho_l}(x) = 0\,,$$

if it is a density. The effective filters $\Phi^{\rho_1 \dots \rho_l}_{\mu_1 \dots \mu_k}$ *are defined by*

$$\Phi^{\rho_1 \dots \rho_l}_{\mu_1 \dots \mu_k}(x) = \frac{(-1)^k}{l!}\, \phi_{\mu_1 \dots \mu_k}(x)\, x^{\rho_1} \dots x^{\rho_l}\,.$$

(Note that a minus sign has been absorbed into the definition of the effective filters for each lower index.) Since the Gaussian family $(-1)^k \phi_{\mu_1 \dots \mu_k}$ $(k \in \mathbb{Z}_0^+)$ is complete, the set of filters $\Phi^{\rho_1 \dots \rho_l}_{\mu_1 \dots \mu_k}$ $(k, l \in \mathbb{Z}_0^+)$ must be *redundant*. Hence all filters can be expressed in terms of Gaussian derivatives (the case $l = 0$), in other words, all Lie derivatives are linear combinations of image derivatives in the sense of scale-space theory; for technical details see Appendix B. Thus we arrive at the conclusion that we have obtained a linear system of equations in the coefficients of interest, the size of which depends upon approximation order M, and the coefficient matrix of which is fully determined by the local image jet of order $2M + 1$ (to see this, consider the highest order term $k = l = M$, and use Result B.2 in Appendix B).

It is more important to note the restriction on the admissible orders: $k \leq M$. We have to take care not to introduce spurious d.o.f.'s by truncating the optic flow field (this would let in a gauge condition through the back-door!). In particular, we should try to maintain gauge invariance for our *approximated* optic flow field v^μ_M. Allowing arbitrary orders of differentiation in Result 6.1 would certainly break invariance in the generic case (i.e. the usual case when $v^\mu \neq v^\mu_M$), and may even yield an *inconsistent* system[3]! Put differently, the approximations of Equations 6.4–6.5 do not permit us to differentiate safely beyond order M. This is why one needs to limit the highest order to $k = M$ in Result 6.1. It can be shown that the resulting linear equations are generically independent and indeed gauge invariant *to the same extent as the exact system* [85]. Thus the M-th order approximation does *not* affect the intrinsic ambiguity of optic flow at all, as it should: approximations should not introduce spurious constraints.

Unfortunately, the literature is abundant of instances in which a unique optic flow field is singled out by shrewd combination of truncation and differentiation. It is this method that has led people to believe that the aperture problem is a "false problem", arguing that it can be circumvented on the basis of higher order image structure. What happens in reality is that the interplay of truncation and differentiation orders implicitly fixes the gauge. One cannot expect this to be a proper way of gauge fixing, except for coincidental cases. As explained before, a spurious gauge merely trades in aperture problem for optic flow ambiguity, making it quite difficult to appreciate *which* optic flow field one eventually ends up with. Gauge fixing, or optic flow disambiguation, should be a matter of *external insight*[4] independent of the image, the OFCE, or anything derived from these.

As opposed to conventional schemes based on M-fold implicit differentiation of the OFCE, *every* k-th order subset of Result 6.1 contains M-th order components of the approximated optic flow field. An important thing to keep in mind is that the degrees of freedom of v^μ_M, i.e. the coefficients $v^\mu_{M;\rho_1 \dots \rho_l}$, depend

---

[3]Such a system is particularly treacherous in combination with "least squares" methods...

[4]E.g. knowledge of how a physical conservation law for sources manifests itself in the image domain.

upon the order M of approximation. In other words, the polynomial approximation v_{M+1}^μ is a refinement of v_M^μ in the sense that *all* coefficients are refined. Hence, it is *not* the Taylor polynomial of v^μ; only in the limiting case we have $\lim_{M\to\infty} v_{M;\rho_1...\rho_l}^\mu = D_{\rho_1...\rho_l} v^\mu(0)$, so that $v_\infty^\mu = v^\mu$, subject to its original, gauge invariant definition. This brings us to another misconception one sometimes encounters in the literature; a polynomial approximation is not necessarily a truncated Taylor expansion.

6.4 Examples

6.4.1 Zeroth, First, and Second Order Systems

The following example illustrates the gauge invariant systems for the approximated zeroth, first, and second order optic flow field, respectively. The source of interest will be interpreted as a scalar field (recall Definition 6.1). The example illustrates the general principle of refinement: the $(M+1)$-st order system has the same form as the M-th order system except for additional terms of order $M+1$ in the flow field's approximation. The transition $M \to M+1$ will generally affect *all* coefficients in the formal expansion of the optic flow field.

Example 6.1 (Lowest Orders)
The zeroth order, gauge invariant optic flow field equation is given by $\mathcal{L}_{v_0} F[\phi] = 0$, or, using Notation 3.1, Page 65:

$$v_0^\rho \, L_\rho = 0 \,.$$

The first order equations are given by $\mathcal{L}_{v_1} F[\phi] = D_\mu \mathcal{L}_{v_1} F[\phi] = 0$, i.e.

$$v_1^\rho \, L_\rho + v_{1;\sigma}^\rho \, L_\rho^\sigma = 0 \,,$$

$$v_1^\rho \, L_{\rho\mu} + v_{1;\sigma}^\rho \, L_{\rho\mu}^\sigma + v_{1;\mu}^\rho \, L_\rho = 0 \,.$$

The second order equations are given by $\mathcal{L}_{v_2} F[\phi] = D_\mu \mathcal{L}_{v_2} F[\phi] = D_{\mu\nu} \mathcal{L}_{v_2} F[\phi] = 0$, or

$$v_2^\rho \, L_\rho + v_{2;\sigma}^\rho \, L_\rho^\sigma + \frac{1}{2} v_{2;\sigma\tau}^\rho \, L_\rho^{\sigma\tau} + \frac{1}{2} \eta^{\sigma\tau} v_{2;\sigma\tau}^\rho \, L_\rho = 0 \,,$$

$$v_2^\rho \, L_{\rho\mu} + v_{2;\sigma}^\rho \, L_{\rho\mu}^\sigma + v_{2;\mu}^\rho \, L_\rho + \frac{1}{2} v_{2;\sigma\tau}^\rho \, L_{\rho\mu}^{\sigma\tau} + \frac{1}{2} \eta^{\sigma\tau} v_{2;\sigma\tau}^\rho \, L_{\rho\mu} + v_{2;\mu\sigma}^\rho \, L_\rho^\sigma = 0 \,,$$

$$v_2^\rho \, L_{\rho\mu\nu} + v_{2;\sigma}^\rho \, L_{\rho\mu\nu}^\sigma + v_{2;\mu}^\rho \, L_{\rho\nu} + v_{2;\nu}^\rho \, L_{\rho\mu} + \frac{1}{2} v_{2;\sigma\tau}^\rho \, L_{\rho\mu\nu}^{\sigma\tau} + \frac{1}{2} \eta^{\sigma\tau} v_{2;\sigma\tau}^\rho \, L_{\rho\mu\nu} +$$
$$+ v_{2;\mu\sigma}^\rho \, L_{\rho\nu}^\sigma + v_{2;\nu\sigma}^\rho \, L_{\rho\mu}^\sigma + v_{2;\mu\nu}^\rho \, L_\rho = 0 \,.$$

6.4.2 Simulation and Verification

In order to test the theory, and at the same time illustrate how one could proceed in specific applications, we define two analytically tractable (2+1)D stimuli, one simulating a density, the other a scalar field. We then derive an exact, closed-form expression for normal flow in the usual temporal gauge, exploiting the scalar paradigm in *both* cases. This is done for the purpose of analysis and empirical verification, although the construction suggests that in the density case one would probably want to use the density paradigm instead. Simulation has the advantage that evaluation will not be hampered by uncertain factors such as reconstruction artifacts, complications that often arise in real image data (and will have to be handled on the basis of modality specific models).

Noise generally causes any idealisation to be violated, and so we study significant noise perturbations as well. There ought to be no problem according to theoretical prediction: away from isolated singularities, everything depends continuously upon the noise.

The raw images consist of oscillating Gaussian blobs (Figure 6.5) that behave as a density and as a scalar, respectively, and the tests are run on discretised floating-point representations of these, as well as on instances perturbed by 50% multiplicative, pixel-uncorrelated Gaussian noise.

6.4.2.1 Density Gaussian

Consider the following stimulus definition in a Cartesian coordinate system:

$$F(x, y, t) = \frac{1}{4\pi s(t)} e^{-\frac{x^2+y^2}{4s(t)}}, \tag{6.6}$$

with the following choice of kinematics:

$$s(t) = A + B \sin(\frac{2\pi t}{T}) \quad (A > B > 0). \tag{6.7}$$

Conservation for this stimulus follows from

$$\int dx dy \, F(x, y, t) = 1 \quad \text{for all } t. \tag{6.8}$$

In other words, each time-slice contains the same amount of "mass". The Lie derivative w.r.t. the vector field $\underline{v} = (v^t; v^x, v^y)$ is

$$\mathcal{L}_{\underline{v}} F = F_x v^x + F_y v^y + F_t v^t \stackrel{\text{def}}{=} 0, \tag{6.9}$$

in which a subscript denotes a partial derivative. This result holds globally. The temporal component can locally be gauged to unity as usual,

$$v^t \stackrel{\text{def}}{=} 1, \tag{6.10}$$

provided topological detail is conserved. For simplicity let us write $(v^x, v^y) = (u, v)$. Straightforward computation yields the following linear equation:

$$xu + yv = \left(\frac{r^2}{2s(t)} - 2\right) \dot{s}(t) \stackrel{\text{def}}{=} \alpha(r, t). \tag{6.11}$$

Normal flow can be obtained by imposing an additional spatial gauge identical to Equation 6.9, but with $\underline{v} = (1; u, v)$ replaced by its spatial dual $*\underline{v} = (0; -v, u)$:

$$xv - yu = 0. \tag{6.12}$$

In other words: the spatial velocity vector $\vec{v}$ is everywhere parallel to $\vec{x} = (x, y)$, which is also obvious from considerations of symmetry. Solving Equations 6.10, 6.11 and 6.12 yields the following solution:

$$\begin{pmatrix} u \\ v \end{pmatrix} = \frac{\alpha(r, t)}{r^2} \begin{pmatrix} x \\ y \end{pmatrix}. \tag{6.13}$$

The origin is problematic in the temporal gauge (why?) even though the example is not at all pathological[5]. Note also that $\vec{v} = \vec{0}$ on the oscillating circle defined by $r^2 = 4s(t)$ (on which F_t vanishes identically), and that the direction of the optical flow vector flips across this circle (which, by the way, runs counter to our visual percept, which gives the impression of alternating expansions and contractions). The situation at time $t \approx 0$ is as follows:

- within the zero-flow radius, $\vec{v}$ points inward, i.e. towards the centre of the blob,

- outside the zero-flow radius, $\vec{v}$ points outward, and

- the symmetry centre is a singularity.

The singularity behaves as

$$\|\vec{v}(r)\| \sim 2\frac{|\dot{s}|}{r} \qquad (r \downarrow 0), \tag{6.14}$$

and arises as an artifact of the scalar flow paradigm, for which the imposed gauge conditions are apparently too strong. In reality there is no fundamental problem, of course; the reader may verify that the singularity vanishes when using the density paradigm subject to a natural gauge instead, and calculating physically meaningful quantities, such as the "mass flux" through a sphere containing the singularity. For example, a radial field $\vec{v}(r) \propto \vec{r}/r^2$ in 3D will yield a constant flux (by the divergence theorem).

6.4.2.2 Scalar Gaussian

We now consider the scalar case. We create a raw image of the form

$$G(x, y, t) = e^{-\frac{x^2 + y^2}{4s(t)}}. \tag{6.15}$$

For mathematical convenience we take (with $\sigma = \sqrt{2s}$)

$$\sigma(t) = A + B\sin(\frac{2\pi t}{T}) \qquad (A > B > 0). \tag{6.16}$$

Equation 6.8 does not hold. The Lie derivative w.r.t. the optical flow vector field $\underline{v} = (v^t; v^x, v^y)$ is given by Equation 6.9, with F replaced by G. Again we assume the temporal gauge to hold: $(v^t; v^x, v^y) = (1; u, v)$. Unlike previously, there ought

---

[5]Note that the singularity cannot be "blurred out"; blurring amounts to an offset in $s(t)$.

to be no problem with this, since the scalar flow paradigm is the appropriate one to use for this stimulus. We get the following linear equation:

$$xu + yv = \frac{r^2}{2s(t)} \dot{s}(t) \stackrel{\text{def}}{=} \beta(r, t).$$
(6.17)

Normal flow can be obtained using the default gauges (Equations 6.10 and 6.12). Combined with Equation 6.17 this yields the following solution:

$$\left(\begin{array}{c} u \\ v \end{array} \right) = \frac{\beta(r, t)}{r^2} \left(\begin{array}{c} x \\ y \end{array} \right).$$
(6.18)

The solution differs qualitatively from Equation 6.13. First of all, $\beta(r, t)$ has a global sign, equal to that of $\dot{s}(t)$; there is no zero-flow spatial circle across which the flow inverts (in agreement with perceptual flow). Secondly, as anticipated, there is no singularity.

6.4.2.3 Numerical Test

The frame of interest is the first one, corresponding to $t = 0$. The spatial symmetry centre is taken as the origin $(x, y) = (0, 0)$. Parameter values are as follows. For Equations 6.6–6.7 we take $A = 52$, $B = 20$, $T = 16$. For Equations 6.15—6.16 we choose $A = 8$, $B = 4$ and $T = 16$. We select the same spatial and temporal scale parameters in both cases: $\sigma = 2$ and $\tau = 1$ (all values relative to pixel scale or frame interval). Both images are of size $16 \times 128 \times 128$, so that they accommodate full periods, and are computed with 4 bytes-per-pixel floating point precision[6].

Figures 6.6–6.17 show the results. They all refer to the normal flow field equations

$$\left\{ \begin{array}{rcl} \mathcal{L}_{\underline{v}} F[\phi] & = & 0 \\ \mathcal{L}_{*\underline{v}} F[\phi] & = & 0, \end{array} \right.$$

for $\underline{v} = (1; u, v)$, whence $*\underline{v} = (0, -v, u)$, using the analytical, zeroth or first order approximating schemes.

The analytical study has been discussed in-depth in the previous sections. We again adhere to Notation 3.1, Page 65, with one subtle but important reinterpretation, viz. that each index μ now denotes a *scaled* derivative w.r.t. x^μ; e.g. $L_{xtt} = -\sigma \tau^2 F[\phi_{xtt}]$, etc. It is important for the discussion that follows to distinguish between the explicit scaling factors σ, τ showing up in the amplitudes, and the inner scale parameters of the Gaussian filter ϕ.

The $k = 0$ approximation is based on a "locally constant" field $u_0(t; x, y) = u$, $v_0(t; x, y) = v$, and is found by inversion of

$$\left\{ \begin{array}{rcl} uL_x + vL_y & = & -L_t \\ -vL_x + uL_y & = & 0, \end{array} \right.$$

---

[6]Floating point precision of data is not required (cf. the results of noise simulation), but is useful in computations.

which is the same as the traditional OFCE for normal flow. For $k=1$ we insert first order polynomials $u_1(t; x, y) = u + u_t t + u_x x + u_y y$, $v_1(t; x, y) = v + v_t t + v_x x + v_y y$, for which we get instead

$$
\begin{cases}
uL_x + vL_y + u_t L_{xt} + v_t L_{yt} + u_x L_{xx} + v_x L_{xy} + u_y L_{xy} + v_y L_{yy} & = -L_t \\
uL_{xt} + vL_{yt} + u_t(L_x + L_{xtt}) + v_t(L_y + L_{ytt}) + u_x L_{xxt} + v_x L_{xyt} + u_y L_{xyt} + v_y L_{yyt} & = -L_{tt} \\
uL_{xx} + vL_{xy} + u_t L_{xxt} + v_t L_{xyt} + u_x(L_x + L_{xxx}) + v_x(L_y + L_{xxy}) + u_y L_{xxy} + v_y L_{xyy} & = -L_{xt} \\
uL_{xy} + vL_{yy} + u_t L_{xyt} + v_t L_{yyt} + u_x L_{xxy} + v_x L_{xyy} + u_y(L_x + L_{xyy}) + v_y(L_y + L_{yyy}) & = -L_{yt} \\
uL_y - vL_x + u_t L_{yt} - v_t L_{xt} + u_x L_{xy} - v_x L_{xx} + u_y L_{yy} - v_y L_{xy} & = 0 \\
uL_{yt} - vL_{xt} + u_t(L_y + L_{ytt}) - v_t(L_x + L_{xtt}) + u_x L_{xyt} - v_x L_{xxt} + u_y L_{yyt} - v_y L_{xyt} & = 0 \\
uL_{xy} - vL_{xx} + u_t L_{xyt} - v_t L_{xxt} + u_x(L_y + L_{xxy}) - v_x(L_x + L_{xxx}) + u_y L_{xyy} - v_y L_{xxy} & = 0 \\
uL_{yy} - vL_{xy} + u_t L_{yyt} - v_t L_{xyt} + u_x L_{xyy} - v_x L_{xxy} + u_y(L_y + L_{yyy}) - v_y(L_x + L_{xyy}) & = 0.
\end{cases}
$$

The lowest order system has $1 + 1$ equations in 2 unknowns, u, v (not to be confused with those of the first order system!), and is determined in terms of the image's first order derivatives (or $1 + 2$ equations in 3 unknowns if temporal gauge is made explicit). The first order system comprises $4 + 4$ equations in $2 + 6$ unknowns, $u, v, u_x, u_y, u_t, v_x, v_y, v_t$, and requires derivatives of orders $1, 2, 3$ (or $4 + 8$ equations in $3 + 9$ unkowns, respectively). Recall that the optic flow parameters are not the field's derivatives; for example, the parameters u, v arising from the latter system generally refine those of the former (the order tag has been left out for notational simplicity).

Both systems above have been solved numerically (pixel-wise) by a standard method called "LU decomposition" as described in Numerical Recipes [238, section 2.3]. The coefficients (in the spatial domain: linear convolutions of the raw image with Gaussian derivative filters) have been computed using FFT in the straightforward way.

6.4.2.4 ✳ Conceptual Comparison with Similar Methods

Upon first glance there is nothing remarkable about the linear systems in Section 6.4.2.3 in comparison to similar ones proposed in the literature. It would lead too far to make a scrutinised comparison here. Suffice it to say that, since our point of departure is different, one naturally expects to find different equations. Still, the similarity with other approaches, such as suggested by Otte and Nagel [234], or by Werkhoven and Koenderink [295], is particularly eminent, and it is instructive to point out how the theory in this chapter differs from theirs (apart from the constraint on order of approximation relative to order of differentiation, which has been argued to be quite essential).

Among others, Otte and Nagel discuss two OFCE-based linear systems, which they refer to as the "Rigorous Condition" (RC) and the "Integrated Condition" (IC), respectively, of which they have presented particular orders by way of example. The gauge invariant part of the RC [234, Equation 8, top four rows] is readily obtained from the first order system above by taking the limit $\sigma, \tau \downarrow 0$ for the explicit scale factors. This means that such factors as $L_x + L_{xtt}$ (more explicitly $\sigma L_x + \sigma \tau^2 L_{xtt}$) will effectively converge to L_x (i.e. σL_x; the overall scale factor will cancel in the equation), etc. If the remaining derivatives are computed at finite inner scales $\sigma, \tau > 0$ (which they are by necessity), we obtain an interpretation of a conservation law that is assumed to hold *at these finite scales*. Recall that in our scheme, conservation applies to the hypothetical level $\sigma = \tau = 0$, while only its *manifestation* is investigated at finite scales. The reader may verify that if the scalar OFCE is applied to the scalar Gaussian of Equation 6.15 after spatial blurring to scale $\sigma_0 > 0$, the flow will develop a singularity at the origin, which behaves (again we write $\sigma_0 = \sqrt{2s_0}$) as

$$\|\vec{v}(r)\| \sim \frac{s_0}{s} \frac{|\dot{s}|}{r} \qquad (r \downarrow 0),$$

which signifies the inadequacy of the RC. The rest of the RC [234, Equation 8, all other rows] has no counterpart in our theory, as we argued this to be part of a specific kind of gauge fixing.

However, more closely related is Otte and Nagel's IC [234, Equations 10–11a]. Again, to be able to compare we must disregard their second order equations, which correspond to gauge conditions. In addition we must include temporal derivatives; altogether we must consider the analogous, underdetermined (gauge invariant) set of 4 equations in 8 unknowns, as in the $k = 1$ system above. In a way the IC is based on duality, which explains the similarity with our method. However, under the IC the (approximate) Lie derivative is integrated over a spatiotemporal cylinder (the temporal extension of the spatial sphere used by Otte and Nagel [234, Equation 9]), and not computed as a smoothly weighted integral. From a duality perspective, the IC-filter is an indicator function on a cylinder of spatial radius and temporal extent proportional to σ and τ, respectively, which prohibits transposition of derivatives. Recall that for the purpose of well-posed and operationally well-defined derivation, filter smoothness is an essential demand. Taking into account the abovementioned modifications to the IC, one finds upon careful inspection that the coefficients resulting from the IC approach are indeed identical in form to the ones of the $k = 1$ system above.

In the context of visual motion detection Werkhoven and Koenderink [295] conjecture that

> "The aperture problem [...] is not an inherent problem in visual motion detection."

This is evidently true, since visual routines must somehow be gauge constrained (we do have a definite, or at least very limited transparent motion percept, even if illusory). Curiously, although the authors do in fact propose an explicit and visually plausible mechanism for gauge fixing, viz. by monitoring the output of a *restricted* set of receptive fields from the Gaussian family [167], they refrain from stating unequivocally that this *ipso facto* brings in an external source of information, witnessing their somewhat confusing remark:

> "Except for image irradiance patterns that vary only in one dimension, it [i.e. the aperture problem] does not arise if higher order spatial derivatives are considered."

Technically, however, Werkhoven and Koenderink's approach clearly comes nearest to the one described in this chapter, as it is in fact a gauge constrained instance based on the same duality principle.

Niessen *et al.* [224, 227] have exploited the deep structure of the proposed OFCE (likewise to first order) by selecting appropriate scales at each pixel so as to optimise a reliability measure for the extracted flow field, and have made a comparison of the result with a few of the best performing algorithms as described by Barron *et al.* [13]. Applied to a standard video sequence, the first order scheme with scale selection already shows competitive or even superior performance, provided *a priori* knowledge is brought in properly in the form of gauge constraints. However, schemes such as the ones illustrated here are more rigorously applicable to medical imagery. Along the same lines, Niessen *et al.* have studied the left ventricular wall motion of the canine heart in cine-MR and cine-CT [223, 224, 227]. For an in-depth investigation of scale selection and a quantitative comparison with other methods the reader is referred to Florack *et al.* [86].

6.5 Summary and Discussion

In this chapter we have presented conceptual and computational aspects of optic flow measurements. It has been argued that, from a syntactical viewpoint, the aperture problem is not at all a false problem, as is sometimes suggested by the *ipse dixit* claiming that its solution can be deduced from higher order image structure. In other words, it cannot be solved on the basis of data evidence without semantical input. One can interpret the intrinsic ambiguity as a gauge invariance proper to image intrinsic optic flow. The codimension of isophotes limits its intrinsic d.o.f. to one per base point and per level of resolution. Normal flow is a "least-committed" representative of gauge invariant optic flow. Tangential flow corresponds to its virtual, or extrinsic d.o.f.'s. A model or gauge condition is required to disambiguate possibilities. In a typical disambiguation scheme, normal flow and tangential flow are inherently coupled.

An operational approach based on a novel interpretation of the OFCE has been shown to yield a linear system of equations for optic flow measurement different from existing schemes. The coefficients are linear combinations of well-posed local image derivatives, computed in accordance with scale-space theory. The resulting system has been formulated in closed-form by exploiting specific properties of the Gaussian family. Depending upon the situation, one can obtain a meaningful solution after adding admissible gauge conditions. The details of this are beyond the scope of this book; rather, the manifest separation of image evidence versus model in the process of extracting optic flow has been emphasized.

An aspect that has not been addressed here is *optimal order*; the optic flow extraction method has been formulated for arbitrary orders of approximation M, involving image derivatives of orders as high as $2M + 1$. Optimality of differential order is therefore a decisive factor; this depends upon various details such as noise, discretisation characteristics, reconstruction quality, resolution of interest, etc. Higher orders are quite likely to improve results even further as far as these quality criteria permit. This is to be expected because the formal expansion v_M converges to the actual optic flow defining generator v as $M \to \infty$, at least in theory. In practice it is useful to consider the Cauchy differences $v_{M+1} - v_M$, since these are actually computable up to some order M. From these one can infer the rate of successive refinement of the various optic flow parameters, and thus get an impression of the rate of convergence of the local optic flow field expansion. In turn this enables predictions concerning which orders are relevant and which can be safely ignored, depending upon image structure, instead of having to commit oneself to a predetermined truncation order, which may either be overdone or inadequate at any given location [3, 217, 218, 234, 272]. Moreover, if the Cauchy differences do not decrease sufficiently fast, so that they become negligible before the optimal order prohibits further refinement, then we effectively have an optic flow *discontinuity*. This is of interest in segmentation tasks. Finally, knowledge of refinement is also important because all optic flow parameters, even the lowest order ones, are only reliably extracted if M-th order provides a sufficiently accurate local approximation.

A second problem that has not been addressed here suggests itself: the formulation of an optic flow theory in the context of real-time processing compatible with the temporally causal scale-spacetime construct of Section 4.3. Such a theory might well lead to natural models for the so-called "Reichardt motion detectors" in biological vision [239].

The theory in this chapter has been demonstrated by means of an analytically tractable stimulus. Numerical simulation of optic flow extraction shows qualitatively acceptable results to lowest order, and quantitatively quite accurate results to first order approximation. It has also been verified that the numerical computation is robust, a property which has explicitly been taken into consideration in the theoretical construction. Application to real data, such as cine-MR, is straightforward, but results must be evaluated in the light of modality specific quality factors and functional goal. The *real* problem is of course of a semantical nature, viz. to establish appropriate gauge conditions on the basis of *a priori* knowledge. The case of normal flow is of interest in its own right and has been extensively discussed.

☞ Box 6.1 (History)

The idea of "flow" can be traced a long way back in history. An early scientific account must be credited to the ancient Greek philosophers. Reasoning led Parmenides (ca. 540–480 B.C.) to the conclusion that "nothing changes". He was willing to sacrifice empirical evidence to support that view, arguing that our senses are deceptive; he was a profound rationalist. Despite his peculiar view on empirics, the core of his idea has survived two and a half thousand years: after all, *invariance* has become the very backbone of state-of-the-art physics! Heraclitus, a contemporary of Parmenides, seemed to hold the opposite view: "παντα ρει, και ουδεν μενει" ("everything flows, nothing abides"). Note his carefully chosen vocabulary; the key word is *flow*, not *change*. Clearly, Heraclitus' view was not at all contradicted by empirical facts. Indeed, he heavily relied on the faithfulness of sensory perception; his inclination was clearly more empirical.

Several ancient philosophers have felt the need to reconcile both views (Empedocles, 490–430 B.C., Anaxagoras, 500–428 B.C., a.o.), and, with some ad hoc-ery, more or less succeeded in doing that. Most interesting is perhaps Democritus (ca. 460–370 B.C.), who postulated the existence of immutable particles of different flavours. He considered these so-called *atoms* to be the elementary building blocks of which everything is made. Indeed, nothing actually changes in this view, but since the atoms are in principle free to flow, their mutual spatial relations and their relative contributions to atomic mixtures may vary. Thus Democritus was able to explain the apparent phenomenon of flow, as well as the very existence of different substances on the basis of a simple and elegant model. After a scientifically interesting epoch, dogmatism started to dominate the scene. It took more than two thousand years before the ancient ideas could be given a rock solid basis. Leibniz and Newton initiated the development of the powerful apparatus of *differential calculus*, providing convenient tools for quantifying "rate of change". Lie introduced Lie groups (recall Box 3.2, Page 75), allowing one to combine the principles of flow and invariance. The notion of a *Lie derivative*, which expresses the rate of change of a quantity when co-moving with some flow field, is in fact the perfect tool for setting up local optic flow models. Schwartz collected a large number of ideas and techniques that had arisen in the mathematical literature and put these into a coherent framework. His theory of distributions has turned the apparatus of differential calculus into something that can actually be used in the context of measurements and computations, as we have seen in previous chapters (earlier, the physical non-entity of an "infinitesimal" had already been abandoned by Cartan in his exterior differential calculus).

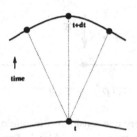

Figure 6.1: There is equal support from the data for all vectors connecting points on corresponding iso-grey-level contours at two successive moments. Gauge transformations—explained in the text—are all diffeomorphisms confined to isophotes; under such transformations a given optic flow field is mapped to one that is *equivalent* in the sense that there is *no observable effect* on the data.

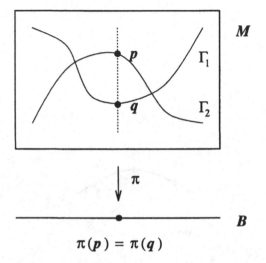

$$\pi(p) = \pi(q)$$

Figure 6.2: In a gauge theory one considers a manifold M of higher dimensionality than the actual base space of physical observables B. By virtue of a symmetry principle postulated for M ("gauge invariance") there is a stratification into "invariant orbits" that induce identical identical observations on B via a "projection map" $\pi(p) = \pi(q)$. This implies that one can impose any gauge condition $\Gamma_1, \Gamma_2, \ldots$ in practice, as long as it is admissible, i.e. transversal to the orbits. The manifold M, together with its inherent symmetry, thus merely enters as a "model space" for B, in which the manifestation of the latter may be significantly simpler than in any non-gauge model confined to B. In the case at hand M will be the vector bundle of n-vectors v^μ representing all possible optic flow vector fields compatible with the image data, while B comprises the actual optic flow evidence in the form of, say, the normal flow field components (more precisely, the 1D homotopy of iso-grey-level contours). The virtue of a gauge theoretical approach to optic flow lies in the fact that one can model B via a simple, viz. linear model for M, at the price of only a mild concession: local redundancy, or gauge invariance. In this case there are $n - 1$ gauge d.o.f.'s (per base point), which can be fixed by imposing equally many gauge conditions, none of which contradicts data evidence ("tangential flow"). These gauge conditions must reflect *a priori* knowledge or a hypothesis about the physical cause of the induced flow.

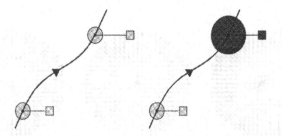

Figure 6.3: In a scalar interpretation of the OFCE (left) it is asserted that two points (or fixed-scale samples taken at these points), lying on one flow line, have the same grey-value attribute. In the density interpretation (right) grey-values are effectively attributed to volumes. In that case local volume elements are susceptible to the divergence of the optic flow field. In these sketches, the indicated patches are intended to represent corresponding volume elements. In the case of a density, the OFCE entails that volumetrically integrated grey-values be the same provided the integration aperture is consistently transformed by the flow. In the scalar case volume elements are not affected by the flow. This is also the case in density images if the flow happens to be divergence-free.

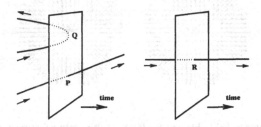

Figure 6.4: Left: a transversal and a non-transversal flow-line (through P and Q, respectively). The latter one is excluded by the transversality requirement. The upper branch of that curve has an anti-causal orientation. It is quite easy to give it a causal interpretation simply by reversing the arrow; in that case, however, Q becomes an annihilation point, thus conflicting the assumption of "conservation of topological detail". Right: if the spatial optic flow field vanishes at a point R, the flow will be parallel to the time axis.

Figure 6.5: First frame of the optic flow test sequence, corresponding to $t = 0$.

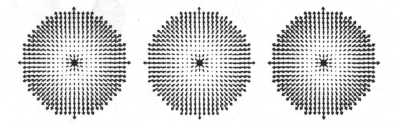

Figure 6.6: Left: Analytically determined velocity field for the density stimulus (6.6). **Middle & Right**: Velocity field obtained in the zeroth and first order approximation ($\sigma_s = 2; \sigma_t = 1$). Vectors are plotted at 50% of their magnitude for clarity.

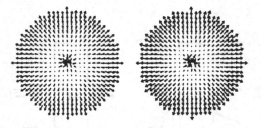

Figure 6.7: Velocity field obtained for (6.6). Multiplicative Gaussian noise (50%) has been added to the original data. Vectors are plotted at 50% of their magnitude for clarity. **Left**: zeroth order approximation. **Right**: first order approximation.

Figure 6.8: Error in velocity field obtained for the density stimulus (6.6) in the noise free case. **Left**: zeroth order approximation. **Right**: first order approximation.

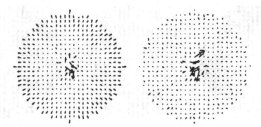

Figure 6.9: Error in velocity field obtained for the density stimulus (6.6) after the original image has been corrupted with multiplicative Gaussian noise (50%). **Left:** zeroth order approximation. **Right:** first order approximation.

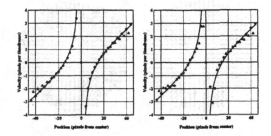

Figure 6.10: Cross-section of the velocity field obtained for the density stimulus (6.6); analytical (solid line), zeroth (triangles) and first order (stars). **Left:** noise free case. **Right:** 50% noise.

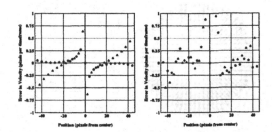

Figure 6.11: Error in cross-section of the velocity field obtained for the density stimulus (6.6); zeroth (triangles) and first order (stars). **Left:** noise free case. **Right:** 50% noise.

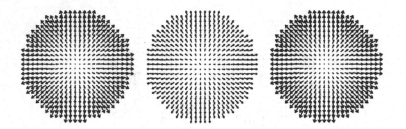

Figure 6.12: **Left:** Analytically determined velocity field for the scalar stimulus (6.15). **Middle & Right:** Velocity field obtained in the zeroth and first order approximation ($\sigma_s = 2; \sigma_t = 1$). Vectors are plotted at 20% of their magnitude for clarity.

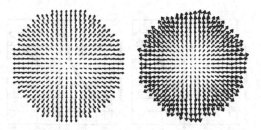

Figure 6.13: Velocity field obtained for the scalar stimulus (6.15). Multiplicative Gaussian noise (50%) has been added to the original data. Vectors are plotted at 20% of their magnitude for clarity. **Left:** zeroth order approximation. **Right:** first order approximation.

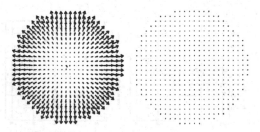

Figure 6.14: Error in velocity field obtained for the scalar stimulus (6.15). Vectors are plotted at 20% of their magnitude for clarity. **Left:** zeroth order approximation. **Right:** first order approximation.

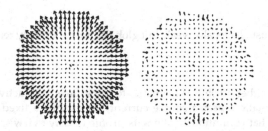

Figure 6.15: Error in velocity field obtained for the scalar stimulus (6.15), after the original image has been corrupted with multiplicative Gaussian noise (50%). Vectors are plotted at 40% of their magnitude for clarity. **Left:** zeroth order approximation. **Right:** first order approximation.

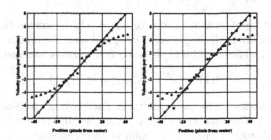

Figure 6.16: Cross-section of the velocity field obtained for the scalar stimulus (6.15); analytical (solid line), zeroth (triangles) and first order (stars). **Left:** noise free case. **Right:** 50% noise.

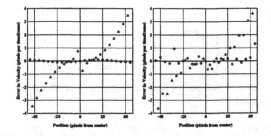

Figure 6.17: Error in cross-section of the velocity field obtained for the scalar stimulus (6.15); zeroth (triangles) and first order (stars). **Left:** noise free case. **Right:** 50% noise.

Problems

6.1

Page 180: "This is a usual—locally weak, but globally not necessarily realistic—assumption..."
Explain.

6.2

By duality we may either attribute kinematics to the source field (active view: dynamical world model; kinematic cause: a source current), or assume a fixed source field being exposed to a filter that continuously alters its profile (passive view: static world model; kinematic cause: a filter current density). In this exercise we consider the latter option. Let M be the symmetric, positive definite $(d+1) \times (d+1)$ matrix, given by the Cartesian components

$$M^{\mu\nu} = \left(\begin{array}{c|c} M^{00} & \emptyset \\ \hline \emptyset & M^{ij} \end{array} \right),$$

with arbitrary positive M^{00} and symmetric, positive definite matrix M^{ij} ($M^{i0} = M^{0j} = 0$). Define

$$\phi_M(x) = \frac{1}{\sqrt{2\pi}^{d+1}} \frac{1}{\sqrt{\det M}} \exp(-\frac{1}{2} x^\mu M^{\text{inv}}_{\mu\nu} x^\nu).$$

In a suitable orthonormal basis we can take $M = \sigma^2 I$ (ct-convention: Definition 3.17, Page 62). In that case we shall simply write $\phi(x)$ and omit the matrix subscript M. Now, suppose that the spatial flow field $v^i(x)$ is linear in x^i relative to some fixation point, say

$$v^\mu(x) = (v^0(x); v^i(x)) = (1; A^i_j x^j).$$

Recall that Latin indices refer to space, not time. Furthermore, consider the parametrised spacetime curve $x(\varepsilon)$, with time as the affine parameter, and with initial condition $x(0) = x = (t; \vec{x})$. A dot denotes *transposed* (i.e. *minus*) differentiation w.r.t. ε; in particular we have $(\dot{t}; \dot{\vec{x}}) = -(1; \vec{v}(\vec{x}))$ at $\varepsilon = 0$.

a. Separating space and time explicitly, $\phi(x) = \phi^{\text{time}}(t) \phi^{\text{space}}(\vec{x})$, show that, in the dual view, the filter transformation $\dot{\phi}(x) = -D_\mu j^\mu(x)$ is given by

$$\begin{cases} \dot{\phi}^{\text{time}}(t) & = & t\,\phi^{\text{time}}(t) & = & -D_t \phi^{\text{time}}(t), \\[2mm] \dot{\phi}^{\text{space}}(\vec{x}) & = & \left(x_i A^i_j x^j - A^i_i \right) \phi^{\text{space}}(\vec{x}) & = & A^i_j D_i D^j \phi^{\text{space}}(\vec{x}). \end{cases}$$

b. Define

$$\begin{cases} \phi^{\text{time}}(t; \varepsilon = 0) & = & \phi^{\text{time}}_I(t), \\[2mm] \phi^{\text{space}}(\vec{x}; \varepsilon = 0) & = & \phi^{\text{space}}_I(\vec{x}), \end{cases}$$

and show that "exponentiation" of the equations in **a** with these initial conditions yield the following solution:

$$\begin{cases} \phi^{\text{time}}(t; \varepsilon) & \overset{\text{def}}{=} & e^{-\varepsilon D_t} \phi^{\text{time}}_I(t) & = & \phi^{\text{time}}_I(t - \varepsilon), \\[2mm] \phi^{\text{space}}(\vec{x}; \varepsilon) & \overset{\text{def}}{=} & e^{\varepsilon A^i_j D_i D^j} \phi^{\text{space}}_I(\vec{x}) & = & \phi^{\text{space}}_{I + \varepsilon(A + A^T)}(\vec{x}). \end{cases}$$

c. Conclude that the filter continues, at least for some finite time, to adapt its profile to the linear flow. Argue that, depending on the details of the flow field, the adaptation process may cease to make sense after a certain finite time due to physical limitations. Consider a "pure, negative divergence", for which $A_j^i = \alpha \delta_j^i$ with $\alpha < 0$ (so $d\alpha = A_i^i = D_i v^i(x)$ is the only degree of freedom in the flow), and show that in this case the whole thing collapses after a time $\varepsilon = -1/(2\alpha)$ if one refrains from reinitialising the system periodically.

6.3

Page 178: "...indeed, quite a number of assumptions..." There are at least five.

Appendix A

Geometry and Tensor Calculus

A.1 Literature

There is plenty of introductory literature on differential geometry and tensor calculus. It depends very much on scientific background which of the following references provides the most suitable starting point. The list is merely a suggestion and is not meant to give a complete overview.

- Y. Choquet-Bruhat, C. DeWitt-Morette, and M. Dillard-Bleick, *Analysis, Manifolds, and Physics. Part I: Basics* [36]: overview of many fundamental notions of analysis, differential calculus, the theory of differentiable manifolds, the theory of distributions, and Morse theory. An excellent summary of standard mathematical techniques of generic applicability. A good reminder for those who have seen some of it already; probably less suited for initiates.

- M. Spivak, *Calculus on Manifolds* [264]: a condensed introduction to basic geometry. A "black piste" that starts off gently at secondary school level. A good introduction to Spivak's more comprehensive work.

- M. Spivak, *Differential Geometry* [265]: a standard reference for mathematicians. Arguably a bit of an overkill; however, the first volume contains virtually all tensor calculus one is likely to run into in the image literature for the next decade or so.

- C. W. Misner, K. S. Thorne, and J. A. Wheeler, *Gravitation*: an excellent course in geometry from a physicist's point of view [211]. Intuitive and practical, even if you have not got the slightest interest in gravitation (in which case you may want to skip the second half of the book).

- J. J. Koenderink, *Solid Shape* [158]: an informal explanation of geometric concepts, helpful in developing a pictorial attitude. The emphasis is on *understanding* geometry. For this reason a useful companion to any tutorial on the subject.

- A. P. Lightman, W. H. Press, R. H. Price, and S. A. Teukolsky, *Problem Book in Relativity and Gravitation* [185]: ±500 problems on relativity and gravitation to practice with; a good way to acquire computational skill in geometry.

- M. Friedman, *Foundations of Space-Time Theories: Relativistic Physics and Philosophy of Science* [92]: a philosophical view on differential geometry and its relation to classical, special relativistic, and general relativistic spacetime theories. Technique is de-emphasized in favour of understanding the role of geometry in physics.

- Other classics: R. Abraham, J. E. Marsden, T. Ratiu, *Manifolds, Tensor Analysis, and Applications* [1]; V. I. Arnold, *Mathematical Methods of Classical Mechanics* [6]; M. P. do Carmo, *Riemannian Geometry* [31] and *Differential Geometry of Curves and Surfaces* [30]; R. L. Bishop and S. I. Goldberg, *Tensor Analysis on Manifolds* [18]; H. Flanders, *Differential Forms* [72].

The next section contains a highly condensed summary of geometric concepts introduced in this book. The reader is referred to the text and to aforementioned literature for detailed explanations.

A.2 Geometric Concepts

A.2.1 Preliminaries

Kronecker symbol or *Kronecker tensor:*

$$\delta^\alpha_\beta = \begin{cases} 1 & \text{if } \alpha = \beta \\ 0 & \text{if } \alpha \neq \beta \end{cases} \tag{A.1}$$

Generalised Kronecker or *permutation symbols:*

$$\delta^{\mu_1 \cdots \mu_k}_{\nu_1 \cdots \nu_k} = \det \begin{pmatrix} \delta^{\mu_1}_{\nu_1} & \cdots & \delta^{\mu_1}_{\nu_k} \\ \vdots & & \vdots \\ \delta^{\mu_k}_{\nu_1} & \cdots & \delta^{\mu_k}_{\nu_k} \end{pmatrix} =$$

$$\begin{cases} +1 & \text{if } (\mu_1, \ldots, \mu_k) \text{ even permutation of } (\nu_1, \ldots, \nu_k), \\ -1 & \text{if } (\mu_1, \ldots, \mu_k) \text{ odd permutation of } (\nu_1, \ldots, \nu_k), \\ 0 & \text{otherwise}. \end{cases} \tag{A.2}$$

Completely antisymmetric symbol in n dimensions:

$$[\mu_1 \ldots \mu_n] = \begin{cases} +1 & \text{if } (\mu_1, \ldots, \mu_n) \text{ is an even permutation of } (1, \ldots, n), \\ -1 & \text{if } (\mu_1, \ldots, \mu_n) \text{ is an odd permutation of } (1, \ldots, n), \\ 0 & \text{otherwise}. \end{cases} \tag{A.3}$$

The *tensor product* of any two $\mathbb{R}$-valued operators: if

$$X : \text{Dom}\, X \to \mathbb{R} \quad : \quad x \mapsto X(x), \tag{A.4}$$

$$Y : \text{Dom}\, Y \to \mathbb{R} \quad : \quad y \mapsto Y(y), \tag{A.5}$$

then

$$X \otimes Y : \text{Dom}\, X \times \text{Dom}\, Y \to \mathbb{R} : (x; y) \mapsto X \otimes Y(x; y) \tag{A.6}$$

is the operator defined by

$$X \otimes Y(x; y) \stackrel{\text{def}}{=} X(x) \cdot Y(y). \tag{A.7}$$

Alternation or *antisymmetrisation* of an operator: if

$$X : \underbrace{V \otimes \ldots \otimes V}_{k} \to \mathbb{R} : (x_1, \ldots, x_k) \mapsto X(x_1, \ldots, x_k), \tag{A.8}$$

then

$$\text{Alt}\, X : \underbrace{V \otimes \ldots \otimes V}_{k} \to \mathbb{R} : (x_1, \ldots, x_k) \mapsto \text{Alt}\, X(x_1, \ldots, x_k)$$

$$\stackrel{\text{def}}{=} \frac{1}{k!} \sum_{\pi} \text{sgn}\, \pi\, X(x_{\pi[1]}, \ldots, x_{\pi[k]}), \tag{A.9}$$

in which the sum extends over all permutations π of $(1, \ldots, k)$, and in which sgn π denotes the sign of π.

Symmetrisation of an operator: if

$$X : \underbrace{V \otimes \ldots \otimes V}_{k} \to \mathbb{R} : (x_1, \ldots, x_k) \mapsto X(x_1, \ldots, x_k), \tag{A.10}$$

then

$$\text{Sym}\, X : \underbrace{V \otimes \ldots \otimes V}_{k} \to \mathbb{R} : (x_1, \ldots, x_k) \mapsto \text{Sym}\, X(x_1, \ldots, x_k)$$

$$\stackrel{\text{def}}{=} \frac{1}{k!} \sum_{\pi} X(x_{\pi[1]}, \ldots, x_{\pi[k]}), \tag{A.11}$$

in which the sum extends over all permutations π of $(1, \ldots, k)$.

Index symmetrisation and *antisymmetrisation*:

$$X_{(\mu_1 \ldots \mu_k)} = \frac{1}{k!} \sum_{\pi} X_{\mu_{\pi(1)} \ldots \mu_{\pi(k)}}, \tag{A.12}$$

$$X_{[\mu_1 \ldots \mu_k]} = \frac{1}{k!} \sum_{\pi} \text{sgn}\, \pi\, X_{\mu_{\pi(1)} \ldots \mu_{\pi(k)}}. \tag{A.13}$$

Einstein summation convention in n dimensions:

$$X_\alpha Y^\alpha \stackrel{\text{def}}{=} \sum_{\alpha=1}^{n} X_\alpha Y^\alpha. \tag{A.14}$$

Einstein summation convention for antisymmetric sums: if $X_{\mu_1 \ldots \mu_k}$ and $Y^{\mu_1 \ldots \mu_k}$ are antisymmetric, then

$$X_{|\mu_1 \ldots \mu_k|} \, Y^{\mu_1 \ldots \mu_k} \overset{\text{def}}{=} \sum_{\mu_1 < \ldots < \mu_k} X_{\mu_1 \ldots \mu_k} \, Y^{\mu_1 \ldots \mu_k} = \frac{1}{k!} X_{\mu_1 \ldots \mu_k} \, Y^{\mu_1 \ldots \mu_k} \,. \qquad \text{(A.15)}$$

A.2.2 Vectors

A *vector* lives in a vector (or linear) space:

$$\vec{v} \in V \,. \qquad \text{(A.16)}$$

Pictorial representation: a vector is an arrow with its tail attached to some base point (Figure A.1). Vectors are *rates*: if f is a $\mathbb{R}$-valued function, then

$$\vec{v}[f] = \vec{v}(df) = df(\vec{v}) = v^\alpha \, \partial_\alpha f \,, \qquad \text{(A.17)}$$

i.e. the rate of change of f along $\vec{v}$ (for definition of df, see Sections A.2.3 and A.2.7). This also shows you how to define polynomials and power series of a vector, e.g.

$$\exp \vec{v}[f] = \sum_{n=0}^{\infty} \frac{1}{n!} v^{\alpha_1} \partial_{\alpha_1} \ldots v^{\alpha_n} \partial_{\alpha_n} f \,. \qquad \text{(A.18)}$$

This is just Taylor's expansion at the (implicit) base point to which $\vec{v}$ is assumed to be attached (see "tangent space").
Leibniz's product rule:

$$\vec{v}[fg] = g \, \vec{v}[f] + f \, \vec{v}[g] \,. \qquad \text{(A.19)}$$

Decomposition relative to basis vectors:

$$\vec{v} = v^\alpha \, \vec{e}_\alpha \,. \qquad \text{(A.20)}$$

Standard coordinate basis:

$$\vec{e}_\alpha = \partial_\alpha \,. \qquad \text{(A.21)}$$

Operational definition: a vector is a linear, $\mathbb{R}$-valued function of covectors (or 1-forms), producing the *contraction* of the vector and the covector:

$$\vec{v} : V^* \to \mathbb{R} : \tilde{\omega} \mapsto \vec{v}(\tilde{\omega}) \,. \qquad \text{(A.22)}$$

Often, one attaches a vector space V to each base point p of a manifold M (Figure A.1):

$$V \sim TM_p \,. \qquad \text{(A.23)}$$

TM_p is the *tangent space at p*, or the "fibre over p" of the manifold's *tangent bundle* (Figure A.1):

$$TM = \bigcup_{p \in M} TM_p \,. \qquad \text{(A.24)}$$

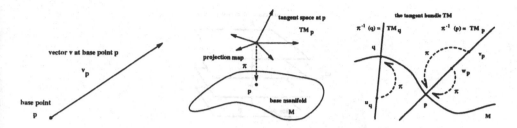

Figure A.1: Vector, tangent space, tangent bundle.

The "vector $\vec{v}$ at base point p":

$$\vec{v}_p \equiv (p; \vec{v}) \in \mathrm{TM}_p .\tag{A.25}$$

One can find out to which base point a vector is attached by means of the *projection map*:

$$\pi : \mathrm{TM} \to \mathrm{M} : \vec{v}_p \mapsto p .\tag{A.26}$$

The tangent space TM_p projects, by definition, to p:

$$\pi^{-1} : \mathrm{M} \to \mathrm{TM} : p \mapsto \mathrm{TM}_p .\tag{A.27}$$

A *vector field* is a smooth *section* of the tangent bundle:

$$s : \mathrm{M} \to \mathrm{TM} : p \mapsto s(p) = \vec{v}_p .\tag{A.28}$$

Any vector field projects to the identity map of the base manifold:

$$\pi \circ s = \mathrm{I_M} .\tag{A.29}$$

A.2.3 Covectors

A *covector* (or *1-form*) lives in a covector space:

$$\tilde{\omega} \in \mathrm{V}^* .\tag{A.30}$$

Pictorial representation: a covector is an oriented stack of planes "attached" to some base point (a "planar wave disregarding phase"; see Figure A.2).
Covectors are *gradients*.
Decomposition relative to basis covectors:

$$\tilde{\omega} = \omega_\alpha \, \tilde{e}^\alpha .\tag{A.31}$$

Standard coordinate basis:

$$\tilde{e}^\alpha = dx^\alpha .\tag{A.32}$$

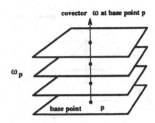

Figure A.2: Covector.

Operational definition: a covector is a linear, $\mathbb{R}$-valued function of vectors, producing the *contraction* of the covector and the vector:

$$\tilde{\omega} : V \to \mathbb{R} : \vec{v} \mapsto \tilde{\omega}\,(\vec{v})\,. \tag{A.33}$$

Often, one attaches a covector space V^* to each base point p of a manifold M:

$$V^* \sim T^*M_p\,. \tag{A.34}$$

T^*M_p is the *cotangent space at* p, or the fibre over p of the manifold's *cotangent bundle*:

$$T^*M = \bigcup_{p \in M} T^*M_p\,. \tag{A.35}$$

The "covector $\tilde{\omega}$ at base point p":

$$\tilde{\omega}_p \equiv (p; \tilde{\omega}) \in T^*M_p\,. \tag{A.36}$$

One can find out to which base point a covector is attached by means of the *projection map*:

$$\pi' : T^*M \to M : \tilde{\omega}_p \mapsto p\,. \tag{A.37}$$

The cotangent space T^*M_p projects, by definition, to p:

$$\pi'^{-1} : M \to T^*M : p \mapsto T^*M_p\,. \tag{A.38}$$

A *covector field* is a smooth *section* of the cotangent bundle:

$$s' : M \to T^*M : p \mapsto s'(p) = \tilde{\omega}_p\,. \tag{A.39}$$

Any covector field projects to the identity map of the base manifold:

$$\pi' \circ s' = I_M\,. \tag{A.40}$$

A.2.4   Dual Bases

Dual bases (no metric required!):

$$\tilde{e}^\alpha\,(\vec{e}_\beta) = \vec{e}_\beta\,(\tilde{e}^\alpha) = \delta^\alpha_\beta\,. \tag{A.41}$$

Figure A.3: The gauge figure.

A.2.5 Riemannian Metric

Riemannian metric tensor:

$$G : V \times V \to \mathbb{R} : (\vec{v}, \vec{w}) \mapsto G(\vec{v}, \vec{w}) . \tag{A.42}$$

A Riemannian metric is a symmetric, bilinear, positive definite, $\mathbb{R}$-valued mapping of two vectors:

$$
\begin{aligned}
\text{linearity:} \quad & G(\lambda \vec{u} + \mu \vec{v}, \vec{w}) &=& \ \lambda G(\vec{u}, \vec{w}) + \mu G(\vec{v}, \vec{w}) , \\
\text{symmetry:} \quad & G(\vec{u}, \vec{v}) &=& \ G(\vec{v}, \vec{u}) , \\
\text{positivity:} \quad & G(\vec{u}, \vec{u}) &>& \ 0 \qquad \forall \vec{u} \neq 0 .
\end{aligned}
\tag{A.43}
$$

The *scalar product* is another name for the metric:

$$\vec{v} \cdot \vec{w} \stackrel{\text{def}}{=} G(\vec{v}, \vec{w}) . \tag{A.44}$$

Pictorial representation: a Riemannian metric is a quadric "centred" at some base point, the *gauge figure*: Figure A.3.
The *sharp operator* converts vectors into covectors:

$$\sharp : V \to V^* : \vec{v} \mapsto \sharp \vec{v} \stackrel{\text{def}}{=} G(\vec{v}, .) \stackrel{\text{def}}{=} v_\alpha \, \tilde{e}^\alpha . \tag{A.45}$$

The $\sharp$-operator is invertible, $\sharp^{-1} \equiv \flat$ (the *flat operator*):

$$\flat : V^* \to V : \tilde{v} \mapsto \flat \tilde{v} \stackrel{\text{def}}{=} H(\tilde{v}, .) \stackrel{\text{def}}{=} v^\alpha \, \vec{e}_\alpha , \tag{A.46}$$

with

$$H(\sharp \vec{v}, \sharp \vec{w}) \stackrel{\text{def}}{=} G(\vec{v}, \vec{w}) . \tag{A.47}$$

In particular, $\sharp$ converts a basis vector into a corresponding covector:

$$\sharp \vec{e}_\alpha = g_{\alpha\beta} \, \tilde{e}^\beta . \tag{A.48}$$

Similarly, $\flat$ converts a basis covector into a corresponding vector:

$$\flat \tilde{e}^\alpha = g^{\alpha\beta} \, \vec{e}_\beta . \tag{A.49}$$

If $\vec{e}_\alpha$ and $\tilde{e}^\beta$ are dual, then $\flat \tilde{e}^\beta$ and $\sharp \vec{e}_\alpha$ are dual, too.
The $\sharp$-operator and its inverse motivate the *lowering* and *raising of indices*:

$$
\begin{aligned}
\downarrow : \ v_\alpha &= \ g_{\alpha\beta} \, v^\beta , \tag{A.50} \\
\uparrow : \ v^\alpha &= \ g^{\alpha\beta} \, v_\beta . \tag{A.51}
\end{aligned}
$$

Pictorial representation: $\sharp$ ($\flat$) represents an inversion in the gauge figure, mapping the tip of a vector to its "polar plane" (vice versa).

Equivalent metric representations:

$$
\begin{aligned}
H &: V^* \times V^* \rightarrow \mathbb{R} : (\tilde{v},\tilde{w}) \mapsto H(\tilde{v},\tilde{w}) \overset{\text{def}}{=} G(\flat\tilde{v},\flat\tilde{w}), \\
I &: V \times V^* \rightarrow \mathbb{R} : (\vec{v},\tilde{w}) \mapsto I(\vec{v},\tilde{w}) \overset{\text{def}}{=} G(\vec{v},\flat\tilde{w}).
\end{aligned}
\tag{A.52}
$$

Decompositions relative to basis:

$$
\begin{aligned}
G &= g_{\alpha\beta}\,\tilde{e}^\alpha \otimes \tilde{e}^\beta = \sharp\,\vec{e}_\alpha \otimes \tilde{e}^\alpha, \\
H &= g^{\alpha\beta}\,\vec{e}_\alpha \otimes \vec{e}_\beta = \flat\,\tilde{e}^\alpha \otimes \vec{e}_\alpha, \\
I &= \delta^\beta_\alpha\,\tilde{e}^\alpha \otimes \vec{e}_\beta = \tilde{e}^\alpha \otimes \vec{e}_\alpha.
\end{aligned}
\tag{A.53}
$$

Corresponding metric components:

$$
g_{\alpha\beta} = G(\vec{e}_\alpha,\vec{e}_\beta), \quad g^{\alpha\beta} = H(\tilde{e}^\alpha,\tilde{e}^\beta), \quad \delta^\beta_\alpha = I(\vec{e}_\alpha,\tilde{e}^\beta).
\tag{A.54}
$$

Relation between various metric components:

$$
g^{\alpha\mu}g_{\mu\beta} = \delta^\alpha_\beta.
\tag{A.55}
$$

A.2.6   Tensors

Operational definition: a (mixed) *tensor* is a multilinear, $\mathbb{R}$-valued function of vectors and covectors:

$$
T : \underbrace{V \otimes \ldots \otimes V}_{k} \otimes \underbrace{V^* \otimes \ldots \otimes V^*}_{l} \rightarrow \mathbb{R} :
$$

$$
(\vec{v}_1,\ldots,\vec{v}_k,\tilde{\omega}_1,\ldots,\tilde{\omega}_l) \mapsto T(\vec{v}_1,\ldots,\vec{v}_k,\tilde{\omega}_1,\ldots,\tilde{\omega}_l).
\tag{A.56}
$$

Decomposition relative to basis vectors and covectors:

$$
T = T^{\beta_1\ldots\beta_l}_{\alpha_1\ldots\alpha_k}\,\tilde{e}^{\alpha_1} \otimes \ldots \otimes \tilde{e}^{\alpha_k} \otimes \vec{e}_{\beta_1} \otimes \ldots \otimes \vec{e}_{\beta_l}.
\tag{A.57}
$$

Components:

$$
T^{\beta_1\ldots\beta_l}_{\alpha_1\ldots\alpha_k} = T(\vec{e}_{\alpha_1},\ldots,\vec{e}_{\alpha_k},\tilde{e}^{\beta_1},\ldots,\tilde{e}^{\beta_l}).
\tag{A.58}
$$

Spivak's convention (tag "S"):

- T is a *tensor* of type $\binom{k}{l}_{\text{S}}$.

- T is said to be *covariant of rank k* and *contravariant of rank l*.

Misner, Thorne & Wheeler's convention (tag "MTW"):

- $T^{\beta_1\ldots\beta_l}_{\alpha_1\ldots\alpha_k}$ are the *tensor components* of type $\binom{l}{k}_{\text{MTW}}$ (take care!).

- $T^{\beta_1\ldots\beta_l}_{\alpha_1\ldots\alpha_k}$ are said to be *covariant of rank k* and *contravariant of rank l*.

- In a space without metric, this terminology carries over to the tensor T itself.

- In a Riemannian space, a tensor can be generalised so as to incorporate "metrical knowledge": see Equation A.68 below.

- If $T = T^{\beta_1\ldots\beta_l}_{\alpha_1\ldots\alpha_k}\, \vec{e}_{\beta_1} \otimes \ldots \otimes \vec{e}_{\beta_l} \otimes \tilde{e}^{\alpha_1} \otimes \ldots \otimes \tilde{e}^{\alpha_k}$, its components are written as $T^{\beta_1\ldots\beta_l}{}_{\alpha_1\ldots\alpha_k}$, so as to avoid confusion about the ordering of basis vectors and covectors.

Every bundle V gives rise to a *tensor bundle*:

$$T \in \mathcal{J}^k_l(V). \tag{A.59}$$

Dimension:

$$\dim \mathcal{J}^k_l(V) = n^{k+l}. \tag{A.60}$$

Special cases:

$$\mathcal{J}_l(V) \;=\; \mathcal{J}^0_l(V) \quad \text{("contravariant tensors of rank } l\text{")}, \tag{A.61}$$
$$\mathcal{J}^k(V) \;=\; \mathcal{J}^k_0(V) \quad \text{("covariant tensors of rank } k\text{")}. \tag{A.62}$$

A tensor bundle, generalising a vector or covector bundle, comprises fibres and base points:

$$\mathcal{J}^k_l(TM) = \bigcup_{p \in M} \mathcal{J}^k_l(TM_p). \tag{A.63}$$

The base point of a tensor is revealed by the *projection map*:

$$\Pi : \mathcal{J}^k_l(TM) \to M : T_p \mapsto p. \tag{A.64}$$

The fibre $\mathcal{J}^k_l(TM_p)$ projects, by definition, to p:

$$\Pi^{-1} : M \to \mathcal{J}^k_l(TM) : p \mapsto \mathcal{J}^k_l(TM_p). \tag{A.65}$$

A *tensor field* is a smooth *section* of a tensor bundle:

$$\Sigma : M \to \mathcal{J}^k_l(TM) : p \mapsto \Sigma(p) = T_p. \tag{A.66}$$

Any tensor field projects to the identity map of the base manifold:

$$\Pi \circ \Sigma = I_M. \tag{A.67}$$

In a Riemannian space, a tensor can be generalised so as to incorporate built-in conversion rules based on the $\sharp$ and $\flat$-operators:

$$T \sim \{T; \sharp, \flat\} \tag{A.68}$$

having the following interpretation:

- $T \in \mathcal{J}_l^k(TM)$ is *any* "prototype tensor" of fixed total rank $k + l$ as defined above.

- The cast operator $\sharp$ (or its inverse, $\flat$) is implicitly invoked whenever an argument (a vector or covector) passed into a slot of T does not match the prototype T.

- The type of T is characterised by its *total rank* $m = k + l$.

- The partial, covariant and contravariant ranks k and l refer to any representative prototype T (and its components) used in a particular "realization" T of T.

The $\sharp$-operator can be used to convert between different representations of a rank-m tensor T:

$$T \in \mathcal{J}_l^k(V) \rightleftharpoons T' \in \mathcal{J}_{l'}^{k'}(V) \quad \text{with} \quad k' + l' = k + l = m. \tag{A.69}$$

In particular, you can define $\updownarrow T \in \mathcal{J}_k^l(V)$ by applying $\sharp$ and $\flat$ to all basis vectors and forms in the decomposition of T (component representation: raising all lower and lowering all upper indices).

You may similarly define $\downarrow T \in \mathcal{J}^{k+l}(V)$ and $\uparrow T \in \mathcal{J}_{k+l}(V)$ by applying only $\sharp$ or $\flat$ to either basis vectors or covectors.

The *norm* of a tensor: if $T \in \mathcal{J}_l^k(V)$, then $\updownarrow T \in \mathcal{J}_k^l(V)$, so you can "contract T onto itself" to get its squared norm:

$$\| T \|^2 \stackrel{\text{def}}{=} \updownarrow T(T) = \uparrow T(\downarrow T). \tag{A.70}$$

Components:

$$\| T \|^2 = T_{\mu_1 \ldots \mu_l}{}^{\mu_{l+1} \ldots \mu_{k+l}} T^{\mu_1 \ldots \mu_l}{}_{\mu_{l+1} \ldots \mu_{k+l}} = T^{\mu_1 \ldots \mu_{k+l}} T_{\mu_1 \ldots \mu_{k+l}}. \tag{A.71}$$

But: see also Equation A.88 below.

Example of $T \sim \{T; \sharp, \flat\}$: the metric tensor $G : V \times V \to \mathbb{R}$ as defined previously (A.42–A.43) is a prototype for the generalised metric tensor G; depending on what you feed it, G behaves either as the prototype G (2 vectors), H (2 covectors), or I (1 vector and 1 covector, respectively).

Tensors of the same type $\begin{pmatrix} l \\ k \end{pmatrix}_{\text{MTW}} = \begin{pmatrix} k \\ l \end{pmatrix}_{\text{S}}$ can be linearly combined:

$$(\lambda S + \mu T)(\vec{v}_1, \ldots, \vec{v}_k, \tilde{\omega}_1, \ldots, \tilde{\omega}_l) =$$
$$\lambda S(\vec{v}_1, \ldots, \vec{v}_k, \tilde{\omega}_1, \ldots, \tilde{\omega}_l) + \mu T(\vec{v}_1, \ldots, \vec{v}_k, \tilde{\omega}_1, \ldots, \tilde{\omega}_l). \tag{A.72}$$

Tensors of arbitrary types $\begin{pmatrix} l \\ k \end{pmatrix}_{\text{MTW}}$ and $\begin{pmatrix} n \\ m \end{pmatrix}_{\text{MTW}}$ can be multiplied; if

$$\vec{V}_k \equiv (\vec{v}_1, \ldots, \vec{v}_k), \quad \vec{W}_m \equiv (\vec{w}_1, \ldots, \vec{w}_m), \quad \tilde{\Omega}_l \equiv (\tilde{\omega}_1, \ldots, \tilde{\omega}_l), \quad \tilde{\Sigma}_n \equiv (\tilde{\sigma}_1, \ldots, \tilde{\sigma}_n), \tag{A.73}$$

then

$$(S \otimes T)(\vec{V}_k, \tilde{\Omega}_l, \vec{W}_m, \tilde{\Sigma}_n) = S(\vec{V}_k, \tilde{\Omega}_l) \cdot T(\vec{W}_m, \tilde{\Sigma}_n), \tag{A.74}$$

yielding a product tensor of type $\left(\begin{smallmatrix} l+n \\ k+m \end{smallmatrix}\right)_{MTW}$.

Symmetrising tensors: if $S \in \mathcal{J}^k(V)$, then

$$\mathrm{Sym}\, S \in \Sigma^k(V). \tag{A.75}$$

Components:

$$(\mathrm{Sym}\, S)_{\mu_1 \ldots \mu_k} = S_{(\mu_1 \ldots \mu_k)}. \tag{A.76}$$

Similar definitions apply to contravariant and mixed tensors (symmetrisation of mixed tensors is carried out as an independent symmetrisation of covariant and/or contravariant parts): if $T \in \mathcal{J}_l(V)$ and $U \in \mathcal{J}_l^k(V)$, then

$$\mathrm{Sym}\, T \;\in\; \Sigma_l(V), \tag{A.77}$$
$$\mathrm{Sym}\, U \;\in\; \Sigma_l^k(V). \tag{A.78}$$

Alternating or *antisymmetrising tensors*: if $S \in \mathcal{J}^k(V)$, then

$$\mathrm{Alt}\, S \in \Lambda^k(V). \tag{A.79}$$

Components:

$$(\mathrm{Alt}\, S)_{\mu_1 \ldots \mu_k} = S_{[\mu_1 \ldots \mu_k]}. \tag{A.80}$$

Similar definitions apply to contravariant and mixed tensors (alternation of mixed tensors is carried out as an independent alternation of covariant and contravariant parts): if $T \in \mathcal{J}_l(V)$ and $U \in \mathcal{J}_l^k(V)$, then

$$\mathrm{Alt}\, T \;\in\; \Lambda_l(V), \tag{A.81}$$
$$\mathrm{Alt}\, U \;\in\; \Lambda_l^k(V). \tag{A.82}$$

The Alt-operator is idempotent:

$$\mathrm{Alt} \circ \mathrm{Alt} = \mathrm{Alt}. \tag{A.83}$$

The Alt-operator is linear, hence can be viewed as a *constant tensor*. When acting on a rank-k cotensor S with components $S_{\mu_1 \ldots \mu_k}$, Alt has components

$$\mathrm{Alt}_{\nu_1 \ldots \nu_k}{}^{\mu_1 \ldots \mu_k} = \frac{1}{k!}\, \delta^{\mu_1 \ldots \mu_k}_{\nu_1 \ldots \nu_k}. \tag{A.84}$$

Decomposition relative to basis: if $S \in \Lambda^k(V)$, then

$$S = S_{\mu_1 \ldots \mu_k}\, \tilde{e}^{\mu_1} \otimes \ldots \otimes \tilde{e}^{\mu_k} = S_{|\mu_1 \ldots \mu_k|}\, \tilde{e}^{\mu_1} \wedge \ldots \wedge \tilde{e}^{\mu_k}, \tag{A.85}$$

with $S_{\mu_1 \ldots \mu_k} = S_{[\mu_1 \ldots \mu_k]}$ antisymmetric; see also Equation A.93 below.
Similar definitions apply to the alternation of contratensors and mixed tensors (the latter corresponding to the tensor product of two permutation tensors).
The *wedge product* is an antisymmetrised tensor product: if $S_i \in \Lambda^{k_i}(V)$ then

$$S_1 \wedge \ldots \wedge S_p = \frac{(k_1 + \ldots + k_p)!}{k_1! \ldots k_p!}\, \mathrm{Alt}\, (S_1 \otimes \ldots \otimes S_p). \tag{A.86}$$

Components: if $\kappa_j = \sum_{i=1}^{j} k_i$, then

$$(S_1 \wedge \ldots \wedge S_p)_{\mu_1 \ldots \mu_{\kappa_p}} = \delta_{\mu_1 \ldots \mu_{\kappa_p}}^{\rho_1 \ldots \rho_{\kappa_p}} S_{1|\rho_1 \ldots \rho_{\kappa_1}|} \cdots S_{p|\rho_{\kappa_{p-1}+1} \ldots \rho_{\kappa_p}|} . \tag{A.87}$$

Rule of thumb: $S_1 \wedge \ldots \wedge S_p$ equals $S_1 \otimes \ldots \otimes S_p$ plus terms that guarantee complete antisymmetry. Likewise for $S_i \in \Lambda_{k_i}(V)$. In the general case of mixed tensors you are supposed to "wedge" the covariant and contravariant parts independently. Special cases: *k-forms* and *k-vectors*: if $\tilde{\omega}_1, \ldots, \tilde{\omega}_k$ and $\vec{v}_1, \ldots, \vec{v}_k$ are k covectors and k vectors, respectively, then $\tilde{\omega}_1 \wedge \ldots \wedge \tilde{\omega}_k \in \Lambda^k(V)$ is a k-form, and $\vec{v}_1 \wedge \ldots \wedge \vec{v}_k \in \Lambda_k(V)$ is a k-vector.

When defining the *norm of an antisymmetric tensor* one often incorporates a combinatorial factor to avoid "overcounting": if $\mathbf{A}$ is an antisymmetric tensor of rank k (typically prototyped by some k-form $\mathbf{A} \in \Lambda^k(V)$ or k-vector $\mathbf{A} \in \Lambda_k(V)$), then

$$\| \mathbf{A} \|^2 = A_{|\mu_1 \ldots \mu_k|} A^{\mu_1 \ldots \mu_k} \tag{A.88}$$

Physical interpretations of n-forms and n-vectors (in n dimensions):

- n-forms are *densities*;

- n-vectors are *capacities*.

Basis k-forms: the set of all

$$\tilde{e}^{\mu_1} \wedge \ldots \wedge \tilde{e}^{\mu_k} \quad 1 \leq \mu_1 < \ldots < \mu_k \leq n \tag{A.89}$$

is a basis for $\Lambda^k(V)$.

Basis k-vectors: the set of all

$$\vec{e}_{\mu_1} \wedge \ldots \wedge \vec{e}_{\mu_k} \quad 1 \leq \mu_1 < \ldots < \mu_k \leq n \tag{A.90}$$

is a basis for $\Lambda_k(V)$.
Consequently:

$$\dim \Lambda^k(V) = \dim \Lambda_k(V) = \binom{n}{k} . \tag{A.91}$$

A unit k-form is designed to exactly fit its dual unit k-vector:

$$\tilde{e}^{\mu_1} \wedge \ldots \wedge \tilde{e}^{\mu_k} (\vec{e}_{\nu_1} \wedge \ldots \wedge \vec{e}_{\nu_k}) = \vec{e}_{\nu_1} \wedge \ldots \wedge \vec{e}_{\nu_k} (\tilde{e}^{\mu_1} \wedge \ldots \wedge \tilde{e}^{\mu_k}) = \delta_{\nu_1 \ldots \nu_k}^{\mu_1 \ldots \mu_k} . \tag{A.92}$$

It is more natural to use the wedge product basis than the ordinary tensor product basis for decomposing an antisymmetric tensor: if $S \in \Lambda^k(V)$, then

$$S = S_{|\mu_1 \ldots \mu_k|} \tilde{e}^{\mu_1} \wedge \ldots \wedge \tilde{e}^{\mu_k} \tag{A.93}$$

Reminder: tensor coefficients $S_{\mu_1 \ldots \mu_k}$ are usually defined in terms of the ordinary tensor product basis.

The *Levi-Civita tensor* in n dimensions is the unique unit n-form of positive orientation:

$$\varepsilon \stackrel{\text{def}}{=} \sqrt{g}\, \tilde{e}^1 \wedge \ldots \wedge \tilde{e}^n = \varepsilon_{\mu_1 \ldots \mu_n} \tilde{e}^{\mu_1} \wedge \ldots \wedge \tilde{e}^{\mu_n} , \tag{A.94}$$

with $g \stackrel{\text{def}}{=} \det g_{ij}$.

Covariant and contravariant components:

$$\varepsilon_{\mu_1 \ldots \mu_n} = \sqrt{g} \, [\mu_1 \ldots \mu_n] \, , \tag{A.95}$$

$$\varepsilon^{\mu_1 \ldots \mu_n} = \frac{1}{\sqrt{g}} \, [\mu_1 \ldots \mu_n] \, . \tag{A.96}$$

The *Hodge star operator* converts k-forms (k-vectors) into $(n - k)$-forms ($(n - k)$-vectors, respectively):

$$* : \Lambda^k(V) \to \Lambda^{n-k}(V) : A \mapsto *A . \tag{A.97}$$

It can be defined recursively: if $S \in \Lambda^k(V)$ is a k-form and $\tilde{\omega} \in \Lambda^1(V)$ is a 1-form, then

$$* : \left\{ \begin{array}{ll} *1 & = \varepsilon , \\ *(S \wedge \tilde{\omega}) & = (*S) \, (\sharp \tilde{\omega}) . \end{array} \right. \tag{A.98}$$

In general, if

$$S = S_{|\mu_1 \ldots \mu_k|} \, \tilde{e}^{\mu_1} \wedge \ldots \wedge \tilde{e}^{\mu_k} , \tag{A.99}$$

then

$$*S = [*S]_{|\mu_{k+1} \ldots \mu_n|} \, \tilde{e}^{\mu_{k+1}} \wedge \ldots \wedge \tilde{e}^{\mu_n} , \tag{A.100}$$

where

$$[*S]_{\mu_{k+1} \ldots \mu_n} \stackrel{\text{def}}{=} S^{|\mu_1 \ldots \mu_k|} \varepsilon_{\mu_1 \ldots \mu_n} . \tag{A.101}$$

A.2.7 Push Forward, Pull Back, Derivative Map

Push forward and pull back:

$$\left\{ \begin{array}{l} f_* : V \to W \\ f^* : W^* \to V^* , \end{array} \right. \tag{A.102}$$

in which f_* is any linear map and

$$f^* \tilde{\omega} \, (\vec{v}) = \tilde{\omega} \, (f_* \vec{v}) \qquad \text{for all } \vec{v} \in V \text{ and } \tilde{\omega} \in W^* . \tag{A.103}$$

Diagrammatically:

$$
\begin{array}{ccc}
f^* \tilde{\omega} \in T^*M & \xleftarrow{\;f^*\;} & \tilde{\omega} \in T^*N \\
\downarrow & & \downarrow \\
f^* \tilde{\omega} \, (\vec{v}) & = & \tilde{\omega} \, (f_* \vec{v}) \\
\uparrow & & \uparrow \\
\vec{v} \in TM & \xrightarrow{\;f_*\;} & f_* \vec{v} \in TN
\end{array}
\tag{A.104}
$$

In general you can pull back any kind of tensors:

$$\begin{cases} f_* : V \to W \\ f^* : \mathcal{J}^k(W) \to \mathcal{J}^k(V), \end{cases} \tag{A.105}$$

with, by definition,

$$f^* T(\vec{v}_1, \ldots, \vec{v}_k) = T(f_*(\vec{v}_1), \ldots, f_*(\vec{v}_k)) \quad \text{for all } \vec{v}_1, \ldots, \vec{v}_k \in V \text{ and } T \in \mathcal{J}^k(W). \tag{A.106}$$

Typically, any function on a manifold $f : M \to N$ induces a linear map, its *differential*:

$$f_* \equiv df. \tag{A.107}$$

The differential gives you the difference between "tip and tail" of a vector (see Section A.2.2):

$$df(\vec{v}) = \vec{v}[f]. \tag{A.108}$$

Decomposition relative to coordinate basis:

$$df = \partial_\mu f \, dx^\mu. \tag{A.109}$$

Commutative diagram:

$$\begin{array}{ccc} TM & \xrightarrow{f_*} & TN \\ \pi \downarrow & & \downarrow \pi \qquad \pi \circ f_* = f \circ \pi \\ M & \xrightarrow{f} & N \end{array} \tag{A.110}$$

The chain rule,

$$d(g \circ f)(p) = dg(f(p)) \circ df(p), \tag{A.111}$$

can be restated as

$$(g \circ f)_* = g_* \circ f_*. \tag{A.112}$$

Appendix B

The Filters $\Phi^{\rho_1 \dots \rho_l}_{\mu_1 \dots \mu_k}$

Consider the first case of Result 6.1, Page 184. Using the following lemma we can get rid of the derivative ∂_μ in the integrand (we use Cartesian coordinates).

Lemma B.1
Using parentheses to denote index symmetrisation, we have

$$\partial_\mu \Phi^{\rho_1 \dots \rho_l}_{\mu_1 \dots \mu_k} (x) = -\Phi^{\rho_1 \dots \rho_l}_{\mu_1 \dots \mu_k \mu} (x) + \delta^{(\rho_l}_\mu \Phi^{\rho_1 \dots \rho_{l-1})}_{\mu_1 \dots \mu_k} (x).$$

It is understood that $\Phi^{\rho_1 \dots \rho_{l-1}}_{\mu_1 \dots \mu_k} \equiv 0$ *if* $l = 0$.

The proof of this lemma is straightforward and will be omitted. Using this lemma we can rewrite Result 6.1:

Result B.1
See Result 6.1.

$$\partial_{\mu_1 \dots \mu_k} \mathcal{L}_{v_M} F[\phi] = \sum_{l=0}^{M} v^\mu_{M;\rho_1 \dots \rho_l} \int dx \, f(x) \left[\Phi^{\rho_1 \dots \rho_l}_{\mu_1 \dots \mu_k \mu}(x) - \Phi^{\rho_1 \dots \rho_{l-1}}_{\mu_1 \dots \mu_k}(x) \delta^{\rho_l}_\mu \right].$$

(Note that index symmetrisation w.r.t. $\rho_1, \dots, \rho_l$ is automatic by virtue of symmetry of the coefficients $v^\mu_{M;\rho_1 \dots \rho_l}$.) One may compare this result to the second case of Result 6.1.

The essence is now to express the overcomplete set of filters $\Phi^{\rho_1 \dots \rho_l}_{\mu_1 \dots \mu_k}$ in terms of Gaussian derivative filters $\phi_{\mu_1 \dots \mu_m}$. Consider the following diagram.

$$
\begin{array}{ccc}
\Phi^{\rho_1 \dots \rho_l}_{\mu_1 \dots \mu_k}(x) & \xrightarrow{\;\; \mathbf{F} \;\;} & \hat{\Phi}^{\rho_1 \dots \rho_l}_{\mu_1 \dots \mu_k}(\omega) \\
{\scriptstyle \star} \downarrow & & \downarrow {\scriptstyle \star\star} \\
\Phi^{\rho_1 \dots \rho_l}_{\mu_1 \dots \mu_k}(x) & \xleftarrow{\;\; \mathbf{F}^{\mathrm{inv}} \;\;} & \hat{\Phi}^{\rho_1 \dots \rho_l}_{\mu_1 \dots \mu_k}(\omega)
\end{array}
$$

Instead of simplifying directly in the spatial domain (the arrow marked by a $\star$ indicates a simplification step), we take the Fourier route ($\mathbf{F} \to \star\star \to \mathbf{F}^{\mathrm{inv}}$), and simplify in Fourier space. According to Definition 2.8, we can make the following formal identifications of operators (the l.h.s. in the spatial domain, the r.h.s. in the Fourier domain):

$$x^\rho \equiv i\frac{\partial}{\partial\omega_\rho}\,, \qquad \frac{\partial}{\partial x^\rho} \equiv i\omega_\rho\,. \qquad\qquad (B.1)$$

We need one more definition.

Definition B.1 (Hermite Polynomials)
The Hermite polynomial of order k, H_k, is defined by

$$\frac{d^k}{dx^k}e^{-\frac{1}{2}x^2} = (-1)^k H_k(x)\,e^{-\frac{1}{2}x^2}\,.$$

This is appropriate for the 1-dimensional case. Let us define the n-dimensional analogue of the Hermite polynomials as follows.

Definition B.2 (Hermite Polynomials in n Dimensions)
The n-dimensional Hermite polynomial of order k, $\mathcal{H}_{i_1...i_k}$, is defined by

$$\frac{\partial^k}{\partial x^{i_1}\dots\partial x^{i_k}}e^{-\frac{1}{2}x^2} = (-1)^k \mathcal{H}_{i_1...i_k}(x)\,e^{-\frac{1}{2}x^2}\,.$$

These n-dimensional Hermite polynomials are related to the standard ones in the following way.

Lemma B.2 (Relation to Standard Definition)
The n-dimensional Hermite polynomials as defined according to Definition B.2 are related to the standard definition, Definition B.1, as follows.

$$\mathcal{H}_{i_1...i_k}(x) = \prod_{j=1}^{n} H_{\alpha_j^{i_1...i_k}}(x^j)\,,$$

in which $\alpha_j^{i_1...i_k}$ denotes the number of indices in $i_1,\dots,i_k$ equal to j.

Clearly we have $\sum_{j=1}^{n}\alpha_j^{i_1...i_k} = k$, since this simply sums up all indices. The separability property of Lemma B.2 follows straightforwardly from Definition B.1, when applied to a multidimensional Gaussian.

Having established all basic ingredients and notational matters, we can now relate the overcomplete family of filters $\Phi^{\rho_1...\rho_l}_{\mu_1...\mu_k}$ to the Gaussian family. This is easy, since all we need to do is to use Leibniz's product rule for differentiation in

$$\hat{\Phi}^{\rho_1...\rho_l}_{\mu_1...\mu_k}(\omega) = \frac{(-1)^k}{l!}i\frac{\partial}{\partial\omega_{\rho_1}}\dots i\frac{\partial}{\partial\omega_{\rho_l}}\left(i\omega_{\mu_1}\dots i\omega_{\mu_k}\hat{\phi}(\omega)\right)\,, \qquad (B.2)$$

(see Formula B.1 and the definition of the filters in Result 6.1). Then, each time we have to take a derivative of $\hat{\phi}(\omega)$, we use the explicit property of the Gaussian stated in Definition B.2. In this way we arrive at

Result B.2 (The Filters $\Phi^{\rho_1\cdots\rho_l}_{\mu_1\cdots\mu_k}$ and the Gaussian Family)

Let S denote the index symmetrisation operator (applying separately to upper and lower indices), then we have

$$\hat{\Phi}^{\rho_1\cdots\rho_l}_{\mu_1\cdots\mu_k}(\omega) =$$

$$\frac{(-1)^k}{l!} S \left\{ \sum_{m=0}^{\min(k,l)} \binom{l}{m} \frac{(-1)^m k!}{(k-m)!} \delta^{\rho_1}_{\mu_1} \cdots \delta^{\rho_m}_{\mu_m} i\omega_{\mu_{m+1}} \cdots i\omega_{\mu_k} (-i)^{l-m} \mathcal{H}^{\rho_{m+1}\cdots\rho_l}(\omega)\, \hat{\phi}(\omega) \right\}.$$

Fourier inversion yields

$$\Phi^{\rho_1\cdots\rho_l}_{\mu_1\cdots\mu_k}(x) =$$

$$\frac{(-1)^k}{l!} S \left\{ \sum_{m=0}^{\min(k,l)} \binom{l}{m} \frac{(-1)^m k!}{(k-m)!} \delta^{\rho_1}_{\mu_1} \cdots \delta^{\rho_m}_{\mu_m} \partial_{\mu_{m+1}} \cdots \partial_{\mu_k} i^{l-m} \mathcal{H}^{\rho_{m+1}\cdots\rho_l}(i\nabla)\, \phi(x) \right\}.$$

Note that this expression is real in the spatial domain, since $i^p \mathcal{H}^{\rho_1\cdots\rho_p}(i\nabla)$ is a real differential operator for any $p \in \mathbb{Z}_0^+$. To see this, look at the explicit form of a Hermite polynomial:

$$H_k(x) = \sum_{m=0}^{[k/2]} (-1)^m \binom{k}{2m} (2m-1)!!\, x^{k-2m}, \tag{B.3}$$

in which $[x]$ denotes the *entier* of $x \in \mathbb{R}$, i.e. the largest integer less than or equal to x, and in which the double factorial $(2m-1)!!$ indicates the product $1 \times 3 \times \ldots \times (2m-1)$. Consequently,

$$i^k H_k(i\frac{d}{dx}) = (-1)^k \sum_{m=0}^{[k/2]} \binom{k}{2m} (2m-1)!! \frac{d^{k-2m}}{dx^{k-2m}}, \tag{B.4}$$

very real indeed. The general n-dimensional case follows from this observation. Note also that the r.h.s. of Result B.2 is a linear combination of Gaussian derivatives of the type $\phi_{\mu_1\cdots\mu_p}$, with $p = 0,\ldots,k+l$. Thus we have indeed proven overcompleteness of the (apparently $(k+l)$-th order) filters $\Phi^{\rho_1\cdots\rho_l}_{\mu_1\cdots\mu_k}$ by explicitly rewriting them in terms of Gaussian derivatives.

Appendix C

Proof of Proposition 5.4

To proof Proposition 5.4 we embed the $k \times k$ matrix $A^{\nu_1 \ldots \nu_k}_{\mu_1 \ldots \mu_k}$, introduced in Definition 5.8, into a $d \times d$-matrix $\tilde{A}^{\nu_1 \ldots \nu_k}_{\mu_1 \ldots \mu_k}$, which has the same determinant, by adding a $(d-k) \times (d-k)$ identity block, as follows:

$$\tilde{A}^{\nu_1 \ldots \nu_k}_{\mu_1 \ldots \mu_k} = \left(\begin{array}{c|c} A^{\nu_1 \ldots \nu_k}_{\mu_1 \ldots \mu_k} & 0_{k \times (d-k)} \\ \hline 0_{(d-k) \times k} & 1_{(d-k) \times (d-k)} \end{array} \right) .$$

We can write the determinant of the matrix $\tilde{A}^{\nu_1 \ldots \nu_k}_{\mu_1 \ldots \mu_k}$ with the use of a double ε-product:

$$\det \tilde{A}^{\nu_1 \ldots \nu_k}_{\mu_1 \ldots \mu_k} = \frac{1}{d!} \varepsilon_{\alpha_1 \ldots \alpha_d} \varepsilon^{\beta_1 \ldots \beta_d} (\tilde{A}^{\nu_1 \ldots \nu_k}_{\mu_1 \ldots \mu_k})^{\alpha_1}_{\beta_1} \ldots (\tilde{A}^{\nu_1 \ldots \nu_k}_{\mu_1 \ldots \mu_k})^{\alpha_d}_{\beta_d} .$$

Despite its d^{2d} terms, this is actually a rather sparse sum: only those terms for which the indices $(\alpha_1, \ldots, \alpha_d)$ and $(\beta_1, \ldots, \beta_d)$ are permutations of $(1, \ldots, d)$ survive, and we may reorder them in such a way that the first k indices address the actual matrix elements of $A^{\nu_1 \ldots \nu_k}_{\mu_1 \ldots \mu_k}$. To this end we consider all permutations $(k_1, \ldots, k_d)$ of $(1, \ldots, d)$, such that the k-tuple $(\alpha_{k_1}, \ldots, \alpha_{k_k})$ and the $(d-k)$-tuple $(\alpha_{k_{k+1}}, \ldots \alpha_{k_d})$ are permutations of $(1, \ldots, k)$ and $(k+1, \ldots, d)$, respectively. If we take into account a combinatorial factor, counting the various possibilities to choose this separation, we may assume that the indices $\alpha_{k_1}, \ldots, \alpha_{k_k}$ and $\beta_{k_1}, \ldots, \beta_{k_k}$ index the actual matrix elements of the block $A^{\nu_1 \ldots \nu_k}_{\mu_1 \ldots \mu_k}$, whereas $\alpha_{k_{k+1}}, \ldots, \alpha_{k_d}$ and $\beta_{k_{k+1}}, \ldots, \beta_{k_d}$ index the identity block. In other words,

$$(\tilde{A}^{\nu_1 \ldots \nu_k}_{\mu_1 \ldots \mu_k})^{\alpha_{k_i}}_{\beta_{k_i}} = (A^{\nu_1 \ldots \nu_k}_{\mu_1 \ldots \mu_k})^{\alpha_{k_i}}_{\beta_{k_i}} = \delta^{\nu_{\alpha_{k_i}}}_{\mu_{\beta_{k_i}}} ,$$

for the first k subindices $i = 1, \ldots, k$, and

$$(\tilde{A}^{\nu_1 \ldots \nu_k}_{\mu_1 \ldots \mu_k})^{\alpha_{k_i}}_{\beta_{k_i}} = \delta^{\alpha_{k_i}}_{\beta_{k_i}} ,$$

for the last $d-k$ indices $i = k+1, \ldots, d$. This leads to the following expression:

$$\det \tilde{A}^{\nu_1 \ldots \nu_k}_{\mu_1 \ldots \mu_k} = \binom{d}{k} \frac{1}{d!} \varepsilon_{\alpha_{k_1} \ldots \alpha_{k_d}} \varepsilon^{\beta_{k_1} \ldots \beta_{k_d}} \delta^{\nu_{\alpha_{k_1}}}_{\mu_{\beta_{k_1}}} \ldots \delta^{\nu_{\alpha_{k_k}}}_{\mu_{\beta_{k_k}}} \delta^{\alpha_{k_{k+1}}}_{\beta_{k_{k+1}}} \ldots \delta^{\alpha_{k_d}}_{\beta_{k_d}} ,$$

or, upon a relabelling of dummy contraction indices:

$$\det \tilde{A}^{\nu_1...\nu_k}_{\mu_1...\mu_k} = \frac{1}{k!(d-k)!} \, \varepsilon_{\alpha_{k_1}...\alpha_{k_k}\lambda_1...\lambda_{d-k}} \varepsilon^{\beta_{k_1}...\beta_{k_k}\lambda_1...\lambda_{d-k}} \, \delta^{\nu_{\alpha_{k_1}}}_{\mu_{\beta_{k_1}}} \cdots \delta^{\nu_{\alpha_{k_k}}}_{\mu_{\beta_{k_k}}} .$$

For a given k-tuple $(k_1, \ldots, k_k)$ there are $k!$ equally contributing terms in this expression (corresponding to all permutations of this k-tuple), and so we may finally rewrite this expression into:

$$\det \tilde{A}^{\nu_1...\nu_k}_{\mu_1...\mu_k} = \frac{1}{(d-k)!} \, \varepsilon_{\mu_1...\mu_k\lambda_1...\lambda_{d-k}} \varepsilon^{\nu_1...\nu_k\lambda_1...\lambda_{d-k}} .$$

Since $\det A^{\nu_1...\nu_k}_{\mu_1...\mu_k} = \det \tilde{A}^{\nu_1...\nu_k}_{\mu_1...\mu_k}$, we have completed the proof.

Appendix D

Proof of Proposition 5.5

Write $\{L, S_0, \ldots, S_{d-1}, I_1, \ldots, I_d\}$, with $S_k = L_{i_1} L_{i_1 i_2} L_{i_2 i_3} \ldots L_{i_k i_{k+1}} L_{i_{k+1}}$ ($k = 0, 1, 2, \ldots$) and $I_k = L_{i_1 i_2} L_{i_2 i_3} \ldots L_{i_k i_1}$ ($k = 1, 2, 3, \ldots$; both S_k and I_k contain k 2-vertices), and observe that all connected polynomial diagrams are of the form L, S_k or I_{k+1} for some $k \geq 0$.

We will first consider a system with 2-vertices only, and show that all I_k for $k > d$ are reducible. Then we will turn to the general case, first by including the S_k, showing their reducibility for $k > d - 1$, and then by extending it with the trivial zeroth order member L.

D.1   Irreducible System for $\{L_{ij}\}$

We concentrate on the second order system $\{L_{ij}\}$.

Definition D.1
Recall Definition 5.8, and define

$$L_{ij}^{[k]} \overset{\text{def}}{=} \delta_{i_1 \ldots i_k i; j_1 \ldots j_k j} L_{i_1 j_1} \ldots L_{i_k j_k},$$

$$X^{[k]} \overset{\text{def}}{=} \delta_{i_1 \ldots i_k; j_1 \ldots j_k} L_{i_1 j_1} \ldots L_{i_k j_k}.$$

By developing the $(k + 1) \times (k + 1)$-determinant underlying the generalised Kronecker tensor of rank $k + 1$ in the definition of $L_{ij}^{[k]}$ w.r.t. the last column into $k \times k$-determinants one may derive the following identity:

$$L_{ij}^{[k]} = \delta_{ij} X^{[k]} - k L_{i\alpha}^{[k-1]} L_{\alpha j}. \tag{D.1}$$

By induction we then have (L_{ij}^k denotes $L_{i i_1} L_{i_1 i_2} \ldots L_{i_{k-2} i_{k-1}} L_{i_{k-1} j}$):

$$L_{ij}^{[k]} = (-1)^k k! \, L_{ij}^k + \sum_{i=1}^{k} \binom{k}{i} X^{[i]} L_{ij}^{[k-i]}, \tag{D.2}$$

and since, by construction, the generalised Kronecker tensor of rank $d+1$ vanishes identically, we obtain the so-called *Cayley-Hamilton theorem* [1]:

$$L_{ij}^{[d]} = 0. \tag{D.3}$$

This is a polynomial of order d in the Hessian L_{ij}. This so-called "characteristic polynomial" has d (generally distinct) roots λ_k, $k = 1, \ldots, d$, the eigenvalues of the Hessian, which are functions of the invariants $X^{[k]}$. These d invariants correspond to the d independent degrees of freedom of L_{ij}. Instead of $X^{[k]}$, or λ_k, we may use the traces I_k, which completes the proof of the irreducibility of the set $\{I_1, \ldots, I_d\}$.

D.2   Irreducible System for $\{L, L_i, L_{ij}\}$

The irreducibility of the set $\{S_0, \ldots, S_{d-1}, I_1, \ldots, I_d\}$ associated with the first and second order tensors $\{L_i, L_{ij}\}$ could be proved in a similar way as for the second order case. However, it is more economical to proceed differently, using previous results.

We introduce two independent parameters λ and μ and consider the following, symmetric 2-tensor:

Definition D.2

$$H_{ij}(\lambda, \mu) \overset{\text{def}}{=} \lambda L_i L_j + \mu L_{ij} \qquad (\lambda, \mu \in \mathbb{R}).$$

We can use $H_{ij}(\lambda, \mu)$ to form similar polynomial invariants as we did for the Hessian L_{ij}.

Definition D.3

$$\tilde{I}_k(\lambda, \mu) \overset{\text{def}}{=} H_{i_1 i_2}(\lambda, \mu) H_{i_2 i_3}(\lambda, \mu) \ldots H_{i_k i_1}(\lambda, \mu).$$

Upon expanding this product we find:

$$\tilde{I}_k(\lambda, \mu) = \mu^k I_k + \sum_{i=0}^{k-1} \binom{k}{i} \lambda^{k-i} \mu^i S_0^{k-i-1} S_i. \tag{D.4}$$

Since $\left\{\tilde{I}_1(\lambda, \mu), \ldots, \tilde{I}_d(\lambda, \mu)\right\}$ is an irreducible system for the system $\{H_{ij}(\lambda, \mu)\}$ involving only the sets $\{S_0, \ldots, S_{d-1}\}$ and $\{I_1, \ldots, I_d\}$ we conclude that these two sets are *sufficient* for constructing any mixed first and second order polynomial invariant. That they are also *necessary* follows by a simple counting argument: there are exactly $2d$ independent invariant degrees of freedom in $\{L_i, L_{ij}\}$ (which is obvious in a coordinate system in which the Hessian is diagonal).

By including the independent, zeroth order image value L we have finally proved the irreducibility of $\{L, S_0, \ldots, S_{d-1}, I_1, \ldots, I_d\}$ for the case of the 2-jet tensors $\{L, L_i, L_{ij}\}$, which completes the proof of Proposition 5.5.

Solutions to Problems

Solution 2.2. $F \in S'(\mathbb{R}^n)$ is continuous: if $\lim_{k \to \infty} \phi_k = \phi$ in $S(\mathbb{R}^n)$, then $\lim_{k \to \infty} F[\phi_k] = F[\phi]$. The claim that $S(\mathbb{R}^n)$ has "a very strong topology" means that the construction of a convergent sequence in $S(\mathbb{R}^n)$ is a very hard job. Once you have accomplished that it will be difficult to construct a linear map that does *not* converge (in the usual sense of convergence in $\mathbb{R}$ as explained above). Continuity is desirable in the context of measurements, because it is a necessary condition for stability.

This suggests that, in order to construct a discontinuous functional, one has to weaken the topology. Here is an example of a discontinuous functional from Rudin's book on functional analysis [246, Chapter 1, Exercise 13]. Let C be the vector space of all complex functions on $[0, 1]$. Let (C, σ) be C with the topology induced by the metric

$$d(f, g) = \int_0^1 dx \, \frac{|f(x) - g(x)|}{1 + |f(x) - g(x)|}.$$

Moreover, let (C, τ) be the topological vector space defined by the seminorms

$$p_x(f) = |f(x)| \qquad 0 \le x \le 1.$$

Then the identity operator from (C, τ) to (C, σ) fails to be continuous.

Solution 2.3.

a. We have (in n dimensions)

$$\chi_{\beta(a;\sigma)}[\phi] = \frac{\int_{\|x-a\| < \sigma} dx \, \{\phi(a) + \mathcal{O}(\|x - a\|)\}}{\int_{\|x-a\| < \sigma} dx} = \phi(a) + \mathcal{O}(\sigma^n),$$

for all $\phi \in S(\mathbb{R}^n)$, so that $\lim_{\sigma \downarrow 0} \chi_{\beta(a;\sigma)} = \delta_a$.

b. Note that the integral of $\gamma_{a;\sigma}$ depends neither on a nor on σ. Define $\gamma(x) \equiv \gamma_{a=0;\sigma=1}(x)$. The standard integral

$$\int dx \, \gamma(x) = 1$$

can be easily proven for $n = 2$ using polar coordinates, and from there for any $n \in \mathbb{Z}^+$ by virtue of separability.

c. Directly in the spatial domain we can switch from "pull back" to "push forward" representation, as follows. Reparametrise $x \equiv a + \sigma\xi$ and define $\phi_{a;\sigma}(\xi) \equiv \phi(a + \sigma\xi)$, then

$$\lim_{\sigma\downarrow 0} \gamma_{a;\sigma}[\phi] = \lim_{\sigma\downarrow 0} \gamma[\phi_{a;\sigma}] = \gamma[\lim_{\sigma\downarrow 0} \phi_{a;\sigma}] = \phi(a),$$

in which we have made use of the previous result. This holds for all $\phi \in S(\mathbb{R}^n)$, in other words, $\lim_{\sigma\downarrow 0} \gamma_{a;\sigma} = \delta_a$. Alternatively one can look at the Fourier representation of the function $\gamma_{a;\sigma}(x)$:

$$\widehat{\gamma}_{a;\sigma}(\omega) = e^{ia\omega - \frac{1}{2}\sigma^2\omega^2},$$

which in the limit of vanishing scale becomes $e^{ia\omega} = \widehat{\delta}_a(\omega)$.

Solution 2.4.

a. Plain substitution; note that you can put $a = 0$ by virtue of shift invariance.

b. The r.h.s. satisfies the diffusion equation and has the right limit for $\sigma \downarrow 0$; since the solution in $S(\mathbb{R}^n)$ is unique, it must be another notation for the l.h.s.

See also Solution 2.12.

Solution 2.5.

a. $H \in S'(\mathbb{R})$ because $H[\phi] \stackrel{\text{def}}{=} \int_0^\infty dx\, \phi(x)$ is always well-defined and continuous if $\phi \in S(\mathbb{R})$.
$H'[\phi] \stackrel{\text{def}}{=} - \int_0^\infty dx\, \phi'(x) = - [\phi(x)]_{x=0}^{x\to\infty} = \phi(0) = \delta[\phi]$, so $H' = \delta$.

b. $|x|\,[\phi] \stackrel{\text{def}}{=} \int dx\, |x|\, \phi(x) = - \int_{-\infty}^0 dx\, x\, \phi(x) + \int_0^\infty dx\, x\, \phi(x)$, both terms are well-defined and well-behaved if $\phi \in S(\mathbb{R})$.
$|x|'\,[\phi] \stackrel{\text{def}}{=} - \int dx\, |x|\, \phi'(x) = \int_{-\infty}^0 dx\, x\, \phi'(x) - \int_0^\infty dx\, x\, \phi'(x) \stackrel{\text{p.i.}}{=} - \int_{-\infty}^0 dx\, \phi(x) + \int_0^\infty dx\, \phi(x) = \int dx\, [H(x) - H(-x)]\, \phi(x)$, whence $|x|' = \mathrm{sgn}\,(x)$.
Similarly, $|x|''\,[\phi] \stackrel{\text{def}}{=} -|x|'\,[\phi'] = -\int dx\, [H(x) - H(-x)]\, \phi'(x) \stackrel{\text{p.i.}}{=} \int dx\, [H'(x) + H'(-x)]\, \phi(x) = \int dx\, [\delta(x) + \delta(-x)]\, \phi(x) = 2\delta[\phi]$, so $|x|'' = 2\delta(x)$.

Solution 2.6.

a. First observation: $\phi(x) \leq m!\, x^m$ for all $x \in \mathbb{R}$ and $m \in \mathbb{Z}_0^+$; this follows from the series expansion of the exponential (consider the expansion of $1/\phi(u)$ in terms of $u = 1/x$). Secondly, it follows by induction that the k-th order derivative is of the form $\phi^{(k)}(x) = p_k(1/x)\, \phi(x)$, in which $p_k(u)$ is a $(2k)$-th order polynomial. This implies that there exists a constant M_k such that $|p_k(u)| \leq M_k u^{2k}$ if $u \geq 1$, which in turn implies that $\phi^{(k)}(x) \leq M_k m!\, x^{m-2k}$. We are free to choose $m > 2k$, so that we must have $\phi \in C^k(\mathbb{R})$, and $\phi^{(k)}(0) = 0$.

b. In 1D you can take $\psi_1(x) = \phi(x - a)\, \phi(b - x)$; in nD, $\psi_n(x^1, \ldots, x^n) = \prod_{i=1}^n \psi_1(x^i)$.

c. By smoothness and compact support there exists a point $x_0 \in \mathbb{R}$ where all derivatives vanish, $\alpha^{(k)}(x_0) = 0$ for all $k \in \mathbb{Z}_0^+$. By analyticity we have a convergent Taylor series $\alpha(x) = \sum_{k=0}^\infty \frac{\alpha^{(k)}(x_0)}{k!} (x - x_0)^k = 0$ for all $x \in \mathbb{R}$.

Solution 2.8.

a. Take $n, \alpha = 1$, the general case is straightforward. Conventional differentiation is defined for the subclass $f \in C^1(\mathbb{R}) \cap \mathcal{P}(\mathbb{R}) \subset S'(\mathbb{R})$, in which case we have $-F[\phi'] \stackrel{\text{def}}{=} -\int dx\, f(x)\, \phi'(x) \stackrel{\text{p.i.}}{=} \int dx\, f'(x)\, \phi(x) - [f(x)\, \phi(x)]_{x \to -\infty}^{x \to +\infty}$. The boundary term vanishes by construction of $S(\mathbb{R})$, and the remainder equals $F'[\phi]$, in which F' is the RTD corresponding to the classical derivative $f'(x)$. Moreover, if $F'[\phi] = G[\phi]$ for some RTD G corresponding to a function $g \in C^0(\mathbb{R}) \cap \mathcal{P}(\mathbb{R})$, then, by a claim somewhere after Definition 2.12, $f' = g$, so in the appropriate subspace the RTD F' and the classical function f' are one-to-one related.

b. From partial integration in the appropriate subspace it follows that $\nabla^{\dagger} = -\nabla$. From the definition of the complex scalar product (cf. Problem 2.10) it follows that $c^{\dagger} = c^*$ for any complex multiplier $c \in \mathbb{C}$. For any two—not necessarily commuting—linear operators A and B we have $(AB)^{\dagger} = B^{\dagger} A^{\dagger}$ (this follows immediately from the definition of conjugation, $A^{\dagger} v \cdot w \stackrel{\text{def}}{=} v \cdot Aw$, a "last-in first-out" process). Minus signs cancel: $(i\nabla)^{\dagger} = \nabla^{\dagger} i^{\dagger} = (-i)(-\nabla) = i\nabla$.

c. By the same token we have $P^{\dagger}(\lambda, \nabla) = \left(\sum_{|\alpha| \le m} \lambda_{\alpha}(x) \nabla_{\alpha} \right)^{\dagger} = \sum_{|\alpha| \le m} (-1)^{|\alpha|} \nabla_{\alpha} \circ \lambda_{\alpha}^*(x)$. Note the reversed order of operators!

d. The eigenvalue equation for eigenvalue ω is $i\nabla\phi = \omega\phi$, a nonzero solution of which exists for all $\omega \in \mathbb{R}$, given by $\phi_{\omega}(x) = e^{-i\omega x}$ up to arbitrary normalisation. In other words, the eigenvalue spectrum is just the Fourier frequency spectrum, while the corresponding eigenvectors correspond to the Fourier basis of planar waves.

Solution 3.2.

a. Note that the base point is $\pi^{\mu}[\phi_{a;\Lambda}] = a^{\mu}$. The rest follows from shift invariance of the Lebesgue measure $dx = d(x - a)$; the integrand depends only on $x - a$.

b. Results: $\sigma_0[\phi_{a,\Lambda}] = 1$ (by filter normalisation), $\sigma_1^{\mu}[\phi_{a,\Lambda}] = 0$ (by definition of central momenta), $\sigma_2^{\mu\nu}[\phi_{a,\Lambda}] = \Lambda^{\mu\nu}$. The nontrivial ones are easily computed in a Cartesian frame by reparametrising $x = R\xi$, in which R is a rotation matrix such that $R^{\mathrm{T}} \Lambda^{-1} R = \text{diag}\{s_1^{-1}, \ldots, s_n^{-1}\}$, after putting $a = 0$.

c. The case for odd orders $2k + 1$ follows from anti-symmetry of the integrand. The expression given for even orders should be obvious from the definition of momenta (each λ-derivative brings in a factor x^2).

d. If we introduce the symbol $\alpha_j^{\mu_1 \cdots \mu_k}$ that counts the number of indices among the $\mu_1, \ldots, \mu_k$ that equal j, then we can factorise the n-dimensional momentum into n 1-dimensional momenta:

$$\sigma_k^{\mu_1 \cdots \mu_k}[\phi_{a;\Lambda}] = \prod_{j=1}^{n} \sigma_{\alpha_j^{\mu_1 \cdots \mu_k}}[\phi_{a^j; \sigma_j^2}].$$

Note that $\sum_{j=1}^{n} \alpha_j^{\mu_1 \cdots \mu_k} = k$, because this is just the total number of indices. Consequently all we need to do is evaluate the 1-dimensional momenta $\sigma_{\alpha_j^{\mu_1 \cdots \mu_k}}[\phi_{a^j; \sigma_j^2}]$ for each dimension j. From the previous representation we conclude that

$$\sigma_{2k}[\phi_{a;\sigma^2}] = (-2\sigma^2)^k \left\{ \frac{d^k}{d\lambda^k} \frac{1}{\sqrt{\lambda}} \right\}_{\lambda=1} = (2k-1)!! \, \sigma^{2k}.$$

Solution 3.4. For $\theta(x) = Mx + a$ we get $\theta_* \phi(x) = |\det M^{\mathrm{inv}}| \phi(M^{\mathrm{inv}}(x - a))$, which is identical to the Gaussian of Problem 3.2 if we take $\Lambda = MM^\top$. Note that this is always possible by the constraints on Λ, and in particular that $|\det M| = \sqrt{\det \Lambda}$.

Solution 3.5.

a. $(\eta \circ \theta)_* \phi = J_{(\eta \circ \theta)\mathrm{inv}} \phi \circ (\eta \circ \theta)^{\mathrm{inv}} = J_{\eta \mathrm{inv}}(J_{\theta \mathrm{inv}} \phi \circ \theta^{\mathrm{inv}}) \circ \eta^{\mathrm{inv}} = (\eta_* \circ \theta_*)\phi$. From this it follows that for all admissible ϕ, $(\eta \circ \theta)^* F[\phi] = F[(\eta \circ \theta)_* \phi] = F[(\eta_* \circ \theta_*)\phi] = \eta^* F[\theta_* \phi] = (\theta^* \circ \eta^*)F[\phi]$.

b. $\eta^* F[\theta_* \phi] = F[(\eta \circ \theta)_* \phi] = F[\mathrm{id}_* \phi] = F[\phi]$.

c. $f'(y) = \eta^* f(y) = f(\eta(y)) = f(x)$.

d. $\phi'(y) = \theta_* \phi(y) = J_\eta(y) \phi(\eta(y)) = |\det \frac{\partial x}{\partial y}| \phi(x)$.

e. By substitution of variables $x = \eta(y)$ we can rewrite $\lambda' = F'[\phi'] = \int dy \, f'(y) \, \phi'(y) = \int dy \, |\det \frac{\partial \eta}{\partial y}(y)| \, f(\eta(y)) \, \phi(\eta(y)) = \int dx \, f(x) \, \phi(x) = F[\phi] = \lambda$, showing the parametrisation independence of a local sample.

Solution 3.7.

a. One has $\frac{1}{2}(n-1)(n-2)$ Euler angles for rotations, 2 scale parameters, n shift parameters, and $n - 1$ velocity parameters.

b. Subtract $n - 1$ from the previous count.

c. One has $\frac{1}{2}d(d-1)$ Euler angles for rotations, but this time only 1 scale parameter and d shift parameters; in addition one has d velocity parameters when adding Galilean boosts.

Solution 3.8. The eigenvalues λ are the solutions of $\det(X - \lambda I) = (1 - \lambda)^n = 0$. Apparently we have only a single eigenvalue $\lambda = 1$, independent of $\vec{v}$. Eigenvectors are the nontrivial elements of the null space of $X - \lambda I$, which is of dimension $(n-1)$, spanned by the basis vectors $e_i = (0; \vec{e}_i)$, where $\vec{e}_i$ is the standard basis vector with components $e_i^j = \delta_i^j$, $i, j = 1, \ldots, n-1$.

Solution 3.9. If we transform all M quantities $x_i \rightarrow \lambda^{\alpha_i} x_i$ $(i = 1, \ldots, M)$ under a rescaling of a single hidden scale parameter, then $F(x_1, \ldots, x_M) = 0$ will be recasted into $F(\lambda^{\alpha_1} x_1, \ldots, \lambda^{\alpha_M} x_M) = 0$. Differentiation w.r.t. λ at $\lambda = 1$ yields the 1-st order p.d.e. $\alpha_1 x_1 \frac{\partial F}{\partial x_1} + \ldots + \alpha_M x_M \frac{\partial F}{\partial x_M} = 0$. For N independent hidden scale parameters labelled by an index $\mu = 1, \ldots, N$ we thus get N such equations in N independent scale parameters λ_μ, and we likewise obtain a system of N p.d.e.'s by differentiation: $\alpha_{\mu 1} x_1 \frac{\partial F}{\partial x_1} + \ldots + \alpha_{\mu M} x_M \frac{\partial F}{\partial x_M} = 0$.

Solution 3.10.

a. For $\theta_x : z \mapsto z + x$ we get $\theta_x^* f(z) = f(z + x)$ and consequently $\theta_x^* F[\phi] = \int dz \, f(z + x) \, \phi(z) = F \star \phi(x)$. $\Theta^* F$ is symbolic notation for the entire collection $\{\theta^* F \mid \theta \in \Theta\}$, i.e. we can identify $\Theta_x^* F$ with the "function-valued" distribution $F\star$, with independent variable x.

b. Differentiation commutes with translation: $\nabla \circ \theta_x = \theta_x \circ \nabla$. Therefore $\nabla_\alpha F \star \phi = \nabla_\alpha \Theta^* F[\phi] = \Theta^* \nabla_\alpha F[\phi] = \Theta^* F[\nabla_\alpha^\dagger \phi] = F \star \nabla_\alpha^\dagger \phi$. The equality $\nabla_\alpha F \star \phi = F \star \nabla_\alpha \phi$ follows directly from the two equivalent integral representations for $F \star \phi$: $\int dz \, f(z) \, \phi(x - z) = \int dz \, f(x - z) \, \phi(z)$, or, less sloppy, by noting that $F \star \phi = F \star \tilde{\phi}$ (where $\tilde{\phi}(x) = \phi(-x)$).

Solution 3.13.

a. Δ is, by definition, a linear space, so we can add and scalar-multiply filters in the proper way, and in addition we have the $*$ operation, which satisfies all criteria of Equations 3.2 and 3.5.

b. This is Lemma 2.1. We can rewrite $x^\alpha \nabla_\beta (\phi * \psi)(x)$, using $x^\alpha = \sum_{\gamma \le \alpha} \binom{\alpha}{\gamma} (x - y)^{\alpha - \gamma} y^\gamma$, as $\sum_{\gamma \le \alpha} \binom{\alpha}{\gamma} (x^{\alpha - \gamma} \phi) * (x^\gamma \nabla_\beta \psi)$, which is a finite sum of convolution products, in each term of which one of the factors is integrable, and the other one bounded. Integrability and boundedness of the two respective factors in a convolution product suffice to make it bounded. Thus $x^\alpha \nabla_\beta (\phi * \psi)(x)$ is bounded for all multi-indices α and β, in other words, $\phi * \psi \in S(\mathbb{R}^n)$.

c. There is no identity element in $S(\mathbb{R}^n)$, and one cannot invert the filters. Convolution is commutative (abelian).

Solution 3.19. Since $m_2[\phi_t] = 0$ for $\alpha > 2$, these cases cannot correspond to positive filters. For $\alpha \ne 2$ the momenta do not all exist, hence the filter cannot be in $S(\mathbb{R})$ (recall that spatial momenta correspond to derivatives at zero frequency in Fourier space).

Solution 3.23.

a. $N^\dagger = (A^\dagger A)^\dagger = (A^\dagger A) = N$. This implies that N has only real eigenvalues, and an orthonormal basis of associated eigenfunctions. For any ψ we have $\psi \cdot N\psi = \|A\psi\|^2 \ge 0$. In particular, if ψ is itself a normalised eigenfunction ($\psi \cdot \psi = 1$) with eigenvalue λ, then $\lambda = \psi \cdot N\psi \ge 0$.

b. This follows from $[\frac{d}{d\xi}, \xi] = 1$.

c. By induction we find $[A, A^{\dagger k+1}] = A^{\dagger k}[A, A^\dagger] + [A, A^{\dagger k}]A^\dagger = (k+1)A^{\dagger k}$, in which we have used the previous result. For a general, analytical function f, we can use its Taylor series expansion in combination with the present result to find $[A, f(A^\dagger)] = f'(A^\dagger)$.

d. Let ψ_0 be the normalised eigenfunction corresponding to the smallest eigenvalue n_0, $\|\psi_0\|^2 = 1$. Noting that $N = AA^\dagger - 1$, we get $NA\psi_0 = A(A^\dagger A - 1)\psi_0 = (n_0 - 1)A\psi_0$. However, $n_0 - 1$ cannot be an eigenvalue, so $A\psi_0 = 0$. Furthermore we obtain $n_0 = \psi_0 \cdot N\psi_0 = \|A\psi_0\|^2 = 0$.

e. Once we have solved for the initial eigenfunction ψ_0, we can create a sequence of eigenfunctions by successive application of $A^\dagger$. To see this, note that ψ_k is indeed an eigenfunction of N with eigenvalue k, because $N\psi_k = AA^\dagger(\frac{1}{\sqrt{k!}}A^{\dagger k}\psi_0) = \frac{1}{\sqrt{k!}}([A, A^{\dagger k}] + A^\dagger A)\psi_0 = k\psi_0$. Using $\psi_k = \frac{1}{\sqrt{k}}A^\dagger \psi_{k-1}$ we see that it is also normalised: $\frac{1}{k}\|A^\dagger \psi_{k-1}\|^2 = \frac{1}{k}\psi_{k-1} \cdot (N+1)\psi_{k-1} = 1$ (last step by induction).

f. $H = N + \frac{1}{2}I$.

g. This should be evident from the previous results.

h. We only prove orthonormality. Assume $l \ge k$, and evaluate $\psi_k \cdot \psi_l$ with the help of d and e: $\psi_k \cdot \psi_l = \sqrt{\frac{k!}{l!}}\psi_0 \cdot A^{l-k}\psi_0 = \delta_{kl}$. The notation for completeness is explained as follows: for any $f(x)$ we can make the eigenfunction decomposition $f(x) = \sum_{k \in \mathbb{Z}_0^+} f \cdot \psi_k \, \psi_k(x) = \int dy \left\{ \sum_{k \in \mathbb{Z}_0^+} \psi_k(x)\psi_k(y) \right\} f(y)$. Thus we can make the identification $\sum_{k \in \mathbb{Z}_0^+} \psi_k(x)\psi_k(y) = \delta(x - y)$.

i. Left to the reader (separation of variables).

Solution 4.3. If the orders in a term $\int dx\, \phi^{(j)}(x)\, \phi^{(k)}(x)$ add up to an odd integer, say $j + k = 2i + 1$, then repeated partial integration can be used to cast it into the form $(-1)^i \int dx\, \phi^{(i)}(x)\, \phi^{(i+1)}(x)$. Thus the integrand is a total derivative, so that the integral vanishes by virtue of rapid decay. Otherwise, if the orders add up to an even integer, say $j + k = 2i$, then terms can be put in the form $(-1)^i \int dx\, \phi^{(i)}(x)\, \phi^{(i)}(x)$ by the same trick.

Solution 4.7. Consider the gradient of the second order scale-space polynomial,

$$\left\{ \begin{array}{rcl} L_x(x, y) & = & L_x + L_{xx}\, x + L_{xy}\, y \\ L_y(x, y) & = & L_y + L_{xy}\, x + L_{yy}\, y \,. \end{array} \right.$$

Disregard all pixels except candidates that might be hiding critical points, i.e. pixels at which we measure $0 \le L_x^2 + L_y^2 \le \varepsilon^2 M^2$, in which εM is a suitable threshold for the gradient magnitude. If M is the average or maximum gradient magnitude in the image, we may assume that $0 < \varepsilon \ll 1$ and make a perturbation expansion to find critical points. Say the origin (the pixel of interest) is such a candidate (note that it may be part of a cluster, depending on the threshold). Define $L_x = \varepsilon G_x$, $L_y = \varepsilon G_y$, with $(G_x, G_y) = \mathcal{O}(M)$. If $\varepsilon = 0$ then the origin is a critical point, but this won't happen in the typical case. We can nevertheless expect a critical point at sub-pixel position near the origin, so we make a perturbative expansion in ε: $x = \varepsilon X + \mathcal{O}(\varepsilon^2)$, $y = \varepsilon Y + \mathcal{O}(\varepsilon^2)$. Substitution into the equation $\nabla L = 0$ yields

$$\left\{ \begin{array}{rcl} L_{xx}\, X + L_{xy}\, Y & = & -G_x \\ L_{xy}\, X + L_{yy}\, Y & = & -G_y \,, \end{array} \right.$$

which is readily solved for (X, Y). Accept the solution if, say, $-\frac{1}{2} \le \varepsilon X, \varepsilon Y \le \frac{1}{2}$ (pixel units). Repeat the same procedure for every candidate pixel. Obviously the method fails near (but could be adapted to account for) degeneracies of the Hessian. For more details, cf. [76].

Solution 4.13. Using a "bra" to denote a source we may write, with the help of closure, $\mathrm{id} = \int dx\, |x\rangle\, \langle x|$,

$$\langle F| = \int dx\, \langle F|x\rangle\, \langle x| \,.$$

Solution 4.15.

a. Note that $F[\chi] = \int dt'\, \theta(a - t')\, f(t')\, \chi(a - t')$, in which $\theta(x)$ is the Heaviside function (1 if $x > 0$, 0 otherwise), and in which the integration is over the entire real axis. Differentiation w.r.t. a yields two contributions, which can be identified with the two given terms (recall $\theta'(x) = \delta(x)$).

b. $\delta_a F[\chi] = \int du\, \delta(u)\, f(a - u)\, \chi(u) = f(a)\, \chi(0) = 0$.

c. Trivial, since $u = a - t'$.

d. **b** states that $\delta_a F$ is the null functional, so we can disregard it, while **c** explains how the minus signs show up in odd orders.

e. If $f \in C^\infty(\mathbb{R})$, then we can perform an n-fold partial integration starting from the r.h.s. of **d**.

f. The information for signal updates is acquired, as with any measurement, through integration over intervals of finite measure. The output is not instantaneous: $\exists \tau > 0$ such that at base points $t \in (a - \tau, a)$ the active system is "busy" producing the sample at base point $t = a - \tau$ (or at least reaches the asymptotic level after which this sample does not significantly change). The system "sees" with a delay, like astronomical observations reveal only the history of galactical objects due to finite speed of light; the more recent history is still "on its way". Put differently, the information for signal updates is already hidden inside the active system, i.c. in the interval $(a - \tau, a)$.

g. Instead of t-derivatives we consider s-derivatives in Conclusion 4.1. We have $(a - t)\partial_t = \partial_s$ every time we take a derivative (s and t refer to dummy filter arguments); recall Problem 4.12. This brings in a factor $a - t$ for each order. The filters of Conclusion 4.1 are more natural in the context of scale-spacetime theory, since they obey the convolution-algebraic closure property.

Solution 4.17. It is natural to discretise on an equidistant grid in s-time, say $\Delta s = 1$ after suitable scaling of grid constant. Accordingly we may replace $ds = \frac{dt}{a-t}$ by $1 = \frac{\Delta t}{a-t}$, i.e. $\Delta t = (a - t)$ using the same natural unit. In other words, take sampling intervals proportional to elapsed time.

Solution 4.19.

a. Interpreting $\phi_{\mu_1 \ldots \mu_k}$ as a correlation filter, we must show that $\mathfrak{F}_\omega \star \phi_{\mu_1 \ldots \mu_k}$ equals $\lambda_{\mu_1 \ldots \mu_k} \mathfrak{F}_\omega$. Indeed, from the integral representation we see that

$$\mathfrak{f}_\omega \star \phi_{\mu_1 \ldots \mu_k}(x) = \int dz\, \mathfrak{f}_\omega(x+z)\, \phi_{\mu_1 \ldots \mu_k}(z) = \widehat{\phi}_{\mu_1 \ldots \mu_k}(\omega)\, \mathfrak{f}_\omega(x) = \lambda_{\mu_1 \ldots \mu_k}(\omega; \sigma)\, \mathfrak{f}_\omega(x),$$

where we have made use of the Lemmas 3.4, 3.5 and 3.6.

b. Separability follows directly from that of the zeroth order Gaussian point operator. The temporal component equals $\mathfrak{f}_\nu(s)$, which yields the desired result after substitution $s = -\log \frac{a-t'}{a-t}$.

c. If $t' \approx t$ then $-i\nu \log \frac{a-t'}{a-t} \approx \frac{\nu}{a-t} t'$. If the approximation holds, then we can interpret $\frac{\nu}{a-t}$ as ordinary temporal frequency.

d. Eigenfunctions, albeit source fields, are in fact detector properties, not independent sources living in the world.

Solution 4.20.

a. By the assumption of uniform distribution in the stochastic variable s, $\log_{10}(k+1)$ gives the cumulative probability for first digits $1, \ldots, k$. Subtract the cumulative probability for $1, \ldots, k - 1$, and the result is P_k. (Put differently: 10^s is a number that starts with digit $k = 1, \ldots, 9$ iff $s \in [\log_{10} k, \log_{10}(k + 1))$ *modulo* 1.)

b. A scaling in ordinary time yields a shift in canonical time. But in canonical time the distribution is uniform, hence shift invariant.

Solution 5.5. In a Cartesian coordinate system we have $L_{i_1 \ldots i_k} = \partial_{i_1 \ldots i_k} L = D_{i_1 \ldots i_k} L$. If we omit the middle part we have a tensor equation, which is valid in any coordinate system.

Solution 5.7.

a. Straightforward.

b. All $\Gamma^k_{ij} = 0$ except $\Gamma^\theta_{\phi\phi} = -\sin\theta\cos\theta$, $\Gamma^\phi_{\theta\phi} = \Gamma^\phi_{\phi\theta} = \cot\theta$.

c. All $R^i_{jkl} = 0$ except $R^\theta_{\phi\theta\phi} = -R^\theta_{\phi\phi\theta} = \sin^2\theta$, $R^\phi_{\theta\phi\theta} = -R^\phi_{\theta\theta\phi} = 1$. All $R_{ij} = 0$ except $R_{\theta\theta} = -1$ and $R_{\phi\phi} = -\sin^2\theta$. Alternatively: $R^i_j = \mathrm{diag}\{-1,-1\}$. The two equal eigenvalues correspond to the principal curvatures of the umbilical surface. Gaussian curvature is an overall constant: $\det R^i_j = 1$. $R = -2$ (i.e. the sum of principal curvatures).

d. Since there exists no coordinate system in which the Riemann tensor vanishes identically, it is apparently impossible to find flat coordinates for a sphere.

Solution 5.8.

a. See Equation 5.6 for the notation. Taking the determinant of $g'_{ij} = B_i{}^k B_j{}^l g_{kl}$ readily yields $g' = (\det B)^2 g$, in other words, $\sqrt{g'} = |\det B|\sqrt{g}$.

b. We have to check whether the transformation rule $[i_1\ldots i_d]' = (\det B)^{-1} B_{i_1}{}^{q_1}\ldots B_{i_d}{}^{q_d} [q_1\ldots q_d]$ is consistent with the coordinate independent definition of the permutation symbol. Since $B_{i_1}{}^{q_1}\ldots B_{i_d}{}^{q_d}[q_1\ldots q_d]$ equals $[i_1\ldots i_d]\det B$ (the direct consequence of the definition of a determinant), we see that the r.h.s. is indeed equal to the l.h.s.

c. Since $\varepsilon_{i_1\ldots i_d} = \sqrt{g}[i_1\ldots i_d]$, it follows from a and b that the determinant factors cancel up to a possible minus sign, $\varepsilon'_{i_1\ldots i_d} = \mathrm{sgn}(\det B) B_{i_1}{}^{q_1}\ldots B_{i_d}{}^{q_d}\varepsilon_{q_1\ldots q_d}$.

Solution 5.9. A relative pseudo-tensor of weight w (cf. Equation 5.6 for the notation):
$$P'^{i_1\ldots i_l}{}_{j_1\ldots j_k} \stackrel{\mathrm{def}}{=} \mathrm{sgn}(\det B)(\det B)^w A^{i_1}{}_{p_1}\ldots A^{i_l}{}_{p_l} B_{j_1}{}^{q_1}\ldots B_{j_k}{}^{q_k} P^{p_1\ldots p_l}{}_{q_1\ldots q_k}.$$

Solution 5.11. We have $\sqrt{g'} = |\det B|\sqrt{g}$, and $dx'^1\ldots dx'^d = (\det A)\,dx^1\ldots dx^d$ in the notation of Equation 5.6. Jacobians cancel up to a possible minus sign: $\sqrt{g'}\,dx'^1\ldots dx'^d = \mathrm{sgn}(\det A)\sqrt{g}\,dx^1\ldots dx^d$. (Spivak considers the oriented d-form $\sqrt{g}\,dx^1\wedge\ldots\wedge dx^d$ rather than Equation 5.18 [265, Volume 1, Chapter 7, Corollory 8], which manifestly incorporates the orientation of the coordinate basis, and in which case one naturally obtains a transformation of the coefficient of the d-form as an *even* scalar density.)

Solution 5.18. By a method similar to that used in Example 5.12 one shows that the gauge $L_u = L_v = L_{uv} = 0$ corresponds to $w_u = w_v = w_{uv} = 0$, where $w(u,v)$ is a Monge patch parametrisation of the isophote surface near the point of interest $(u,v,w) = (0,0,0)$. From $w(u,v) = \frac{1}{2}(\kappa_1 u^2 + \kappa_2 v^2) + \mathcal{O}(\|(u,v)\|^3)$ it shows that the u and v axes indeed correspond to the principal directions (the method also produces the formulas for $\kappa_{1,2}$ in gradient gauge).

Solution 5.21.

a. This is a special case of Proposition 5.4.

b. This is the direct consequence of the previous result.

c. In gradient gauge we have

$$h = -\frac{1}{2}\frac{L_{uu} + L_{vv}}{L_w} \quad \text{and} \quad k = \frac{L_{uu} L_{vv}}{L_w^2}.$$

d. In a Cartesian coordinate frame we have

$$
h = \frac{1}{2} \frac{2L_x L_y L_{xy} - L_x^2 L_{yy} - L_y^2 L_{xx} + \text{cycl.}(x, y, z)}{(L_x^2 + L_y^2 + L_z^2)^{3/2}},
$$

$$
k = \frac{-L_x^2 L_{yz}^2 + L_x^2 L_{yy} L_{zz} + 2L_x L_y L_{xz} L_{yz} - 2L_x L_z L_{xz} L_{yy} + \text{cycl.}(x, y, z)}{(L_x^2 + L_y^2 + L_z^2)^2},
$$

in which cycl.(x, y, z) stands for all terms obtained from the previous ones by the two cyclic permutations $(x, y, z) \rightarrow (y, z, x) \rightarrow (z, x, y)$.

Symbols

The following list contains only the most frequently used symbols together with their default meanings. It will be clear from the context if a symbol has a different interpretation.

$\mathbb{C}, \mathbb{R}, \mathbb{Q}, \mathbb{Z}, \mathbb{N}$	complex, real, rational, integer, natural (positive integer) numbers
$d, n = d + 1$	dimension of space, dimension of spacetime
$x, y, z, \ldots$	spacetime coordinates
$\delta x, \delta y, \delta z, \ldots$	small spatiotemporal variations
$f, g, h, \ldots$	spacetime functions
$\delta f, \delta g, \delta h, \ldots$	small function perturbations
$\Omega \subset \mathbf{M}, \Omega \subset \mathbb{R}^n,$	connected region in spacetime, respectively in $\mathbb{R}^n$
$\mathbf{M}$	spacetime manifold
$\|f\|_p$	p-norm of f $(p \geq 1)$
$L^p(\Omega)$	p-normed functions $(p \geq 1)$ on Ω
$C^k(\Omega)$	k-fold continuously differentiable functions on Ω $(k \in \mathbb{Z}_0^+ \cup \{\infty\})$
$C^\omega(\Omega)$	analytical functions on Ω
$\approx$	approximately equal
$\stackrel{\text{def}}{=}, \equiv$	equal by definition
Δ	device space
Σ	state space
$\phi, \psi, \ldots \in \Delta$	detectors (filters)
$F, G, \ldots \in \Sigma$	sources (raw images)
sup, inf, max, min	supremum, infimum, maximum, minimum
ess sup, ess inf	essential supremum, essential infimum
$'$	topological dual, or 1D derivative
$*$	algebraic dual, complex conjugate, convolution, or Hodge star operator
$\star$	correlation operator, occasionally a label
$\circ$	composition operator
$\mathcal{O}$	Landau order symbol
$\mathcal{D}(\Omega)$	compact smooth test functions on Ω
$\mathcal{E}(\Omega)$	smooth test functions on Ω
$\mathcal{G}(\mathbb{R}^n)$	Gaussian family
$\mathcal{S}(\mathbb{R}^n)$	smooth test functions of rapid decay (Schwartz space)
$\alpha, \beta, \gamma, \ldots$	multi-indices
$i, j, k, l, \ldots$	spatial tensor-indices (range $1, \ldots, d$), or integer labels
$\mu, \nu, \rho, \sigma, \ldots$	spatiotemporal tensor-indices (range $0, \ldots, d$)
$\nabla, \Delta, \Box$	gradient, Laplacian, d'Alembertian (or wave) operator
$\partial_{\mu_1 \ldots \mu_k}, D_{\mu_1 \ldots \mu_k}$	partial derivative, covariant derivative

Glossary

A

Algebra A set of similar objects with a lot of structure, enabling operations commonly referred to as "addition", "multiplication" and "scalar multiplication". The precise criteria are stated in this book. A trivial example is $\mathbb{R}$. Less trivial and most important in this book is that of a convolution algebra, such as Schwartz space $S(\mathbb{R}^n)$. A convolution algebra is the skeleton of linear image processing.

Analysis In a broad context the act of breaking something down into constituent parts for a detailed examination. Trivial prerequisite: *synthesis*.

C

Covector A covector is first of all a vector in the sense of an element of a linear space. The prefix indicates that this is not a standalone space, but one that goes along with another linear space. For this reason one distinguishes between a "vector space" and its associated "covector space". The essence is that vectors and covectors can "communicate": by convention a covector has a slot in which one can drop a vector (it might just as well have been the other way around) so as to obtain a number, which depends linearly on both the covector and its vector argument. This bilinear procedure is called "contraction".

Vectors and covectors are instances of contravariant, respectively covariant *tensors* of rank 1. If you have a *metric* you can convert vectors into covectors, *vice versa*, in a one-to-one way.

The linear spaces involved may be of infinite dimension. For example, smooth functions of rapid decay $S(\mathbb{R}^n)$ (Schwartz space) define an infinite dimensional vector space; the corresponding covector space consists of all tempered distributions $S'(\mathbb{R}^n)$ and is likewise of infinite dimension. In the context of function spaces covectors are usually called *linear functionals* or *distributions*.

D

Density A density is a quantity attributed to a point that produces a numeric value when integrated over some finite volumetric neighbourhood (most frequently in space, time, or spacetime). Such quantities are inherently ambiguous unless you specify the volume; even if you aim for a point measurement you need to specify a finite *inner scale* (the "radius" of your point). An integrated density is a *scalar*.

Mathematicians like to think of a density (in n dimensions) as an n-form, a geometric object living in a 1-dimensional space. Thus densities can be expressed relative to a single basis vector, which—if suitably normalised— is just the Levi-Civita tensor. Thus both scalars as well as densities are 1-dimensional things, and in fact there exists a one-to-one relationship, known as "Hodge duality". The so-called "Hodge-star" is the operator that takes you from one space to the other; in particular one defines $\varepsilon = *1$.

Derivative A derivative is a linear operator satisfying Leibniz's product rule. The classical way of introducing it by means of an infinitesimal limiting procedure complies with this definition, but is a rather unfortunate one, since no such things as infinitesimals exist as physical entities. In practice derivatives can be defined in terms of integral operators. A common misconception is that such integral operators are mere approximations of derivatives.

Diagrammar Set of rules for representing *tensors* by line drawings. A tensor is depicted as a vertex, the number of emanating branches of which equals its rank. Dangling branches represent empty slots. A contraction corresponds to a connection of two branches. Full contraction will result in a closed diagram, i.e. a (possibly relative) scalar. The vertex point may itself be a diagram with internal contractions.

F

Functional integral Also known as *path integral*. Functional integrals were introduced by Feynman in the sixties in the context of quantum fields, and express his heuristic idea of summing complex phase contributions associated with every conceivable particle trajectory over all possible configurations. The result of such a summation leads to the kind of quantum interference that ought to explain particle interaction phenomena. However, functional integrals are more widely applicable. A real ("Euclidean") counterpart is used in this book to show that if one integrates a certain functional, which classically (i.e. by the Euler-Lagrange principle) corresponds to uniform motion, over all possible paths between two fixed end-points, the Gaussian propagator emerges as a result. Unfortunately, general mathematical definitions of functional integrals exist only in rare cases, and numerical methods are typically prohibitively expensive.

I

Ill-posedness A problem usually caused by an unfortunate choice of *topology*, and closely related to lack of continuity of operators. *Regularisation* aims to turn an ill-posed problem into one that is *well-posed*, but sometimes obscures the core of the problem. For instance, conventional differentiation is ill-posed *not* because of lack of regularity of operands (a restriction to smooth functions is of no help) but as a consequence of a weak function topology. Differentiation in distributional sense is well-posed despite virtually no demands on regularity.

M

Measurement The assessment of physical evidence by means of observation. A measurement without theory is unthinkable: apart from the detector interface at which it is produced it is necessary that one possesses knowledge, or at least a hypothesis, of the inner workings of the detector.

Metric The most fundamental *tensor* in a "Riemannian space". Represented as a *covariant tensor* it accepts any pair of vectors and converts them into a number according to the standard rules underlying a scalar product. The metric is symmetric, so ordering is irrelevant. If you refrain from inserting a second vector argument, it apparently turns your first one into a *covector*; in this form the metric is also known as the "sharp operator", the coordinate representation of which is usually referred to as "index lowering". This procedure is invertible, and the inverse is known as the "flat operator", or, in terms of coordinates, as "index raising". The corresponding slot machine, a *contravariant tensor*, is called the "dual" or "inverse metric".

Important examples of metric spaces are classical 3-space and time in the classical Newtonian model. It is equally important to appreciate that certain spaces are *not* metrical, such as spacetime and scale-space, at least in the context of the Newtonian model.

O

Operational... An operational concept—be it a definition, a representation, or whatever—is basically a computation, or an unambiguous recipe for this. For example, if one defines an "edge" as the output of a predefined "edge detector" (notice the circularity) one actually has an operational definition. If we now dispense with the detector, "edges" cease to exist in operational sense. Neither do they exist as the same entities relative to another "edge detector". A theory developed exclusively in terms of operational concepts is not likely to cause confusion as it forces us to distinguish between different realizations of objects despite identical name tags. Different objects can still be compared relative to an operational criterion, such as the extent to which they successfully subserve a given task.

OFCE Acronym for "Optic Flow Constraint Equation", the image scientist's buzzword for a conservation principle in the form of a vanishing Lie derivative. Vector fields satisfying this equation are called "optic flow" fields.

P

Path integral See *functional integral*.

R

Resolution Inverse *scale*.

S

Sample A measurement extracted from a continuum, or a resampling of discrete data. In image analysis a sample means nothing without a model; that a pixel has a value of 134 is a totally void statement. In addition, viable models should always account for the fact that a sample has a finite tolerance. Despite the fact that well-posedness is an essential prerequisite, ill-posedness has always had epidemic tendencies in image analysis.

Scale In physics, scale means size or weight (in general sense), and is defined relative to a fiducial reference unit, such as a yardstick or a counteracting weight in a balance. Asking for the scale of this reference unit in turn ("absolute" or "hidden scale") is bound to lead to infinite recursion. By virtue of the universal law of scale invariance this has no dramatic implications; a consistent use of dimensional units will guarantee this in practice.

In this book scale is primarily used as a spatial or temporal measure. Several scales are of interest.

- Pixel scale defines the graininess of a digital image and poses an obvious technical limitation. Neighbouring pixels may be correlated due to noise and blur.

- Inner scale is the smallest scale of interest. The ratio of pixel (or correlation) scale and inner scale limits data quality. However, one can arbitrarily lower the inner scale in a model, which may be a useful mental procedure.

- Outer scale pertains to the size of the region of interest.

- Image scale is the typical scope or field of view captured by the image. The ratio of outer scale and scope likewise limits data quality.

The inverse of scale is called *resolution*. In the Fourier domain, frequency scale equals spatiotemporal resolution, *vice versa*.

Scalar A scalar is a quantity with a magnitude but no direction or orientation. Particularly important is the kind of scalar that arises from the integration (measurement) of a *density*.

Semantics The branch of logic pertaining to interpretation or meaning of information. Semantics is what it needs to turn raw data into "evidence". In his famous book [120], Douglas Hofstadter argues that one should actually distinguish between "interpretation" and "meaning". Cf. *syntax*. As we have de-emphasized semantics altogether the matter has not been scrutinised in this book.

Syntax Literally an "arrangement". A set of rules for providing structure to a collection of elements, "syntagmata". A syntax defines data formats suitable for subsequent *semantics*.

Synthesis The act of composing separate elements into a coherent system, for instance the construction of basic axioms; a *conditio sine qua non* for *analysis*.

T

Tensor A tensor is a slot machine designed to convert an ordered list of vectors and covectors into a real number in a multilinear fashion. There exist three distinct types: a covariant tensor maps only vectors, a contravariant tensor maps only covectors, and a mixed tensor handles both types.

In a metric space a tensor may be generalised so as to incorporate metrical relations. This means that each input parameter may be either of *"vector"* or of *"covector"* type; the *metric* will convert data types, if necessary, into the format compatible with the actual implementation of the slot machine.

Topology Topology and the notion of tolerance are close-knit. A space endowed with a topology is one with proximity relations. Examples of topological spaces are the basic spacetime manifold and the various function and functional spaces used in this book.

V

Vision Visually guided behaviour, i.e. not passive observation of, but (optically guided) active participation in the world. Mechanical models of vision ("machine vision") are unlike models in image analysis insofar that input and output spaces are essentially disjunct in the latter.

W

Well-posedness A problem is well-posed if it is not *ill-posed*. Well-posedness is an essential demand in image analysis.

Bibliography

[1] R. Abraham, J. E. Marsden, and T. Ratiu. *Manifolds, Tensor Analysis, and Applications*, volume 75 of *Applied Mathematical Sciences*. Springer-Verlag, New York, second edition, 1988.

[2] M. Abramowitz and I. A. Stegun, editors. *Handbook of Mathematical Functions with Formulas, Graphs, and Mathematical Tables*. Dover Publications, Inc., New York, 1965. Originally published by the National Bureau of Standards in 1964.

[3] S. C. Amartur and H. J. Vesselle. A new approach to study cardiac motion: the optical flow of cine MR images. *Magnetic Resonance in Medicine*, 29(1):59–67, 1993.

[4] A. A. Amini. A scalar function formulation for optical flow. In Eklundh [62], pages 125–131.

[5] C. M. Anderson, R. R. Edelman, and P. A. Turski. *Clinical Magnetic Resonance Angiography*, chapter 3. Phase Contrast Angiography, pages 43–72. Raven Press, New York, 1993.

[6] V. I. Arnold. *Mathematical Methods of Classical Mechanics*, volume 60 of *Graduate Texts in Mathematics*. Springer-Verlag, New York, second edition, 1989.

[7] J. Arnspang. Notes on local determination of smooth optic flow and the translational property of first order optic flow. Technical Report 88/1, Department of Computer Science, University of Copenhagen, 1988.

[8] J. Arnspang. Optic acceleration. Local determination of absolute depth and velocity, time to contact and geometry of an accelerating surface. Technical Report 88/2, Department of Computer Science, University of Copenhagen, 1988.

[9] J. Arnspang. Motion constraint equations in vision calculus, April 1991. Dissertation for the Danish Dr. scient. degree, University of Utrecht, Utrecht, The Netherlands.

[10] J. Arnspang. Motion constraint equations based on constant image irradiance. *Image and Vision Computing*, 11(9):577–587, November 1993.

[11] J. Babaud, A. P. Witkin, M. Baudin, and R. O. Duda. Uniqueness of the gaussian kernel for scale-space filtering. *IEEE Transactions on Pattern Analysis and Machine Intelligence*, 8(1):26–33, 1986.

[12] A. B. Bakushinsky and Goncharsky A. V. *Ill-Posed Problems: Theory and Applications*. Kluwer Academic Publishers, Dordrecht, The Netherlands, 1994.

[13] J. L. Barron, D. J. Fleet, and S. S. Beauchemin. Performance of optical flow techniques. *International Journal of Computer Vision*, 12(1):43–77, 1994.

[14] E. Bayro-Corrochano, J. Lasenby, and G. Sommer. Geometric algebra: A framework for computing point and line correspondences and projective structure using n uncalibrated cameras. In *Proceedings of the 13th International Conference on Pattern Recognition (Vienna, Austria, August 1996)* [129], pages 334–338.

[15] F. J. Beekman. *Fully 3D SPECT Reconstruction with Object Shape Dependent Scatter Compensation*. PhD thesis, University of Utrecht, Department of Medicine, Utrecht, The Netherlands, March 8 1995.

[16] R. Benedetti and C. Petronio. *Lectures on Hyperbolic Geometry*. Springer-Verlag, Berlin, 1987.

[17] F. Bergholm. Edge focusing. *IEEE Transactions on Pattern Analysis and Machine Intelligence*, 9:726–741, 1987.

[18] R. L. Bishop and S. I. Goldberg, editors. *Tensor Analysis on Manifolds*. Dover Publications, Inc., New York, 1980. Originally published by The Macmillan Company in 1968.

[19] J. Blom. *Topological and Geometrical Aspects of Image Structure*. PhD thesis, University of Utrecht, Department of Medical and Physiological Physics, Utrecht, The Netherlands, 1992.

[20] J. Blom, B. M. ter Haar Romeny, A. Bel, and J. J. Koenderink. Spatial derivatives and the propagation of noise in Gaussian scale-space. *Journal of Visual Communication and Image Representation*, 4(1):1–13, March 1993.

[21] R. van den Boomgaard. *Mathematical Morphology: Extensions towards Computer Vision*. PhD thesis, University of Amsterdam, March 23 1992.

[22] R. van den Boomgaard. The morphological equivalent of the Gauss convolution. *Nieuw Archief voor Wiskunde*, 10(3):219–236, November 1992.

[23] R. van den Boomgaard and A. W. M. Smeulders. Morphological multi-scale image analysis. In J. Serra and P. Salembier, editors, *International Workshop on Mathematical Morphology and its Applications to Signal Processing (Barcelona, Spain, May 1993)*, pages 180–185. Universitat Politècnica de Catalunya, 1993.

[24] R. van den Boomgaard and A. W. M. Smeulders. The morphological structure of images, the differential equations of morphological scale-space. *IEEE Transactions on Pattern Analysis and Machine Intelligence*, 16(11):1101–1113, November 1994.

[25] N. Bourbaki. *Éléments de Mathématique, Livre V: Espaces Vectoriels Topologiques*. Hermann, Paris, 1964.

[26] V. Braitenberg. *Vehicles*. MIT Press, Cambridge, 1984.

[27] R. D. Brandt and L. Feng. Representations that uniquely characterize images modulo translation, rotation, and scaling. *Pattern Recognition Letters*, 17(9):1001–1015, August 1996.

[28] M. Brill, E. Barrett, and P. Payton. Projective invariants for curves in two and three dimensions. In Mundy and Zisserman [216], chapter 9, pages 193–214.

[29] B. Buck, A. C. Merchant, and S. M. Perez. An illustration of Benford's first digit law using alpha decay half lives. *European Journal of Physics*, 14:59–63, 1993.

[30] M. P. do Carmo. *Differential Geometry of Curves and Surfaces*. Mathematics: Theory & Applications. Prentice-Hall, Englewood Cliffs, New Jersey, 1976.

[31] M. P. do Carmo. *Riemannian Geometry*. Mathematics: Theory & Applications. Birkhäuser, Boston, second edition, 1993.

[32] É. Cartan. Sur les variétés à connexion affine et la théorie de la relativité generalisée (première partie). *Ann. École Norm. Sup.*, 40:325–412, 1923.

[33] E. Cartan. *Leçons sur la Géométrie des Espaces de Riemann*. Gauthiers-Villars, Paris, second edition, 1963.

[34] S. D. Casey and D. F. Walnut. Systems of convolution equations, deconvolution, Shannon sampling, and the wavelet and Gabor transforms. *SIAM Review*, 36(4):537–577, 1994.

[35] H. A. Cerdeira, S. O. Lundqvist, D. Mugnai, A. Ranfagni, V. Sa-yakanit, and L. S. Schulman, editors. *Lectures on Path Integration: Trieste 1991*, Singapore, 1993. World Scientific. Proceedings of a workshop on path integration held in Trieste in 1991.

[36] Y. Choquet-Bruhat, C. DeWitt-Morette, and M. Dillard-Bleick. *Analysis, Manifolds, and Physics. Part I: Basics*. Elsevier Science Publishers B.V. (North-Holland), Amsterdam, 1991.

[37] C. K. Chui. *An Introduction to Wavelets*. Academic Press, San Diego, 1992.

[38] C. K. Chui, editor. *Wavelets: a Tutorial in Theory and Applications*. Academic Press, San Diego, 1992.

[39] W. K. Clifford. Applications of Grassmann's extensive algebra. *Am. J. Math.*, 1:350–358, 1878.

[40] W. T. Cochran et al. What is the Fast Fourier Transform? In *Proceedings of the IEEE,* pages 1664–1674, October 1967.

[41] J. Damon. Local Morse theory for solutions to the heat equation and Gaussian blurring. *Journal of Differential Equations,* 115(2):368–401, January 1995.

[42] J. Damon. Local Morse theory for Gaussian blurred functions. In Sporring et al. [266], chapter 11, pages 147–163.

[43] Per-Eric Danielsson and Olle Seger. Rotation invariance in gradient and higher order derivative detectors. *Computer Vision, Graphics, and Image Processing,* 49:198–221, 1990.

[44] I. Daubechies. Orthonormal bases of compactly supported wavelets. *Communications on Pure and Applied Mathematics,* 41:909–996, 1988.

[45] I. Daubechies. *Ten Lectures on Wavelets.* Number 61 in CBMS-NSF Series in Applied Mathematics. SIAM, Philadelphia, 1992.

[46] R. Dautray and J.-L. Lions. *Mathematical Analysis and Numerical Methods for Science and Technology: Functional and Variational Methods,* volume 2. Springer-Verlag, Berlin, 1988.

[47] G. C. DeAngelis, I. Ohzawa, and R. D. Freeman. Receptive field dynamics in the central visual pathways. *Trends in Neuroscience,* 18(10):451–457, 1995.

[48] P. Delogne, editor. *Proceedings of the International Conference on Image Processing 1996 (Lausanne, Switzerland, September 1996).* IEEE, 1996.

[49] R. Deriche. Using Canny's criteria to derive a recursively implemented optimal edge detector. *International Journal of Computer Vision,* 1:167–187, 1987.

[50] R. Deriche. Fast algorithms for low-level vision. *IEEE Transactions on Pattern Analysis and Machine Intelligence,* 12(1):78–87, 1990.

[51] R. Deriche. Recursively implementing the gaussian and its derivatives. In V. Srinivasan, Ong Sim Heng, and Ang Yew Hock, editors, *Proceedings of the 2nd Singapore International Conference on Image Processing (Singapore, September 1992),* pages 263–267. World Scientific, Singapore, 1992.

[52] R. Deriche. Recursively implementing the gaussian and its derivatives. Technical Report INRIA-RR-1893, INRIA Sophia-Antipolis, France, May 1993.

[53] V. Devlaminck and J. Dubus. Estimation of compressible or incompressible deformable motions for density images. In Delogne [48], pages 125–128.

[54] B. S. DeWitt. Quantum theory of gravity. II. The manifestly covariant theory. *Physical Review,* 162:1195–1239, 1967.

[55] J. D'Haeyer. Determining motion of image curves from local pattern changes. *Computer Vision, Graphics, and Image Processing,* 34:166–188, 1986.

[56] L. Dorst and R. van den Boomgaard. Morphological signal processing and the slope transform. *Signal Processing,* 38:79–98, 1994.

[57] J. J. Duistermaat. M. Riesz's families of operators. *Nieuw Archief voor Wiskunde,* 9(1):93–101, March 1991.

[58] J. J. Duistermaat. Differentiaalrekening op variëteiten. Course notes published by the Mathematics Institute, University of Utrecht, The Netherlands, 1992.

[59] J. J. Duistermaat. Distributies. Course notes published by the Mathematics Institute, University of Utrecht, The Netherlands, 1992.

[60] D. Eberly. A differential geometric approach to anisotropic diffusion. In Haar Romeny [103], pages 371–392.

[61] A. Einstein. *Investigations on the Theory of the Brownian Movement.* Dover Publications, Inc., New York, 1956. Edited with notes by R. Fürth. Translated by A. D. Cowper. Originally published in 1926.

[62] J.-O. Eklundh, editor. *Proceedings of the Third European Conference on Computer Vision (Stockholm, Sweden, May 1994),* volume 800–801 of *Lecture Notes in Computer Science,* Berlin, 1994. Springer-Verlag.

[63] Petra A. van den Elsen. *Multimodality Matching of Brain Images*. PhD thesis, University of Utrecht, Department of Medicine, The Netherlands, 1993.

[64] O. D. Faugeras. *Three-Dimensional Computer Vision: a Geometric Viewpoint*. MIT Press, Cambridge, 1993.

[65] O. D. Faugeras and T. Papadopoulo. A theory of the motion fields of curves. *International Journal of Computer Vision*, 10(2):125–156, 1993.

[66] A. Feinstein. *Foundations of Information Theory*. McGraw-Hill, New York, 1958.

[67] W. Fenchel. *Elementary Geometry in Hyperbolic Space*, volume 11 of *Studies in Mathematics*. Walter de Gruyter, Berlin, 1989.

[68] C. L. Fennema and W. B. Thompson. Velocity determination in scenes containing several moving objects. *Computer Graphics and Image Processing*, 9:301–315, 1979.

[69] R. P. Feynman and A. R. Hibbs. *Quantum Mechanics and Path Integrals*. McGraw-Hill, New York, 1965.

[70] B. Fischl and L. Schwartz. Learning an integral equation approximation to nonlinear anisotropic diffusion in image processing. *IEEE Transactions on Pattern Analysis and Machine Intelligence*, 19(4):342–352, April 1997.

[71] J. M. Fitzpatrick. The existence of geometrical density-image transformations corresponding to object motion. *CVGIP: Image Understanding*, 44:155–174, 1988.

[72] H. Flanders. *Differential Forms*. Academic Press, New York, 1963.

[73] L. M. J. Florack. Grey-scale images. Technical Report ERCIM-09/95-R039, INESC Aveiro, Portugal, September 1995. URL: http://www-ercim.inria.fr/publication/technical_reports.

[74] L. M. J. Florack. The concept of a functional integral—a potentially interesting method for image processing. Technical Report 96/7, Department of Computer Science, University of Copenhagen, 1996.

[75] L. M. J. Florack. Data, models, and images. In Delogne [48], pages 469–472.

[76] L. M. J. Florack. Detection of critical points and top-points in scale-space. In P. Johansen, editor, *Proceedings fra Den Femte Danske Konference om Mønstergenkendelse og Billedanalyse*, pages 73–81, August 1996. DIKU Tech. Rep. Nr. 96/22.

[77] L. M. J. Florack. The intrinsic structure of optic flow in the context of the scale-space paradigm. In T. Moons, E. Pauwels, and L. Van Gool, editors, *From Segmentation to Interpretation and Back: Mathematical Methods in Computer Vision*, Lecture Notes in Computer Science. Springer-Verlag, Berlin, 1996. Selected papers from the Computer Vision and Applied Geometry workshop in Nordfjordeid, Norway.

[78] L. M. J. Florack. Measurement duality. In C. Erkelens, editor, *Biophysics of Shape*, pages 7–13, April 2–4 1997.

[79] L. M. J. Florack, B. M. ter Haar Romeny, J. J. Koenderink, and M. A. Viergever. Families of tuned scale-space kernels. In G. Sandini, editor, *Proceedings of the Second European Conference on Computer Vision (Santa Margherita Ligure, Italy, May 1992)*, volume 588 of *Lecture Notes in Computer Science*, pages 19–23, Berlin, 1992. Springer-Verlag.

[80] L. M. J. Florack, B. M. ter Haar Romeny, J. J. Koenderink, and M. A. Viergever. General intensity transformations and second order invariants. In Johansen and Olsen [136], pages 22–29. Selected papers from the 7th Scandinavian Conference on Image Analysis.

[81] L. M. J. Florack, B. M. ter Haar Romeny, J. J. Koenderink, and M. A. Viergever. Cartesian differential invariants in scale-space. *Journal of Mathematical Imaging and Vision*, 3(4):327–348, November 1993.

[82] L. M. J. Florack, B. M. ter Haar Romeny, J. J. Koenderink, and M. A. Viergever. General intensity transformations and differential invariants. *Journal of Mathematical Imaging and Vision*, 4(2):171–187, 1994.

[83] L. M. J. Florack, B. M. ter Haar Romeny, J. J. Koenderink, and M. A. Viergever. Linear scale-space. *Journal of Mathematical Imaging and Vision*, 4(4):325–351, 1994.

[84] L. M. J. Florack, B. M. ter Haar Romeny, J. J. Koenderink, and M. A. Viergever. The Gaussian scale-space paradigm and the multiscale local jet. *International Journal of Computer Vision*, 18(1):61–75, April 1996. Erratum: Fig. 3 is upside down.

[85] L. M. J. Florack and M. Nielsen. The intrinsic structure of the optic flow field. Technical Report ERCIM-07/94-R033 or INRIA-RR-2350, INRIA Sophia-Antipolis, France, July 1994. URL: http://www-ercim.inria.fr/publication/technical_reports.

[86] L. M. J. Florack, W. J. Niessen, and M. Nielsen. The intrinsic structure of optic flow incorporating measurement duality. Accepted for publication in the International Journal of Computer Vision, 1996.

[87] L. M. J. Florack, A. H. Salden, B. M. ter Haar Romeny, J. J. Koenderink, and M. A. Viergever. Nonlinear scale-space. *Image and Vision Computing*, 13(4):279–294, May 1995.

[88] O. Fogh Olsen. Multi-scale segmentation of grey-scale images. Master's thesis, Department of Computer Science, University of Copenhagen, August 1996.

[89] D. Forsyth, J. L. Mundy, A. Zisserman, C. Coelho, A. Heller, and C. Rothwell. Invariant descriptors for 3-D object recognition and pose. *IEEE Transactions on Pattern Analysis and Machine Intelligence*, 13(10):971–991, October 1991.

[90] J. Fourier. *The Analytical Theory of Heat*. Dover Publications, New York, 1955. Replication of the English translation that first appeared in 1878 with previous corrigenda incorporated into the text, by Alexander Freeman, M.A. Original work: "Théorie Analytique de la Chaleur", Paris, 1822.

[91] F. G. Friedlander. *Introduction to the Theory of Distributions*. Cambridge University Press, Cambridge, 1982.

[92] M. Friedman. *Foundations of Space-Time Theories: Relativistic Physics and Philosophy of Science*. Princeton University Press, Princeton, 1983.

[93] I. M. Gelfand and G. E. Shilov. *Generalized Functions*, volume 1. Academic Press, New York and London, 1968.

[94] I. M. Gelfand and G. E. Shilov. *Spaces of Fundamental and Generalized Functions*, volume 2. Academic Press, New York and London, 1968.

[95] J. J. Gibson. *The Perception of the Visual World*. Houghton-Mifflin, Boston, 1950.

[96] R. Gilmore. *Catastrophe Theory for Scientists and Engineers*. Dover Publications, Inc., New York, 1993. Originally published by John Wiley & Sons, New York, 1981.

[97] L. J. Van Gool, M. H. Brill, E. B. Barrett, T. Moons, and E. J. Pauwels. Semi-differential invariants for nonplanar curves. In Mundy and Zisserman [216], chapter 11, pages 293–309.

[98] L. J. Van Gool, T. Moons, E. J. Pauwels, and A. Oosterlinck. Semi-differential invariants. In Mundy and Zisserman [216], chapter 8, pages 157–192.

[99] L. J. Van Gool, T. Moons, E. J. Pauwels, and A. Oosterlinck. Vision and Lie's approach to invariance. *Image and Vision Computing*, 13(4):259–277, May 1995.

[100] H. Grassmann. Der Ort der Hamilton'schen Quaternionen in der Ausdehnungslehre. *Mathematische Annalen*, 12:375, 1877.

[101] L. D. Griffin. Critical point events in affine scale-space. In Sporring et al. [266], chapter 12, pages 165–180.

[102] L. D. Griffin and A. C. F. Colchester. Superficial and deep structure in linear diffusion scale space: Isophotes, critical points and separatrices. *Image and Vision Computing*, 13(7):543–557, September 1995.

[103] B. M. ter Haar Romeny, editor. *Geometry-Driven Diffusion in Computer Vision*, volume 1 of *Computational Imaging and Vision Series*. Kluwer Academic Publishers, Dordrecht, 1994.

[104] B. M. ter Haar Romeny, L. M. J. Florack, M. de Swart, J. J. Wilting, and M. A. Viergever. Deblurring Gaussian blur. In *Proceedings Mathematical Methods in Medical Imaging II*, volume 2299, San Diego, CA, July, 25–26 1994. SPIE.

[105] B. M. ter Haar Romeny, L. M. J. Florack, J. J. Koenderink, and M. A. Viergever, editors. *Scale-Space '97: Proceedings of the First International Conference on Scale-Space Theory in Computer Vision*, volume 1252 of *Lecture Notes in Computer Science*. Springer-Verlag, Berlin, 1997.

[106] J. Hadamard. Sur les problèmes aux dérivées partielles et leur signification physique. *Bul. Univ. Princeton*, 13:49–62, 1902.

[107] H. F. Harmuth. *Information Theory Applied to Space-Time Physics*. World Scientific, Singapore, 1992.

[108] T. C. Hasley, M. H. Jensen, L. P. Kadanoff, I. Procaccia, and B. I. Shraiman. Fractal measures and their singularities: the characterization of strange sets. *Physical Review A*, 33:1141–1151, 1986.

[109] Sir Thomas L. Heath. *The Thirteen Books of Euclid's Elements*, volume 1 (Books I and II). Dover Publications, Inc., New York, second edition, 1956.

[110] H. J. A. M. Heijmans. Theoretical aspects of grey-level morphology. *IEEE Transactions on Pattern Analysis and Machine Intelligence*, 13(6):568–592, June 1991.

[111] H. J. A. M. Heijmans. Mathematical morphology: a geometrical approach in image processing. *Nieuw Archief voor Wiskunde*, 10(3):237–276, November 1992.

[112] H. J. A. M. Heijmans. Composing morphological filters. Technical Report BS-R9504, CWI, P.O. Box 94079, 1090 GB Amsterdam, The Netherlands, 1995.

[113] C. E. Heil and D. F. Walnut. Continuous and discrete wavelet transforms. *SIAM Review*, 31(4):628–666, 1989.

[114] H. von Helmholtz. *Handbuch der Physiologischen Optik*. Voss, Leipzig, second edition, 1896.

[115] D. Hestenes and G. Sobezyk. *Clifford Algebra to Geometric Calculus: A Unified Language for Mathematics and Physics*. D. Reidel Publishing Company, Dordrecht, 1984.

[116] D. Hestenes and R. Ziegler. Projective geometry with Clifford algebra. *Acta Applicandae Mathematicae*, 23:25–63, 1991.

[117] D. Hilbert. Ueber die vollen Invariantensystemen. *Mathematische Annalen*, 42:313–373, 1893.

[118] E. C. Hildreth. *The Measurement of Visual Motion*. MIT Press, Cambridge, 1983.

[119] E. C. Hildreth. Computations underlying the measurement of visual motion. *Artificial Intelligence*, 23:309–354, 1984.

[120] D. Hofstadter. *Gödel, Escher, Bach: an Eternal Golden Braid*. Basic Books Inc., New York, 1979.

[121] B. K. P. Horn. *Robot Vision*. MIT Press, Cambridge, 1986.

[122] B. K. P. Horn and M. J. Brooks, editors. *Shape from Shading*. MIT Press, Cambridge, 1989.

[123] B. K. P. Horn and B. G. Schunck. Determining optical flow. *Artificial Intelligence*, 17:185–203, 1981.

[124] B. K. P. Horn and B. G. Schunck. "determining optical flow": a retrospective. *Artificial Intelligence*, 59:81–87, 1993.

[125] D. H. Hubel. *Eye, Brain and Vision*, volume 22 of *Scientific American Library*. Scientific American Press, New York, 1988.

[126] R. A. Hummel, B. B. Kimia, and S. W. Zucker. Deblurring Gaussian blur. *Computer Vision, Graphics, and Image Processing*, 38:66–80, 1987.

[127] N. H. Ibragimov, editor. *CRC Handbook of Lie Group Analysis of Differential Equations. Symmetries, Exact Solutions, and Conservation Laws*, volume 1. CRC Press, Boca Raton, Florida, January 1994.

[128] T. Iijima. Basic theory on normalization of a pattern (in case of typical one-dimensional pattern). *Bulletin of Electrical Laboratory*, 26:368–388, 1962. (in Japanese).

[129] The International Association for Pattern Recognition. *Proceedings of the 13th International Conference on Pattern Recognition (Vienna, Austria, August 1996)*. IEEE Computer Society Press, 1996.

[130] P. T. Jackway and M. Deriche. Scale-space properties of the multiscale morphological dilation-erosion. *IEEE Transactions on Pattern Analysis and Machine Intelligence*, 18(1):38–51, January 1996.

[131] A. K. Jain. *Fundamentals of Digital Image Processing*. Prentice-Hall Information and System Sciences Series. Prentice-Hall, Englewood Cliffs, New Jersey, 1989.

[132] B. Jawerth and W. Sweldens. An overview of wavelet based multiresolution analyses. *SIAM Review*, 36(3):377–412, 1994.

[133] E. T. Jaynes. Prior probabilities. *IEEE Transactions on Systems Science and Cybernetics*, 4(3):227–241, 1968.

[134] P. Johansen. On the classification of toppoints in scale space. *Journal of Mathematical Imaging and Vision*, 4(1):57–67, 1994.

[135] P. Johansen. Local analysis of image scale space. In Sporring et al. [266], chapter 10, pages 139–146.

[136] P. Johansen and S. Olsen, editors. *Theory & Applications of Image Analysis*, volume 2 of *Series in Machine Perception and Artificial Intelligence*. World Scientific, Singapore, 1992. Selected papers from the 7th Scandinavian Conference on Image Analysis.

[137] P. Johansen, S. Skelboe, K. Grue, and J. D. Andersen. Representing signals by their top points in scale-space. In *Proceedings of the 8th International Conference on Pattern Recognition (Paris, France, October 1986)*, pages 215–217. IEEE Computer Society Press, 1986.

[138] S. N. Kalitzin. Topological numbers and singularities. In Sporring et al. [266], chapter 13, pages 181–189.

[139] K. Kanatani. *Group-Theoretical Methods in Image Understanding*, volume 20 of *Series in Information Sciences*. Springer-Verlag, 1990.

[140] I. Kant. *Critik der Reinen Vernunft*. Johann Friedrich Hartknoch, Riga, second edition, 1787.

[141] D. C. Kay. *Tensor Calculus*. Schaum's Outline Series. McGraw-Hill, New York, 1988.

[142] H. J. Keisler. *Foundations of Infinitesimal Calculus*. Prindle, Weber & Schmidt, Boston, 1976.

[143] H. J. Keisler, S. Körner, W. A. J. Luxemburg, and A. D. Young, editors. *Selected Papers of Abraham Robinson*, volume 2: Nonstandard Analysis and Philosophy. North-Holland, Amsterdam, 1979. Edited and with introductions by W. A. J. Luxemburg and S. Körner.

[144] W. K. J. Killing. Über die Grundlagen der Geometrie. *J. Reine Angew. Math.*, 109:121–186, 1892.

[145] W. K. J. Killing. *Einführung in die Grundlagen der Geometrie*. Paderborn, 1893.

[146] B. B. Kimia. Deblurring Gaussian blur, continuous and discrete approaches. Master's thesis, Electrical Engineering Department, McGill University, Montreal, Canada, 1986.

[147] B. B. Kimia and S. W. Zucker. Analytic inverse of discrete Gaussian blur. *Optical Engineering*, 32(1):166–176, 1993.

[148] F. Klein. Vergleichende Betrachtungen über neuere geometrische Forschungen (Erlanger Programm). *Mathematische Annalen*, 43:63–100, 1893.

[149] J. J. Koenderink. The concept of local sign. In A. J. van Doorn, W. A. van de Grind, and J. J. Koenderink, editors, *Limits in Perception*, pages 495–547. VNU Science Press, Utrecht, 1984.

[150] J. J. Koenderink. The structure of images. *Biological Cybernetics*, 50:363–370, 1984.

[151] J. J. Koenderink. Optic flow. *Vision Research*, 26(1):161–180, 1986.

[152] J. J. Koenderink. The structure of the visual field. In W. Güttinger and G. Dangelmayr, editors, *The Physics of Structure Formation: Theory and Simulation*. Proceedings of an International Symposium, Tübingen, Germany, October 27–November 2 1986. Springer-Verlag.

[153] J. J. Koenderink. Design for a sensorium. In W. von Seelen, G. Shaw, and U. M. Leinhos, editors, *Organization of Neural Networks - Structures and Models*, pages 185–207. VCH Verlagsgesellschaft mbH, Weinheim, Germany, 1988.

[154] J. J. Koenderink. Design principles for a front-end visual system. In R. Eckmiller and Ch. v. d. Malsburg, editors, *Neural Computers*, volume 41 of *NATO ASI Series F: Computer and Systems Sciences*, pages 111–118. Springer-Verlag, Berlin, 1988.

[155] J. J. Koenderink. Scale-time. *Biological Cybernetics*, 58:159–162, 1988.

[156] J. J. Koenderink. A hitherto unnoticed singularity of scale-space. *IEEE Transactions on Pattern Analysis and Machine Intelligence*, 11(11):1222–1224, November 1989.

[157] J. J. Koenderink. The brain a geometry engine. *Psychological Research*, 52:122–127, 1990.

[158] J. J. Koenderink. *Solid Shape*. MIT Press, Cambridge, 1990.

[159] J. J. Koenderink. Mapping formal structures on networks. In T. Kohonen, K. Mäkisara, O. Simula, and J. Kangas, editors, *Artificial Neural Networks*, pages 93–98. Elsevier Science Publishers B.V. (North-Holland), 1991.

[160] J. J. Koenderink. Local image structure. In Johansen and Olsen [136], pages 15–21. Selected papers from the 7th Scandinavian Conference on Image Analysis.

[161] J. J. Koenderink. What is a "feature"? *Journal of Intelligent Systems*, 3(1):49–82, 1993.

[162] J. J. Koenderink and A. J. van Doorn. Dynamic shape. *Biological Cybernetics*, 53:383–396, 1986.

[163] J. J. Koenderink and A. J. van Doorn. Logical stratification of organic intelligence. In R. Trappl, editor, *Cybernetics and Systems*, pages 871–878. D. Reidel Publishing Company, 1986.

[164] J. J. Koenderink and A. J. van Doorn. Facts on optic flow. *Biological Cybernetics*, 56:247–254, 1987.

[165] J. J. Koenderink and A. J. van Doorn. Representation of local geometry in the visual system. *Biological Cybernetics*, 55:367–375, 1987.

[166] J. J. Koenderink and A. J. van Doorn. The basic geometry of a vision system. In R. Trappl, editor, *Cybernetics and Systems*, pages 481–485. Kluwer Academic Publishers, 1988.

[167] J. J. Koenderink and A. J. van Doorn. Receptive field families. *Biological Cybernetics*, 63:291–298, 1990.

[168] J. J. Koenderink, A. Kappers, and A. J. van Doorn. Local operations: The embodiment of geometry. In G. A. Orban and H.-H. Nagel, editors, *Artificial and Biological Vision Systems*, Basic Research Series, pages 1–23. Springer Verlag, Berlin, 1992.

[169] J. J. Koenderink and A. J. van Doorn. Invariant properties of the motion parallax field due to the movement of rigid bodies relative to an observer. *Optica Acta*, 22(9):773–791, 1975.

[170] J. J. Koenderink and A. J. van Doorn. The structure of two-dimensional scalar fields with applications to vision. *Biological Cybernetics*, 33:151–158, 1979.

[171] J. J. Koenderink and A. J. van Doorn. Operational significance of receptive field assemblies. *Biological Cybernetics*, 58:163–171, 1988.

[172] J. J. Koenderink and A. J. van Doorn. Receptive field assembly pattern specificity. *Journal of Visual Communication and Image Representation*, 3(1):1–12, 1992.

[173] J. J. Koenderink and A. J. van Doorn. Second-order optic flow. *Journal of the Optical Society of America-A*, 9(4):530–538, April 1992.

[174] T. Kohonen. *Self-Organization and Associative Memory*. Springer-Verlag, 1984.

[175] C. Konstantopoulos, R. Hohlfeld, and G. Sandri. Novel deconvolution of noisy Gaussian filters with a modified Hermite expansion. *CVGIP: Graphical Models and Image Processing*, 56(6):433–441, November 1994.

[176] T. H. Koornwinder, editor. *Wavelets: an Elementary Treatment of Theory and Applications*. Number 1 in Series in Approximations and Decompositions. World Scientific, Singapore, 1993.

[177] A. S. E. Koster. *Linking Models for Multiscale Image Segmentation*. PhD thesis, University of Utrecht, Department of Medicine, Utrecht, The Netherlands, November 22 1995.

[178] M. Kouwenhoven, C. J. G. Bakker, M. J. Hartkamp, and W. P. Th. M. Mali. Current MR angiographic imaging techniques, a survey. In Roesch Lanzer, editor, *Vascular Diagnostics*, pages 375–400. Springer-Verlag, Heidelberg, 1994.

[179] J. Lasenby, E. Bayro-Corrochano, A. N. Lasenby, and G. Sommer. A new framework for the formation of invariants and multiple-view constraints in computer vision. In Delogne [48], pages 313–316.

[180] J. Lasenby, E. Bayro-Corrochano, A. N. Lasenby, and G. Sommer. A new methodology for computing invariants in computer vision. In *Proceedings of the 13th International Conference on Pattern Recognition (Vienna, Austria, August 1996)* [129], pages 393–397.

[181] D. F. Lawden. *An Introduction to Tensor Calculus and Relativity.* Spottiswoode Ballantyne, London and Colchester, 1962.

[182] D. N. Lee. A theory of visual control of braking based on information about time-to-collision. *Perception,* 5:437–457, 1976.

[183] D. N. Lee. The optic flow field: the foundation of vision. *Phil. Trans. R. Soc. Lond.* B, 290:169–179, 1980.

[184] S. Lie and F. Engel. *Theorie der Transformationsgruppen,* volume 1–3. B. G. Teubner, Leipzig, 1888–1893.

[185] A. P. Lightman, W. H. Press, R. H. Price, and S. A. Teukolsky. *Problem Book in Relativity and Gravitation.* Princeton University Press, Princeton, 1975.

[186] T. Lindeberg. Scale-space for discrete signals. *IEEE Transactions on Pattern Analysis and Machine Intelligence,* 12(3):234–245, 1990.

[187] T. Lindeberg. On the behaviour in scale-space of local extrema and blobs. In Johansen and Olsen [136], pages 38–47. Selected papers from the 7th Scandinavian Conference on Image Analysis.

[188] T. Lindeberg. Scale-space behaviour of local extrema and blobs. *Journal of Mathematical Imaging and Vision,* 1(1):65–99, March 1992.

[189] T. Lindeberg. Detecting salient blob-like image structures and their scale with a scale-space primal sketch: A method for focus-of-attention. *International Journal of Computer Vision,* 11(3):283–318, 1993.

[190] T. Lindeberg. Effective scale: A natural unit for measuring scale-space lifetime. *IEEE Transactions on Pattern Analysis and Machine Intelligence,* 15(10):1068–1074, October 1993.

[191] T. Lindeberg. *Scale-Space Theory in Computer Vision.* The Kluwer International Series in Engineering and Computer Science. Kluwer Academic Publishers, 1994.

[192] T. Lindeberg. Edge detection and ridge detection with automatic scale selection. In *Proceedings of the IEEE Computer Society Conference on Computer Vision and Pattern Recognition (San Francisco, California, June 1996),* pages 465–470, Los Alamitos, California, 1996. IEEE Computer Society Press.

[193] T. Lindeberg. Edge detection and ridge detection with automatic scale selection. Technical Report ISRN KTH/NA/P–96/06–SE, Royal Institute of Technology, University of Stockholm, Dept. of Numerical Analysis and Computing Science, Stockholm, Sweden, 1996.

[194] T. Lindeberg. Feature detection with automatic scale selection. Technical Report ISRN KTH/NA/P–96/18–SE, Royal Institute of Technology, University of Stockholm, Dept. of Numerical Analysis and Computing Science, Stockholm, Sweden, 1996.

[195] T. Lindeberg and J.-O. Eklundh. On the computation of a scale-space primal sketch. *Journal of Visual Communication and Image Representation,* 2(1):55–78, 1990.

[196] T. Lindeberg and D. Fagerstrom. Scale-space with causal time direction. In B. Buxton and R. Cipolla, editors, *Proceedings of the Fourth European Conference on Computer Vision (Cambridge, England, April 1996),* volume 1064 of *Lecture Notes in Computer Science,* pages 229–240, Berlin, 1996. Springer-Verlag.

[197] T. Lindeberg and J. Gårding. Shape from texture from a multi-scale perspective. In H. H. Nagel, editor, *Proceedings of the 4th International Conference on Computer Vision (Berlin, Germany, June 20–23, 1993),* pages 683–691. IEEE Computer Society Press, 1993.

[198] M. M. Lipschutz. *Differential Geometry.* Schaum's Outline Series. McGraw-Hill, New York, 1969.

[199] J. Llacer, B. M. ter Haar Romeny, L. M. J. Florack, and M. A. Viergever. The representation of medical images by visual response functions. *IEEE Engineering in Medicine and Biology,* 3(93):40–47, 1993.

[200] Lord Rayleigh. The principle of similitude. *Nature*, XCV:66–68, 644, March 1915.

[201] H. Lotze. *Mikrokosmos*. Hirzel, Leipzig, 1884.

[202] Y. Lu and R. C. Jain. Behaviour of edges in scale space. *IEEE Transactions on Pattern Analysis and Machine Intelligence*, 11(4):337–357, 1989.

[203] S. O. Lundqvist, A. Ranfagni, V. Sa-yakanit, and L. S. Schulman, editors. *Path Integration: Achievements and Goals*, Singapore, 1988. World Scientific. Proceedings of a workshop on path integration held in Trieste in 1987.

[204] J. B. Maintz. *Retrospective Registration of Tomographic Brain Images*. PhD thesis, University of Utrecht, Department of Medicine, Utrecht, The Netherlands, December 10 1996.

[205] B. A. Mair, Z. Réti, D. C. Wilson, E. A. Geiser, and B. David. A q-series approach to deblurring the discrete Gaussian. *Computer Vision and Image Understanding*, 66(2):247–254, May 1997.

[206] P. Mansfield and P. G. Morris. *NMR Imaging in Biomedicine*. Academic Press, London and New York, 1982.

[207] D. C. Marr. *Vision*. Freeman, San Francisco, CA, 1982.

[208] J. B. Martens. Deblurring digital images by means of polynomial transforms. *Computer Vision, Graphics, and Image and Stochastic Processing*, 50:157–176, 1990.

[209] Y. Meyer, editor. *Wavelets and Applications*. Number 20 in Research Notes in Applied Mathematics. Springer-Verlag, 1991.

[210] Y. Meyer. *Wavelets: Algorithms and Applications*. SIAM, Philadelphia, 1993.

[211] C. W. Misner, K. S. Thorne, and J. A. Wheeler. *Gravitation*. Freeman, San Francisco, 1973.

[212] O. Monga, N. Ayache, and P. T. Sander. From voxel to intrinsic surface features. *Image and Vision Computing*, 10(6):403–417, July/August 1992.

[213] T. Moons, L. J. Van Gool, M. Van Diest, and E. J. Pauwels. Affine reconstruction from perspective image pairs. In J. Mundy, A. Zisserman, and D. Forsyth, editors, *Applications of Invariance in Vision*, volume 825 of *Lecture Notes in Computer Science*, pages 297–316. Springer-Verlag, Berlin, 1994.

[214] T. Moons, E. J. Pauwels, L. J. Van Gool, M. H. Brill, and E. B. Barrett. Recognizing 3D curves from a stereo pair of images: a semi-differential approach. In O et al. [229], pages 433–442.

[215] J.-C. Müller, J.-P. Lagrange, and R. Weibel, editors. *GIS and Generalization, Methodology and Practice*, volume 1 of *GISDATA Series*. Taylor & Francis Ltd, London, 1995.

[216] J. Mundy and A. Zisserman, editors. *Applications of Invariance in Vision*. MIT Press, 1992.

[217] H.-H. Nagel. On the estimation of optical flow: Relations between different approaches and some new results. *Artificial Intelligence*, 33:299–324, 1987.

[218] H.-H. Nagel. Direct estimation of optical flow and its derivatives. In G. A. Orban and H.-H. Nagel, editors, *Artificial and Biological Vision Systems*, Basic Research Series, pages 193–224. Springer Verlag, Berlin, 1992.

[219] H.-H. Nagel and W. Enkelmann. An investigation of smoothness constraints for the estimation of displacement vector fields from image sequences. *IEEE Transactions on Pattern Analysis and Machine Intelligence*, 8(1):565–593, 1986.

[220] S. K. Nayar, K. Ikeuchi, and T. Kanade. Surface reflection: Physical and geometrical perspectives. *IEEE Transactions on Pattern Analysis and Machine Intelligence*, 13(7):611–634, July 1991.

[221] M. Nielsen, L. Florack, and R. Deriche. Regularization and scale space. Technical Report INRIA-RR-2352, INRIA Sophia-Antipolis, France, September 1994.

[222] M. Nielsen, R. Maas, W. J. Niessen, L. M. J. Florack, and B. M. ter Haar Romeny. Local disparity structure by scale-space operators. Technical Report 96/17, Department of Computer Science, University of Copenhagen, 1996.

[223] W. J. Niessen, J. S. Duncan, L. M. J. Florack, B. M. ter Haar Romeny, and M. A. Viergever. Spatiotemporal operators and optic flow. In T. S. Huang and D. N. Metaxas, editors, *Workshop on Physics-Based Modeling in Computer Vision*, pages 78–84, Cambridge, Massachusetts, June 18–19 1995. IEEE Computer Society Press.

[224] W. J. Niessen, J. S. Duncan, M. Nielsen, L. M. J. Florack, B. M. ter Haar Romeny, and M. A. Viergever. A multiscale approach to image sequence analysis. *Computer Vision and Image Understanding*, 65(2):259–268, February 1997.

[225] W. J. Niessen, B. M. ter Haar Romeny, L. M. J. Florack, and M. A. Viergever. A general framework for geometry-driven evolution equations. *International Journal of Computer Vision*, 21(3):187–205, 1997.

[226] W. J. Niessen and R. Maas. Optic flow and stereo. In Sporring et al. [266], chapter 3, pages 31–42.

[227] W. J. Niessen, M. Nielsen, L. M. J. Florack, R. Maas, B. M. ter Haar Romeny, and M. A. Viergever. Multiscale optic flow using physical constraints. In P. Johansen, editor, *Proceedings of the Copenhagen Workshop on Gaussian Scale-Space Theory*, pages 72–79, May 1996. DIKU Tech. Rep. Nr. 96/19.

[228] L. Nottale. Scale relativity and fractal space-time: Applications to quantum physics, cosmology and chaotic systems. *Chaos, Solitons & Fractals*, 7(6):877–938, 1996.

[229] Y.-L. O, A. Toet, H. J. A. M. Heijmans, D. H. Foster, and P. Meer, editors. *Proceedings of the NATO Advanced Research Workshop Shape in Picture - Mathematical Description of Shape in Greylevel Images*, volume 126 of *NATO ASI Series F: Computer and Systems Sciences*, Berlin, 1994. Springer-Verlag.

[230] P. Olver, G. Sapiro, and A. Tannenbaum. Differential invariant signatures and flows in computer vision: a symmetry group approach. In Haar Romeny [103], pages 255–306.

[231] P. J. Olver. *Applications of Lie Groups to Differential Equations*, volume 107 of *Graduate Texts in Mathematics*. Springer-Verlag, 1986.

[232] W. Ostwald. *Die Farbenlehre*, volume 1–5. Verlag Unesma, Leipzig, 1921.

[233] N. Otsu. *Mathematical Studies on Feature Extraction in Pattern Recognition*. PhD thesis, Electrotechnical Laboratory, Ibaraki, Japan, 1981. (in Japanese).

[234] M. Otte and H.-H. Nagel. Optical flow estimation: Advances and comparisons. In Eklundh [62], pages 51–60.

[235] E. J. Pauwels, L. J. Van Gool, P. Fiddelaers, and T. Moons. An extended class of scale-invariant and recursive scale space filters. *IEEE Transactions on Pattern Analysis and Machine Intelligence*, 17(7):691–701, July 1995.

[236] Y. Pnueli and A. M. Bruckstein. Gridless halftoning: A reincarnation of the old method. *Graphical Models and Image Processing*, 58(1):38–64, January 1996.

[237] T. Poston and I. Steward. *Catastrophe Theory and its Applications*. Pitman, London, 1978.

[238] W. H. Press, B. P. Flannery, S. A. Teukolsky, and W. T. Vetterling. *Numerical Recipes in C; the Art of Scientific Computing*. Cambridge University Press, Cambridge, 1988.

[239] W. Reichardt. Autocorrelation, a principle for the evaluation of sensory information by the central nervous system. In W. A. Rosenblith, editor, *Sensory Communication*, pages 303–317. MIT Press, Cambridge, 1961.

[240] A. Rényi. On the foundations of information theory. In Pál Turán, editor, *Selected Papers of Alfréd Rényi*, volume 2, pages 304–318. Akadémiai Kiadó, Budapest, 1976. (Originally: Rev. Inst. Internat. Stat., 33, 1965, p. 1–14).

[241] A. Rényi. Some fundamental questions of information theory. In Pál Turán, editor, *Selected Papers of Alfréd Rényi*, volume 3, pages 526–552. Akadémiai Kiadó, Budapest, 1976. (Originally: MTA III. Oszt. Közl., 10, 1960, p. 251–282).

[242] B. Riemann. Über die Hypothesen, welche der Geometrie zu Grunde liegen. In H. Weber, editor, *Gesammelte Mathematische Werke*, pages 272–287. Teubner, Leipzig, 1892.

[243] M. Riesz. L'intégrale de Riemann-Liouville et le problème de Cauchy. *Acta Math.*, 81:1–223, 1949.

[244] A. Robinson. *Non-Standard Analysis*. North-Holland, Amsterdam, revised edition, 1974.

[245] A. H. Robinson, J. L. Morrison, P. C. Muehrcke, A. J. Kimerling, and S. C. Guptill. *Elements of Cartography*. John Wiley & Sons, New York, sixth edition, 1995.

[246] W. Rudin. *Functional Analysis*. Tata McGraw-Hill, New Delhi, fifteenth edition, 1992. Originally published by McGraw-Hill in 1973.

[247] M. B. Ruskai, G. Beylkin, R. Coifman, I. Daubechies, S. Mallat, Y. Meyer, and L. Raphael, editors. *Wavelets and their Applications*. Jones and Bartlett, 1992.

[248] B. Russell. *The Problems of Philosophy*. Opus. Oxford University Press, Oxford, ninth edition, 1980.

[249] A. H. Salden. *Dynamic Scale-Space Paradigms*. PhD thesis, University of Utrecht, Department of Medicine, Utrecht, The Netherlands, November 12 1996.

[250] P. T. Saunders. *An Introduction to Catastrophe Theory*. Cambridge University Press, Cambridge, 1980.

[251] L. L. Schumaker and G. Webb, editors. *Recent Advances in Wavelet Analysis*. Academic Press, 1993.

[252] B. G. Schunck. The motion constraint equation for optical flow. In *Proceedings of the 7th International Conference on Pattern Recognition*, pages 20–22, Montreal, Canada, 1984.

[253] L. Schwartz. *Théorie des Distributions*, volume I, II of *Actualités scientifiques et industrielles; 1091,1122*. Publications de l'Institut de Mathématique de l'Université de Strasbourg, Paris, 1950–1951.

[254] L. Schwartz. *Théorie des Distributions*. Hermann, Paris, second edition, 1966.

[255] J. Segman and Y. Y. Zeevi. Image analysis by wavelet-type transforms: Group theoretic approach. *Journal of Mathematical Imaging and Vision*, 3:51–77, 1993.

[256] J. Serra. *Image Analysis and Mathematical Morphology*. Academic Press, London, 1982.

[257] J. Serra and P. Soille, editors. *Mathematical Morphology and its Applications to Image Processing*, volume 2 of *Computational Imaging and Vision Series*. Kluwer Academic Publishers, Dordrecht, 1994.

[258] C. E. Shannon and W. Weaver. *The Mathematical Theory of Communication*. The University of Illinois Press, Urbana, 1949.

[259] J. Shen and S. Castan. An optimal linear operator for step edge detection. *CVGIP: Graphical Models and Image Processing*, 54(2):112–133, March 1992.

[260] J. Shen and S. Castan. Towards the unification of band-limited derivative operators for edge detection. *Signal Processing*, 31(2):103–119, March 1993.

[261] P. Shi, G. Robinson, and J. Duncan. Myocardial motion and function assessment using 4D images. In Richard A. Robb, editor, *Visualization in Biomedical Computing*, Proceedings SPIE 2359, pages 100–109, Rochester, Minnesota, October 4–7 1994. SPIE - The International Society for Optical Engineering.

[262] A. Singh. *Optic Flow Computation: a Unified Perspective*. IEEE Computer Society Press, Los Alamitos, CA, 1991.

[263] J. A. Sorenson and M. E. Phelps. *Physics in Nuclear Medicine*. Grune and Stratton, Orlando, 1987.

[264] M. Spivak. *Calculus on Manifolds*. W. A. Benjamin, New York, 1965.

[265] M. Spivak. *Differential Geometry*, volume 1–5. Publish or Perish, Berkeley, 1975.

[266] J. Sporring, M. Nielsen, L. M. J. Florack, and P. Johansen, editors. *Gaussian Scale-Space Theory*, volume 8 of *Computational Imaging and Vision Series*. Kluwer Academic Publishers, Dordrecht, 1997.

[267] J. Sporring and J. Weickert. On generalized entropies and scale-space. Technical Report 96/37, Department of Computer Science, University of Copenhagen, 1996.

[268] G. Strang. Wavelets and dilation equations: A brief introduction. *SIAM Review*, 31(4):614–627, 1989.

[269] G. Strang. Wavelet transforms versus Fourier transforms. *Bull. Amer. Math. Soc.*, 28(2):288–305, 1993.

[270] R. Thom. *Stabilité Structurelle et Morphogénèse*. Benjamin, Paris, 1972.

[271] A. Tikhonov. Solution of incorrectly formulated problems and the regularization method. *Soviet. Math. Dokl.*, 4:1035–1038, 1963.

[272] M. Tistarelli. Multiple constraints for optical flow. In Eklundh [62], pages 61–70.

[273] M. Tistarelli. Computation of coherent optical flow by using multiple constraints. In *Proceedings of the 5th International Conference on Computer Vision (Boston, Massachusetts, June 20–23, 1995)*, pages 263–268. IEEE Computer Society Press, 1995.

[274] M. Tistarelli and G. Marcenaro. Using optical flow to analyze the motion of human body organs from bioimages. In *Proceedings of the IEEE Workshop on Biomedical Image Analysis*, pages 100–109, Seattle, Washington, June 24–25 1994. IEEE Computer Society Press.

[275] R. Todd Ogden. *Essential Wavelets for Statistical Applications and Data Analysis*. Birkhauser, 1997.

[276] A. Toet. *Visual Perception of Spatial Order*. PhD thesis, University of Utrecht, Utrecht, March 1987.

[277] O. Tretiak and L. Pastor. Velocity estimation from image sequences with second order differential operators. In *Proceedings of the 7th International Conference on Pattern Recognition*, pages 16–19, Montreal, Canada, 1984.

[278] F. Trèves. *Topological Vector Spaces, Distributions and Kernels*. Academic Press, 1969.

[279] B. M. W. Tsui, X. Zhao, E. C. Frey, and W. H. McCartney. Quantitative SPECT: Basics and clinical considerations. *Seminars in Nuclear Medicine*, 24:38–65, 1994.

[280] R. A. Ulichney. *Digital Halftoning*. MIT Press, Cambridge, 1987.

[281] R. A. Ulichney. Dithering with blue noise. *Proceedings of the IEEE*, 76(1):56–79, January 1988.

[282] S. Uras, F. Girosi, A. Verri, and V. Torre. A computational approach to motion perception. *Biological Cybernetics*, 60:79–87, 1988.

[283] A. Verri, F. Girosi, and V. Torre. Differential techniques for optical flow. *Journal of the Optical Society of America-A*, 7(5):912–922, May 1990.

[284] A. Verri and T. Poggio. Motion field and optical flow: Qualitative properties. *IEEE Transactions on Pattern Analysis and Machine Intelligence*, 11(5):490–498, 1989.

[285] K. L. Vincken. *Probabilistic Multiscale Image Segmentation by the Hyperstack*. PhD thesis, University of Utrecht, Department of Medicine, Utrecht, The Netherlands, November 22 1995.

[286] K. L. Vincken, A. S. E. Koster, and M. A. Viergever. Probabilistic segmentation of partial volume voxels. *Pattern Recognition Letters*, 15(5):477–484, 1994.

[287] D. C. C. Wang, A. H. Vagnucci, and C. C. Li. Digital image enhancement: A survey. *Computer Vision, Graphics, and Image Processing*, 24:363–381, 1983.

[288] S. Webb, editor. *The Physics of Medical Imaging*. Medical Science Series. Institute of Physics Publishing, Bristol, 1988.

[289] H. Weber. *Lehrbuch der Algebra*, volume I–III. Chelsea Publishing Company, New York, 1894.

[290] J. Weber and J. Malik. Robust computation of optical flow in a multi-scale differential framework. *International Journal of Computer Vision*, 14(1):67–81, 1995.

[291] J. A. Weickert. *Anisotropic Diffusion in Image Processing*. PhD thesis, University of Kaiserslautern, Department of Mathematics, Kaiserslautern, Germany, January 29 1996.

[292] J. A. Weickert, B. M. ter Haar Romeny, and M. A. Viergever. Conservative image transformations with restoration and scale-space properties. In Delogne [48], pages 465–468.

[293] J. A. Weickert, S. Ishikawa, and A. Imiya. On the history of Gaussian scale-space axiomatics. In Sporring et al. [266], chapter 4, pages 45–59.

[294] P. Werkhoven. *Visual Perception of Successive Order*. PhD thesis, University of Utrecht, Department of Medical Physics, Utrecht, the Netherlands, May 17 1990.

[295] P. Werkhoven and J. J. Koenderink. Extraction of motion parallax structure in the visual system I. *Biological Cybernetics*, 63:185–191, 1990.

[296] H. Weyl. *Gesammelte Abhandlungen*. Springer-Verlag, Berlin, 1968.

[297] R. L. Wheeden and A. Zygmund. *Measure and Integral: an Introduction to Real Analysis*. Marcel Dekker, New York, 1977.

[298] J. A. Wheeler. *At Home in the Universe*, volume 9 of *Masters of Modern Physics*. AIP Press, Woodbury, N.Y., 1992.

[299] R. T. Whitaker. *Geometry-Limited Diffusion*. PhD thesis, University of North Carolina, Department of Computer Science, Chapel Hill, North Carolina, USA, 1993.

[300] R. T. Whitaker. Geometry-limited diffusion in the characterization of geometric patches in images. *CVGIP: Image Understanding*, 57(1):111–120, January 1993.

[301] R. T. Whitaker and S. M. Pizer. A multi-scale approach to nonuniform diffusion. *CVGIP: Image Understanding*, 57(1):99–110, January 1993.

[302] N. Wiener. *Cybernetics*. Wiley, New York, 1948.

[303] A. P. Witkin. Scale-space filtering. In *Proceedings of the International Joint Conference on Artificial Intelligence*, pages 1019–1022, Karlsruhe, Germany, 1983.

[304] S. Wolfram. *Mathematica: A System for doing Mathematics by Computer*. Addison-Wesley, second edition, 1991. Version 2.

[305] R. A. Young. The Gaussian derivative model for machine vision: I. retinal mechanisms. *Spatial Vision*, 2(4):273–293, 1987.

Index

Bold face page numbers indicate pages with important information about the entry, such as a definition or a detailed explanation. A postfix *ff* labels the first page of a section containing information about the entry; *n* refers to a footnote. Page numbers in normal type indicate a textual reference.

A

affine connection, *see also* metric affinity, 145*ff*, **145**
affinity terms, **145**, 146, 147, 150
algebra, **55**, 57, 60, 72, 73, 78, 92, 135
    autoconvolution, 60, 61
    autoconvolution , 60
    commutative, 55, 85
    convolution, 34, 57, 60, 61, 73, 85
    regular, 55, 85
    singular, 55
    with identity, 55, 85
alternation, 156, **207**, 215
analysis, 6*ff*
analyticity, *see* smoothness
anholonomic basis, 139*n*
annihilation operator, **67**, 87
antisymmetrisation, **207**
aperture, *see* sampling aperture
autoconvolution algebra, *see* algebra
automorphism
    confined to isophote, 178
    of spacetime, **41**, 42–45, 47, 47*n*, 75, 83, 134

B

Banach space, **17**, 18, 20, 24, 26
Benford's law, 112, 112*n*, 122, 132
    relation to canonical time, 112
Borel measure, 17
bracket formalism, 116, **126**
Brownian motion, *see* random walk

C

canonical parametrisation, 68*ff*, **69**, **70**, 96, 100
carry along principle, 41
Cartesian coordinate transformation, 44, 137
catastrophe theory, **120**
    for scale-space, 90, 91, 106, **121**
causality
    w.r.t. scale, 101, 109
    w.r.t. time, 16*n*, 108*ff*, **115**, 175
Cayley-Hamilton theorem, 158, **226**
chain rule, 144, 147, 163, **218**
Christoffel symbols, **146**
Clifford algebra, 135
connection coefficients, **144**, 145, 146
conservation, 11, 66, 175, 177, 179, 182, 191
    of topological detail, **180**, 182, 187, 197
contraction, 40, 49, 56, 141–143, 147, 151, 153, 155, 156, **208**, **210**
convolution, 29*ff*, 37, 44, 57, 59, 60, 84, 85, 154, 190
    on $S(\mathbb{R}^n)$, **29**, 72
    on $S'(\mathbb{R}^n)$, **34**
    vs. tangential dilation, 92
convolution algebra, *see* algebra
coordinates, 15, 15*n*, 24, 39, 40, 42–45, 48, 49, 58–60, 62–65, 67–69, 73, 74, 78, 94, 100, 101, 110, 113,

Computational Imaging and Vision

1. B.M. ter Haar Romeny (ed.): *Geometry-Driven Diffusion in Computer Vision.*
    1994   ISBN 0-7923-3087-0
2. J. Serra and P. Soille (eds.): *Mathematical Morphology and Its Applications to
    Image Processing.* 1994   ISBN 0-7923-3093-5
3. Y. Bizais, C. Barillot, and R. Di Paola (eds.): *Information Processing in Med-
    ical Imaging.* 1995   ISBN 0-7923-3593-7
4. P. Grangeat and J.-L. Amans (eds.): *Three-Dimensional Image Reconstruction
    in Radiology and Nuclear Medicine.* 1996   ISBN 0-7923-4129-5
5. P. Maragos, R.W. Schafer and M.A. Butt (eds.): *Mathematical Morphology
    and Its Applications to Image and Signal Processing.* 1996
                                                 ISBN 0-7923-9733-9
6. G. Xu and Z. Zhang: *Epipolar Geometry in Stereo, Motion and Object Recog-
    nition.* A Unified Approach. 1996   ISBN 0-7923-4199-6
7. D. Eberly: *Ridges in Image and Data Analysis.* 1996   ISBN 0-7923-4268-2
8. J. Sporring, M. Nielsen, L. Florack and P. Johansen (eds.): *Gaussian Scale-
    Space Theory.* 1997   ISBN 0-7923-4561-4
9. M. Shah and R. Jain (eds.): *Motion-Based Recognition.* 1997
                                                 ISBN 0-7923-4618-1
10. L. Florack: *Image Structure.* 1997   ISBN 0-7923-4808-7

Kluwer Academic Publishers – Dordrecht / Boston / London